Impactful Technologies Transforming the Food Industry

Şule Aydın
Nevsehir Haci Bektas Veli University, Turkey

Eda Özgül Katlav
Nevşehir Haci Bektaş Veli University, Turkey

Koray Çamlıca
Nevşehir Haci Bektaş Veli University, Turkey

Firdevs Yönet Eren
Nevşehir Haci Bektaş Veli University, Turkey

A volume in the Advances in Environmental Engineering and Green Technologies (AEEGT) Book Series

IGI Global
PUBLISHER of TIMELY KNOWLEDGE

Published in the United States of America by
IGI Global
Engineering Science Reference (an imprint of IGI Global)
701 E. Chocolate Avenue
Hershey PA, USA 17033
Tel: 717-533-8845
Fax: 717-533-8661
E-mail: cust@igi-global.com
Web site: http://www.igi-global.com

Library of Congress Cataloging-in-Publication Data

Names: Aydin, Şule, 1970- editor. | Özgül Katlav, Eda, 1987- editor. | Çamlıca, Koray, 1988- editor. | Yönet Eren, Firdevs, 1994- editor.
Title: Impactful technologies transforming the food industry / edited by Şule Aydın, Eda Özgül Katlav, Koray Çamlıca, Firdevs Yönet Eren.
Description: Hershey, PA : Engineering Science Reference, [2024] | Includes bibliographical references and index. | Summary: "Impactful Technologies Transforming the Food Industry provides an in-depth exploration of the objectives, motivations, and consequences associated with the adoption of technology in the food industry through meticulous analysis and real-world examples. From advanced ordering systems and kitchen automation to virtual reality and interactive dining experiences, readers gain valuable insights into how technology has revolutionized the way food businesses operate and how it has reshaped the relationship between consumers and food. This book also addresses the emergence of food delivery services, the ethical considerations of technology in the industry, and the potential future trajectory of food tech. It is an essential resource for food industry professionals, students, policymakers, and consumers interested in the intersection of gastronomy and technology"-- Provided by publisher.
Identifiers: LCCN 2023025852 (print) | LCCN 2023025853 (ebook) | ISBN 9781668490945 (hardcover) | ISBN 9781668490952 (paperback) | ISBN 9781668490969 (ebook)
Subjects: LCSH: Food industry and trade--Technological innovations.
Classification: LCC TP370.5 .I47 2024 (print) | LCC TP370.5 (ebook) | DDC 664--dc23/eng/20230729
LC record available at https://lccn.loc.gov/2023025852
LC ebook record available at https://lccn.loc.gov/2023025853

This book is published in the IGI Global book series Advances in Environmental Engineering and Green Technologies (AEEGT) (ISSN: 2326-9162; eISSN: 2326-9170)

British Cataloguing in Publication Data
A Cataloguing in Publication record for this book is available from the British Library.

For electronic access to this publication, please contact: eresources@igi-global.com.

Advances in Environmental Engineering and Green Technologies (AEEGT) Book Series

Sang-Bing Tsai
Zhongshan Institute, University of Electronic Science and Technology of China, China & Wuyi University, China

ISSN:2326-9162
EISSN:2326-9170

Mission

Growing awareness and an increased focus on environmental issues such as climate change, energy use, and loss of non-renewable resources have brought about a greater need for research that provides potential solutions to these problems. Research in environmental science and engineering continues to play a vital role in uncovering new opportunities for a "green" future.

The **Advances in Environmental Engineering and Green Technologies (AEEGT)** book series is a mouthpiece for research in all aspects of environmental science, earth science, and green initiatives. This series supports the ongoing research in this field through publishing books that discuss topics within environmental engineering or that deal with the interdisciplinary field of green technologies.

Coverage

- Cleantech
- Policies Involving Green Technologies and Environmental Engineering
- Contaminated Site Remediation
- Renewable Energy
- Alternative Power Sources
- Radioactive Waste Treatment
- Sustainable Communities
- Green Transportation
- Green Technology
- Industrial Waste Management and Minimization

IGI Global is currently accepting manuscripts for publication within this series. To submit a proposal for a volume in this series, please contact our Acquisition Editors at Acquisitions@igi-global.com or visit: http://www.igi-global.com/publish/.

The Advances in Environmental Engineering and Green Technologies (AEEGT) Book Series (ISSN 2326-9162) is published by IGI Global, 701 E. Chocolate Avenue, Hershey, PA 17033-1240, USA, www.igi-global.com. This series is composed of titles available for purchase individually; each title is edited to be contextually exclusive from any other title within the series. For pricing and ordering information please visit http://www.igi-global.com/book-series/advances-environmental-engineering-green-technologies/73679. Postmaster: Send all address changes to above address.

Titles in this Series

For a list of additional titles in this series, please visit: www.igi-global.com/book-series

Global Industrial Impacts of Heavy Metal Pollution in Sub-Saharan Africa
Joan Nyika (University of Johannesburg, South Africa) and Megersa Olumana Dinka (University of Johannesburg, South Africa)
Engineering Science Reference • © 2023 • 406pp • H/C (ISBN: 9781668471166) • US $255.00

Nanopriming Approach to Sustainable Agriculture
Abhishek Singh (Faculty of Biology, Yerevan State University, Armenia) Vishnu D. Rajput (Academy of Biology and Biotechnology, Southern Federal University, Rostov-on-Don, Russia) Karen Ghazaryan (Faculty of Biology, Yerevan State University, Armenia) Santosh Kumar Gupta (National Institute of Plant Genome Research, India) and Tatiana Minkina (Academy of Biology and Biotechnology, Southern Federal University, Rostov-on-Don, Russia)
Engineering Science Reference • © 2023 • 453pp • H/C (ISBN: 9781668472323) • US $245.00

Handbook of Research on AI-Equipped IoT Applications in High-Tech Agriculture
Alex Khang (Global Research Institute of Technology and Engineering, USA)
Engineering Science Reference • © 2023 • 473pp • H/C (ISBN: 9781668492314) • US $335.00

Contemporary Developments in Agricultural Cyber-Physical Systems
G.S. Karthick (Department of Software Systems, PSG College of Arts and Science, India)
Engineering Science Reference • © 2023 • 293pp • H/C (ISBN: 9781668478790) • US $240.00

Perspectives on Ecological Degradation and Technological Progress
Veli Yilanci (Faculty of Political Sciences, Canakkale Onsekiz Mart University, Turkey)
Engineering Science Reference • © 2023 • 342pp • H/C (ISBN: 9781668467275) • US $225.00

Nematode-Plant Interactions and Controlling Infection
Waleed Fouad Abobatta (Horticulture Research Institute, Agriculture Research Center, Egypt) and Rehab Yasin Ghareeb (Plant Protection and Biomolecular diagnosis Dep., Arid Lands Cultivation Research Institute, City of Scientific Research and Technological Applications, Egypt)
Engineering Science Reference • © 2023 • 300pp • H/C (ISBN: 9781668480830) • US $240.00

Perspectives on Global Biodiversity Scenarios and Environmental Services in the 21st Century
Naveen Kumar Chourasia (Government College, Bichhua, India) and Kavita Chahal (Government College, Bichhua, Chhindwara, India)
Engineering Science Reference • © 2023 • 300pp • H/C (ISBN: 9781668490341) • US $250.00

IGI Global
PUBLISHER of TIMELY KNOWLEDGE

701 East Chocolate Avenue, Hershey, PA 17033, USA
Tel: 717-533-8845 x100 • Fax: 717-533-8661
E-Mail: cust@igi-global.com • www.igi-global.com

Table of Contents

Detailed Table of Contents

Conventional nutrition research provides more information on how food promotes health and preventative medicine. Modern lifestyles encourage sustainable packaged food production. Current food technology uses fermentation, encapsulation, extraction, fat replacement, and enzyme processing to create nutritious food products, reduce, or eliminate unnecessary ingredients, add nutrients or fortified foods, alter ingredients, mask artificial flavors, and stabilize ingredients. 3D printing-compatible food ingredients promote these cost-effective, productive, user-friendly, and environmentally friendly food technology breakthroughs.

Sustainable gastronomy is a concept associated with economic, social, and environmental factors. In this book chapter, the social dimension of sustainable gastronomy is examined. In this context, supporting sustainable local agricultural activities, labeling and standardizing local food products, supporting local producers, encouraging consumers to buy local products, preserving and maintaining local food culture, and eliminating or reducing food waste can be considered sustainable gastronomy's main social objectives. For sustainable gastronomy activities to be carried out successfully, all stakeholders (producers, consumers, employees, and businesses) must be involved in the process. For this reason, transparency and information are among the most important concepts in sustainable gastronomy process. It has been determined that websites and social media tools can be effective in achieving the social goals of sustainable gastronomy activities implemented today. The main purpose of these websites is to increase the awareness of the people who visit these sites by providing information about sustainable gastronomy. According to the purposes of creating the websites, these platforms contain information on supporting producers, supporting local women, influencing demand, encouraging sustainable food practices, online training, exams, and various certification processes about sustainable gastronomy. It is seen that various contents related to sustainable gastronomy are shared on the YouTube platform. It has been determined that information about sustainable gastronomy-related activities is shared throughout these posts and that experiences, local practices, and country gastronomy are promoted within the scope

The adoption of robotics and automation tools in the hospitality sector is growing fast with robots programmed to perform a human-operated task. The shift towards the acceptance of robots in hotel kitchens with robot chefs as a part of the chefs team is a reality in some establishments. Robots are preferred by catering establishments as they reduce errors in food preparation and improve productivity; however, the initial investment for robots is high. Chefs do believe that humans could not be replaced by robots as the sensitivity of taste, and flavor may not be replicated by the robots even though standardization of recipes allows them to prepare distinctive food repeatedly. This study is about the customer's response to robots as chefs in various food establishments, and observation of robots in multi-cuisine establishments through secondary resources. The findings of the study will help hospitality chains to be better prepared for robot chefs and automation in the hotel kitchen.

Artificial intelligence (AI) is rapidly reshaping recipe development by creating new recipes, enhancing existing ones, and providing personalized recommendations based on individual tastes. The key advantage of AI is its ability to swiftly generate and evaluate a large number of recipe variants, enabling more efficient experimentation and discovering novel flavor combinations. Additionally, AI's data analysis prowess allows it to identify patterns in ingredient combinations and cooking times for recipe enhancements. AI's influence extends to personalized recipe recommendations, where algorithms analyze culinary preferences, dietary restrictions, and cooking skills to offer tailored recipes, promoting culinary exploration. However, challenges lie in ensuring the practicality and creativity of AI-generated recipes. In conclusion, AI's integration in recipe development has the potential to revolutionize cooking and eating habits, and we can expect further exciting advancements in this field as technology advances.

The food industry is one of the most rapidly evolving sectors, with new technological advancements constantly emerging. Artificial intelligence (AI) has emerged as a crucial component in this industry, revolutionizing various aspects of food production, from recipe generation to quality control. In recent years, AI has been integrated into the food industry in a big way, and its impact on the industry has been profound. An important benefit of AI in the food industry is its capacity to automate monotonous tasks, freeing up human workers for more complex responsibilities. AI also aids in recipe generation, enabling food companies to create innovative products that cater to consumer preferences. Moreover, AI plays a crucial role in maintaining quality control, guaranteeing that food products adhere to safety and quality requirements before they are delivered to consumers. To provide insights into the impact of AI in the food industry, this chapter will present case studies highlighting successful AI implementations.

In this chapter, augmented reality (AR) and its use in the food industry are evaluated under important headings. AR applications that aim to improve the customer or user experience and the contents of these applications are explained. The use and effects of these applications have been examined separately in important steps of the food industry, such as food safety, food cooking stage, food education, food promotion, and user/consumer experience. The aim is to deeply understand the use of AR and to show that the power of the technological opportunities provided by AR can be used at every step of the food industry. In addition, the perspectives of the applications developed and considered to be developed in the literature are examined to present the aspects that should be considered when planting seeds for new applications.

The chapter presents a comprehensive evaluation of the transformative role of blockchain and internet of things (IoT) in augmenting food supply chain transparency, primarily focusing on inventory management and food safety tracking. The global food supply chain, fraught with myriad challenges such as economic adulteration, microbial pathogens, and non-permitted additives, necessitates the integration of cutting-edge technology to ensure safety and sustainability. Blockchain and IoT, with their inherent qualities of traceability and transparency, can revolutionize the food industry by enabling real-time tracking, fostering responsible sourcing, minimizing waste, and bolstering consumer satisfaction. The chapter underscores these advantages while highlighting some existing challenges within the food supply chain and how these technologies can address them.

The potential of 3D printing in the food industry is enormous, but its current limitations hinder widespread adoption. Despite its ability to create intricate designs, mass produce, and reduce waste and costs, large-scale implementation remains a challenge. One major advantage of 3D printing in food is its ability to customize and personalize products, allowing for the creation of unique and complex designs. This opens

proposed methodology adopts a machine learning approach with social media commentaries classified as positive or negative sentiments and then analysed if they relate to the review ratings. The results reveal that although customers from two countries evaluate a restaurant by a similar rating on TripAdvisor, it doesn´t mean that the associated sentiment has the same value as the rating attributed. Sentiment analysis helps understand the motivations of each gastronomic consumer and realise what leads them to create reviews to express their emotions associated with the service.

Preface

In an era defined by rapid advancements and innovations, the culinary landscape finds itself in the midst of an unprecedented transformation. The confluence of technology and gastronomy has given birth to a new realm of possibilities, one where age-old traditions harmonize with cutting-edge inventions to reshape how we experience, consume, and create food. *Impactful Technologies Transforming the Food Industry* stands as a testament to this exhilarating juncture, delving into the remarkable ways in which technology has become an inseparable ingredient in the recipe of culinary evolution.

As editors of this illuminating volume, we have embarked on a captivating journey to uncover the myriad dimensions of this transformation. Our team, hailing from Nevşehir Haci Bektaş Veli University in Turkey, brings together diverse perspectives and expertise that converge to paint a comprehensive picture of the impact of technology on the world of food. We have been motivated not only by our roles as educators and researchers but also by our passion for exploring the dynamic interplay between gastronomy and innovation.

This book is an exploration, a roadmap, and a tribute to the dynamic relationship between technology and the food industry. It goes beyond the conventional boundaries of culinary literature, inviting readers to traverse the exciting terrain where the tangible and the digital meet, interact, and synergize. Our collaborative effort has resulted in a multifaceted compilation that captures the essence of this transformative journey.

Beginning with an exploration of the fundamental concept of food tech, we traverse the landscape of possibilities, from the intricate mechanisms of restaurant operations to the immersive world of dining experiences. We examine how technology has not only streamlined processes but also redefined the very essence of culinary creativity and interaction. Throughout the pages of this book, you will find insights into the rise of food delivery services, the fusion of gastronomy with virtual and augmented reality, and the ethical considerations that accompany the technological tide.

The scope of this volume extends beyond its core content. It is a dynamic resource, a guide, and a source of inspiration for a diverse audience. Whether you are a seasoned chef seeking to infuse your creations with technological flair, an aspiring food scientist eager to explore the next frontiers, a policymaker shaping the industry landscape, or simply a curious individual intrigued by the evolution of food, this book holds something of value for you.

In the following chapters, you will discover not only the ways technology is shaping the present and future of the food industry but also the synergies that emerge when innovation and tradition intertwine. Our aim is to spark conversations, provoke contemplation, and ignite a passion for exploring the ever-evolving gastronomic landscape.

ORGANIZATION OF BOOK

Chapter Overview

Chapter 1

In this chapter, we delve into the realm of modern food technology, exploring how conventional nutrition research intersects with innovative techniques to create nutritious and sustainable food products. From fermentation to 3D printing-compatible ingredients, we uncover the groundbreaking methods such as freeze drying, microwave processing, ultrasound, and more, that are shaping the landscape of food production. This chapter showcases how these technologies promote health, eliminate unnecessary ingredients, and contribute to a more environmentally friendly approach to food production.

Chapter 2

Sustainability in gastronomy goes beyond just the environment—it encompasses economic and social aspects as well. This chapter focuses on the social dimension of sustainable gastronomy, emphasizing the role of stakeholders in successful implementation. Through transparency and information sharing via websites and social media tools, we explore how sustainable gastronomy is being integrated into consumer consciousness. From supporting local producers to promoting sustainable practices, this chapter provides insights into how these digital platforms enhance the social goals of sustainable gastronomy.

Chapter 3

In a world where technology is transforming industries, the food sector is no exception. This chapter takes you on a journey through the impact of various technologies on the food industry. From the internet and social media to Artificial Intelligence (AI) and Augmented Reality (AR), we dissect how these technological advancements shape global food trends, challenges, and innovations. Whether you're a chef, engineer, or economist, this chapter provides a comprehensive overview of how technology is reshaping the future of food.

Chapter 4

The restaurant and food service industry has witnessed a technological revolution, especially in the wake of the pandemic. This chapter dives deep into the fusion of technology with dining experiences. From AI-driven precision cooking to influencer-led food recommendations, we explore how apps and digital innovations have become an integral part of the food service landscape. We also analyze the advantages and drawbacks of these technologies, guiding consumers toward informed food choices.

Chapter 5

This chapter delves into the fascinating world of food delivery apps and their impact on consumer behavior. Through a comprehensive study, we uncover the reasons behind the popularity of these apps and investigate whether they are replacing traditional dining experiences. By examining the factors influencing consumer choices, we gain insights into the simplicity and convenience of ordering food online through these apps.

Chapter 6

Robots are making their mark in the hospitality sector, especially in culinary roles. This chapter explores the dynamic between human chefs and robot counterparts in hotel kitchens. We analyze how robots contribute to efficiency and productivity while maintaining the distinct qualities of taste and flavor that only humans can provide. By studying customer responses and observations in multi-cuisine establishments, we shed light on the evolving relationship between technology and gastronomy.

Chapter 7

Artificial Intelligence is revolutionizing recipe development, from creating new culinary delights to personalized recommendations. This chapter uncovers the potential of AI in the culinary world, showcasing its ability to generate and enhance recipes. We explore how AI's data analysis capabilities unlock novel flavors and ingredient combinations. While celebrating AI's contribution to culinary exploration, we also delve into the practical and creative challenges that come with AI-generated recipes.

Chapter 8

Artificial Intelligence is rapidly becoming an indispensable tool in the food industry. This chapter dives into the profound influence of AI on various aspects of food production, from generating recipes to maintaining quality control. We highlight the automation of repetitive tasks and the role of AI in creating innovative, consumer-centric food products. Through case studies, we illustrate successful AI implementations that are reshaping the future of food.

Chapter 9

Augmented Reality (AR) is changing the way we experience food across the industry, from safety and cooking stages to education and promotion. This chapter explores the diverse applications of AR, emphasizing its potential impact on every step of the food industry. By examining existing and potential AR applications, we provide insights into the fusion of technology and gastronomy, offering a roadmap for future innovations.

Chapter 10

Blockchain and Internet of Things (IoT) are transforming the food supply chain, ensuring transparency, traceability, and responsible sourcing. In this chapter, we delve into the powerful impact of these technologies on inventory management and food safety tracking. By revolutionizing real-time tracking and reducing waste, we uncover how blockchain and IoT are fostering a new era of sustainable food production.

Chapter 11

3D printing is disrupting the food industry with its potential for customization, efficiency, and waste reduction. This chapter explores the transformative role of 3D printing, from creating intricate designs to mass production. We delve into its benefits, including culinary creativity, precise automation, and personalized nutrition. While acknowledging current limitations, we highlight the revolutionary potential of 3D printing in shaping the future of food.

Chapter 12

The emergence of 3D food printers is poised to revolutionize personalized nutrition and address unique dietary needs. In this chapter, we explore the conveniences and possibilities that 3D food printing technology offers. From flexible structures to tailored diets, we discuss how this rapidly evolving technology is set to become a necessity, catering to various nutritional requirements and preferences.

Chapter 13

Food waste is a critical issue affecting ecological, economic, and social aspects of our world. This chapter investigates the connection between food handlers and food waste in various food production settings. We analyze how lack of food safety knowledge and incomplete information contribute to waste. By examining hygiene habits and knowledge levels, we shed light on the complex relationship between food handlers and food waste.

Chapter 14

Blockchain technology holds promise in addressing food waste and promoting responsible practices. This chapter introduces a conceptual model for integrating blockchain with food waste management. We delve into the economic, social, and environmental dimensions of this integration, while also highlighting challenges and barriers. Through the BEST-Worst Method analysis, we provide insights for practitioners and policymakers to harness the potential of blockchain in reducing food waste.

Chapter 15

The rise of nanotechnology has sparked innovation in sustainable food packaging, aiming to reduce plastic consumption and improve barrier properties. This chapter explores the transformative role of nanotechnology in creating environmentally friendly packaging solutions. We discuss advancements in antimicrobial and antioxidant properties while addressing safety concerns associated with nanomaterial consumption.

Chapter 16

Molecularly imprinted polymers (MIPs) offer a promising avenue for selective food additive detection. This review article presents recent advancements in MIP-based sensors, with a focus on detecting formalin and vanillin. We delve into the synthesis, characterization, and performance evaluation of these sensors, highlighting their application in food safety and quality control. By exploring both commercial and real-time applications, we offer a glimpse into the future of MIP-based food additive detection.

Chapter 17

In the age of digital communication, electronic word-of-mouth (eWoM) plays a pivotal role in shaping the restaurant industry. This chapter delves into sentiment analysis of restaurant reviews, uncovering the motivations behind customer feedback and how it impacts competitiveness. By combining machine learning with social media commentary, we dissect the relationship between sentiment and ratings, revealing insights into gastronomic consumer behavior and emotions associated with dining experiences.

We extend our gratitude to the contributors, whose expertise and insights have enriched this volume, and to the readers, who embark on this culinary odyssey with us. As technology continues to evolve, so too will the culinary world. May *Impactful Technologies Transforming the Food Industry* serve as both a reflection of this momentous juncture and a catalyst for the exciting chapters that lie ahead.

Şule Aydın
Nevsehir Haci Bektas Veli University, Turkey

Eda Özgül Katlav
Nevsehir Haci Bektas Veli University, Turkey

Koray Çamlıca
Nevsehir Haci Bektas Veli University, Turkey

Firdevs Yönet Eren
Nevsehir Haci Bektas Veli University, Turkey

Chapter 1
Advances in Innovative and Emerging Technologies in Food Processing

Mahesh Pattabhiramaiah
Bangalore University, India

Shanthala Mallikarjunaiah
https://orcid.org/0000-0002-4699-7809
Bangalore University, India

ABSTRACT

Conventional nutrition research provides more information on how food promotes health and preventative medicine. Modern lifestyles encourage sustainable packaged food production. Current food technology uses fermentation, encapsulation, extraction, fat replacement, and enzyme processing to create nutritious food products, reduce, or eliminate unnecessary ingredients, add nutrients or fortified foods, alter ingredients, mask artificial flavors, and stabilize ingredients. 3D printing-compatible food ingredients promote these cost-effective, productive, user-friendly, and environmentally friendly food technology breakthroughs.

INTRODUCTION

Food sustains life. As early civilizations did not need to preserve food. As demand expanded, food processing and storage limits forced humans to spend more time cultivating and hunting to survive. Because food was plentiful, meal preparation was less crucial when the world population was lower (Wunderlich & Martinez, 2018). Population growth slowed harvesting and foraging (McClements, 2019). Food makers tirelessly enhance global food safety (Compton et al., 2018; Pandiselvam et al.,2019). Microbiological food safety is an issue that should be addressed by both businesses and consumers (He & Shi, 2021). The industrial revolution spawned the food preparation industry to accommodate a growing population. The food supply is essential to economic development because it facilitates trade between urban and

DOI: 10.4018/978-1-6684-9094-5.ch001

rural areas. As farms diversify and become commercial, they increase income, gain access to global markets, and create more jobs.

Food manufacturing extends shelf life. Processing eliminates food deterioration factors such as microbial activity. Molds and yeasts damage food. Nutrients, humidity, pH, oxygen, and pharmaceuticals affect microbial development—weather and pH impact dietary enzymes. Innovative processing technologies improve food safety, nutritional value, physio-chemical characteristics, operational costs, food waste, production, and efficiency (Steinhart, 2006). Green innovations deliver pleasant, nutritious, and safe meals to fulfill high-quality food demand.

Food refining technologies that don't thaw are altering the industry (Sastry, 2008). The innovative, flexible, and efficient food processing strategies include Freeze dryers (FD), Radiation, Microwave Processing (MP), Infrared drying (ID), Ultraviolet Radiation (UV), Radio frequency drying (RF), Osmotic drying (OD), High-Pressure Processing (HPP), Ohmic Heating (OH), Ozone Processing (OP), Plasma technique (PT), Ultrasound (US), Microfluidics technologies (MFT), Modified Atmospheres technology(MAT), Membrane emulsification technology (MET), Robotics technologies (RBT), 3D printing(3DP) or Additive Manufacturing technologies (AMT), Nanotechnologies (NT), Encapsulation technique(ET), Biosensors (BS), etc.

This paper reviews cutting-edge methods for reducing dangerous microbes in food. Eco-friendly technology has examined product safety. This study highlights how food science and production may produce innovative food processing and formulation techniques.

INNOVATIVE TECHNIQUES IN FOOD PROCESSING

Freeze Dryers

Freeze-drying (FD), known as lyophilization, is a standard method for making high-quality food powders and solids (Karam et al., 2016). This approach is suitable for drying thermally fragile and oxidation-prone foods since it stops most microbiological processes (Falade & Igbeka, 2007). The approach protects sensitive, non-evaporable material. Material is in a high-vacuum container on racks or belts. Food is mainly frozen before filling the dryer. Vacuum pumps condense vapor and heats the food. Rapid freeze-drying employs extended metal sheets between items and heated plates to increase heat transfer to rough regions and moisture removal. The expanding metal and plates are constructed for maximum heat transmission. Refrigerated condensers compress water. Apples, asparagus, blackberries, cocoa, garlic, guava, ginger, maple syrup, pumpkins, strawberries, tea, tomatoes, and others have been FD-treated. Vitamins and other bio-compounds may be lost during freeze-drying. This makes soup, quick drinks, desserts, and other meals with consistency requirements like freeze-dried food rehydration and porosity. FD intensification using new technologies or pretreatments has addressed various processing challenges (Bhatia et al., 2020). Freeze-drying initially dehydrates fruits by 90%. Salt absorption, water desorption, membrane pectin cell degeneration, and water crystal pore size impact freezing dried fruits' rehydration. Frozen-dried food rehydrates fast and tastes fresh. Freezing dry products reduces volume, has little chemical transition, reduces volatile component loss, lasts long, and has antioxidant qualities (Abbasi & Azari, 2009). Freeze drying is costly and energy-intensive. Due to extended freezing drying time, the product may dissolve, lose smell, and become stiff and low-rehydratable (Harnkarnsujarit & Charoenrein, 2011).

Radiation

A food preservation process deliberately exposes the product to radiation energy to modify it. Modern food storage involves irradiation. Most nations are learning that food irradiation reduces post-harvest food losses, meets quarantine requirements, boosts exports, and ensures produce hygiene (Loaharanu & Ahmad, 1991). Radiation drying employs X-rays, Gamma, microwave, infrared, and radio frequency. Pre-packaged or bulk food is irradiated with gamma, x-rays, or electrons. Electron accelerator-produced radioisotope radiation, electrons, or x-rays irradiate food (Mahapatra et al., 2007). Irradiating food requires cobalt-60 and cesium-137 gamma rays, X-rays under 5 MeV, and electrons under 10 MeV.

Fruit and vegetables last longer and taste better after irradiation. Sau et al., (2018) tested low-dose gamma rays on indoor-stored off-season guava. Shelf life, fruit percent, weight loss, sugar content, and all sensory off-season guava metrics maintained in an environment-based storage state have improved. Increasing gamma irradiation (0.5, 1.0, 1.5, and 2 kGy) decreased sesame seed IVPD, according to Hassan et al., 2017. Radiation of potatoes and onions, delays in mango and papaya maturity and senescence, mango fruit fly and seed weevil infections, and fruit surface pasteurization have been researched. Fruit and vegetables' technical properties have improved by X-rays, gamma, and electron beams. In several nations, irradiation is replacing chemical fumigation for quarantine.

Microwave Processing

Food processing and industrial operations employ microwave technology for cooking, heating, and food safety. Microwaves are used in food production for drying, pasteurization, sterilizing, thawing, tempering, and numerous other purposes (Gupta & Wong, 2007). Microwave heating is popular in food processing owing to its quick heating, substantial reduction in cooking time, reasonably uniform heating, safe handling, easy operation, and maintenance-free operation (Zhang et al., 2006; Salazar Gonzalez et al., 2012). Microwaves operate at 300 MHz–300 GHz. Commercial microwave systems employ 915 MHz and 2.45 GHz frequencies and serve users in enclosed cavities (Datta & Anantheswaran, 2000). Microwave heating increases as materials absorb and convert microwave energy into thermal energy. Continually polarised dipolar molecules adjust an electro-oscillating field on liquid molecules. Evaporating electromagnetic field energy cooks food in the microwave cavity. This quickly elevates food product temperature. Polar solvent dipole rotation and dissolved ion conductive mobility transmit microwave energy through molecular friction (Oliveira & Franca, 2002).

High surface temperatures make microwaves unreliable and deplete nutrients (Kodandaram et al., 2020). Microwaves are used in homes to heat and thaw food quickly and in industries to dry, bake, pre-prepare, and cook cereals, meat, and meat products, defrost food, bleach vegetables, meals, pasteurize and sterilize fast food/liquid food, and other food products (Oliveira & Franca, 2002; Liu & Lanier, 2016). The microwave's heating and vacuum's low temperature and rapid transport phenomena dry the product, increasing energy consumption and product value (Duan et al., 2010). Microwave food manufacturing advantages include baking, concentrating, curing, cooking, drying, enzyme inactivation (blanching), finish drying, freeze drying, precooking, pasteurizing, heating, puffing and foaming, solvent removal, sterilizing, thawing, and tempering. Giri and Prasad (2007) demonstrated that microwave heat drying carrot cubes shortened drying time by 25-90% and improved quality. Sharma and Prasad (2001) reported 91% faster microwave convective drying. Kasuriya and Atong (2004) examined continuous microwave and traditional oven drying power for bread making.

Infrared Drying

Infrared radiation (IR) is electromagnetic energy between 700 nm and 1 mm. Infrared rays are shorter than radio waves. IR bands from 300 GHz to 400 THz are higher than microwaves but lower than visible light. Food is dehumidified using infrared heating. IR energy falls between UV and radiowaves. IR radiation's spectrum includes near-infrared (NIR) (700–1400 nm), mid-infrared (MIR) (1400–3000 nm), and far-infrared (FIR) (3000–10,000 nm). Food production is optimal in the FIR because most food components absorb radiation (Wang et al., 2014; Rastogi, 2015). (Wang et al., 2014; Kodandaram et al., 2020). Near-IR is used for industrial heating due to its higher temperatures. IR light accelerates and spins food molecules. Relaxing atoms emit heat. IR radiation impacts food and water the most. IR impacts protein, lipid, and polar carbohydrate groups. Drying, roasting, baking, and microbial inactivation require IR thermal treatment. IR heating boosts food production and shelf life. Unlike typical drying processes, infrared drying only heats what has to be heated. Infrared exposure charges food atomically and molecularly in electronic, vibrational, and rotational states. Carrots, sweet potatoes, and tomatoes have been infrared-dried (Bai et al., 2012). Food companies utilize IR heating to inactivate lipoxygenase, lipases, α-amylases, and enzyme products that provide fragrance in fruit and vegetables and bacteria, spores, yeast, and mold in solid and liquid food (Huang et al., 2013). IR heating can dehydrate, boil, bake, and freeze-dry (Huang et al., 2014). Infrared drying has been utilized for food thawing, baking, blanching, roasting, and heating (Ozkoc et al., 2009). Infrared-assisted drying can address traditional challenges (Nathakaranakule et al., 2010). Most research is on solid foods, but liquid food processing needs to be investigated (Rastogi, 2012). Instant meals cannot employ IR-dried vegetables because they rehydrate poorly (Krishnamurthy et al., 2008). Apple slices were dried by infrared and convection. Infrared drying is cheaper and warms material 50% quicker. Infrared radiation increased banana porosity (Leonard et al., 2008), onion rehydration, pineapple, and potato color changes despite high drying rates, dried blueberry firmness, and did not affect product consistency when used with freeze drying, as in sweet potatoes, yam, and strawberries (Shi et al., 2008).

Ultraviolet Radiation (UV)

UV radiation has long been the sun's major bactericide. UV light may disinfect microbes without heat (Sastry et al., 2008). Air and thin fluid films can be sterilized due to the limited penetration depth. Higher dosages diminish taste and odor till disinfection. Low-dose radiation extends food shelf life without harming quality (Barkworth, 1938). UV-A (315–400 nm), UV-B (280–315), and UV-C (280–400) are ultraviolet radiation bands in the electromagnetic spectrum (200–280 nm). UV-C is germicidal. Deactivation peaks at 254–264 nm on a bell-shaped curve. UV inactivation commonly occurs at 254 nm. Several studies suggest that UV-C radiation penetrates cell membranes and damages DNA owing to thymine dimers, inhibiting microbe transcription and replication and killing cells during disinfection (Bintsis et al., 2000). UV technology produces safer, higher-quality food than heat processing (Ramesh et al., 2016). UV sterilizes without decreasing nutrients.

Titanium oxide (TiO_2) appears safe and inert. TiO_2 kills airborne germs, fights ambient air pollution for cancer therapy, protects plants, purifies water, cleans drinking water, and sterilizes food packaging content and surfaces. It colors candy, transparent sauces, bandages, and powdered meals (Ammendolia et al., 2014). TiO_2 photocatalysts activate or kill bacteria, fungi, algae, and viruses. ROS damages cell processes and structures by chemically altering or cleaving genetic material and cell membranes (Kim et al., 2011). UV-assisted TiO_2 photocatalysis (TUV) of bacterial cells damaged plasma membranes, DNA, and internal organelles. It releases ions and kills cells by disrupting cell walls (Hitkova et al., 2012).

Radio Frequency Drying

Radiofrequency (RF) heating is volumetric and allows fast and improved heat transmission into food products. RF heating, a high-energy radio wave, is produced by a generator heating dielectric material. Heat is created fast in the food center (Maloney & Harrison, 2016). Electromagnetic 1–100 MHz waves penetrate food and react with ions, atoms, and molecules to produce heat energy—electromagnetic energy from 3 kHz to 300 GHz powers RF heating. Commercial, scientific, and medical use is limited to 1356, 27.12, and 40.68 MHz (Marra et al., 2009). RF sterilizes, thaws, disinfests, and dries (Bedane et al., 2017). A rotating electrical field heats food product ions and dipolar molecules. Dipolar substances move dipolar, whereas positive and negative ions move ionically. RF heating requires dipole spinning and ionic conduction. Food manufacturers employ RF heating and dry food instead of using conventional techniques. Due to its speed, affordability, deep thermal penetration, and quality control, RF heating is ideal for heating novel foods. Due to its deep penetration, heat dispersion, and energy efficiency, RF heating is a cutting-edge food application technology. Thus, RF joined the food industry to sterilize, pasteurize, and disinfect fruits, dry nuts, and other agricultural products after baking. RF cooling was investigated in the 1960s. Medical RF heating began in 1895. It was initially used to bleach, heat, and dehydrate food (Moyer & Stotz, 1947). Eggs, vegetables, fruit, fish, and meat (with and without bones) were tested with RF. RF food preparation uses dielectric heating. 3 kHz–300 GHz are radio frequencies. US RF bands include 13.56, 27.12, and 40.68 MHz. RF processing alters dielectric material electrical fields. Electrodes transmit RF energy from triode valves. Food molecule friction generates dielectric heat from alternating high-frequency electric fields. Fast, bulk dielectric heating. RF processing fixes most foods and dielectric materials. Water's dipole moment attracts an electric field. Field adjustments rotate and align polar molecules. The heat from molecules boosts food temperature. Ionic conduction warms food via pulsing dissociative ions.

Viscosity, temperature, chemical composition, and other physiological aspects impact food component dielectric characteristics (Uyar et al., 2015). This is used for continuous and batch heat processing in the food industry (Huang et al., 2016); fast defrosting of frozen fish, beef, and other processed foodstuffs; disinfecting, baking, and sanitizing dry food (seeds; cereals; dried fruits; legumes); and sterilizing solid or viscous packaging products. RF waves stir bound water, making it perfect for spices and milk powders. RF systems may be activated and deactivated instantly using low dynamic feed technology (Chen et al., 2019). Since most packaging is plastic bags or packets, food should be pasteurized using RF waves after processing (Huang et al., 2016).

Osmotic Drying (OD)

Osmotic dehydration (OD) dehydrates biological structures like fruits and vegetables. This approach does not impact product transformation since chemical composition changes (dry content rises, and water behavior alters considerably (Escriche et al., 2000). This approach increases food product quality, notably for plant raw materials (Rastogi et al., 2002; Kowalska, 2006). OD is suggested for air-drying or freezing medium-moisture food (Lewicki & Lenart, 2015). Talens et al., (2002) reported that the pretreatment improves kiwi slice color uniformity, mechanical characteristics, and drip loss. Because it is inexpensive, easy, and utilizes low temperatures, OD is a good food preservation solution (Alfaro et al., 2018). OD encompasses air pressure, vacuum osmosis, high-pressure osmotic vacuum pulse dehydration, microwave vacuum, vacuum-ohmic heating, and ultrasound-assisted drying (Lagnika et al., 2018). Dehydration stabilizes the membrane with a semipermeable membrane by removing water from a lower salt content to a higher concentration (Tiwari, 2005). Food retention is enhanced by OD's reduction of fruit and vegetable water activity. OD is favored for color, smell, nutritional, and flavor retention. This drying process cannot ensure shelf-stable low-humidity material. After air, freeze, or vacuum-drying, the osmotic material should be further processed to make a stable product or dehydrated for proper canning, freezing, and decreased processing. OD decreases water by soaking cellular water-containing particles in concentrated water solutions. Water removal begins with a solution-intracellular fluid concentration gradient. The membrane's semi-permeability blocks liquid from entering cells. Osmotic drying involves soaking food in a hypertonic solution to remove water (Figure 1). Thus, vegetable and fruit tissues absorb solution solutes (Fito et al., 2001). Osmosis changes a product's physical, chemical, nutritional, taste, and structure (Flink, 1979). Since their osmotic pressures differ, the meal solution is mixed with water. Food absorbs osmotic fluid. Feed composition, solid sizes and configuration, mass-exchange area, system pressure, and ODs affect water loss (Herman et al., 2013).

Figure 1.

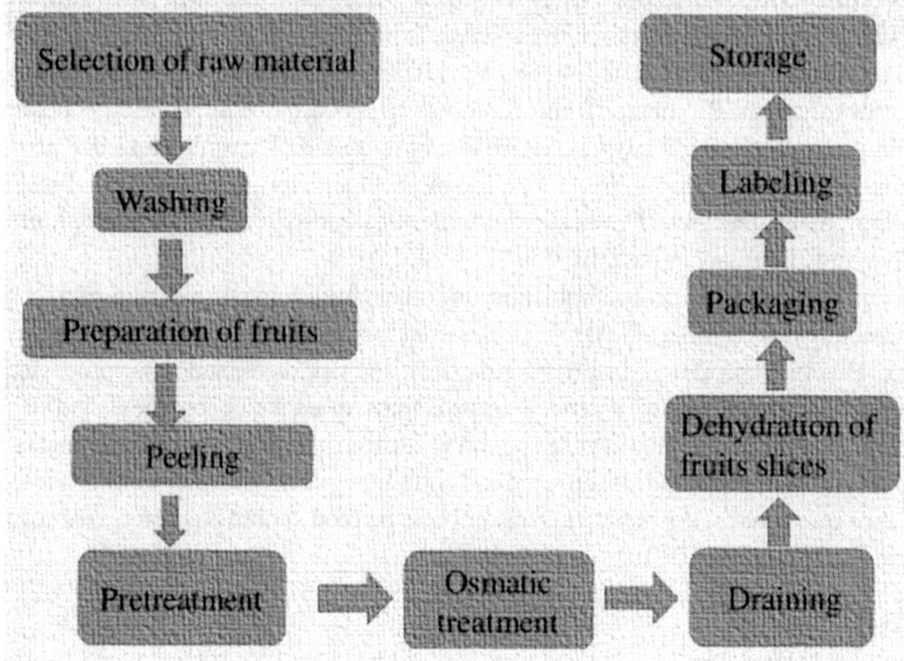

Most osmotically active solutes are monosaccharides, disaccharides, and sodium chloride. Fruits use sugar, and vegetables, fish, and meat use salt chloride. Glucose, lactose, fructose, dextrose, maltose, polysaccharide, maltodextrin, maize starch syrup, milk, sorbitol, ascorbic acid, citric acid, and calcium chloride are other osmotic agents. Hawkes and Flink 1978 and Islam and Flink 1982 employed sucrose to minimize solvent costs and enhance osmosis efficacy using mutual solute combinations. Cost, organoleptic compatibility, and preservative activity determine the mixture's choice. Osmotic dehydration combines sugar syrup with fruit or cubes and salt or vegetable salt (sodium chloride). Water, nutrients, vitamins, fruit acids, etc., pass from fruit or vegetables to solution. Fruit and vegetables absorb sugar and salt.

High Pressure Processing (HPP)

High-pressure processing (HPP) employs 600 MPa / 87,000 psi hydrostatic pressure to cook food in a water-resistant packaging for a few seconds to a few minutes at room temperature (Banerjee & Verma, 2015). High hydrostatic pressure (HHP), ultra-high pressure (UHP), pascalization, or cold pasteurization inactivates numerous enzymes without degrading flavor or nutrients (Rastogi, 2013). Cold pasteurization preserves food and prevents rotting and harmful bacteria. Food processing uses high pressure to improve food safety and taste. HPP can be utilized in food processing to make high-quality food (Bansal et al., 2015). This application has successfully tested beverages, juices, vegetables, nuts, meat, seafood, and ready-to-eat meals (Tribst, 2016). The approach has been used to preserve meat, yogurt, and cheese, minimize milk allergy, and hasten maturation (Pingen et al., 2016). HPP firmly seals flexible packages using the Isostatic rule (Pascal principle) and the Le Chatelier-Braun concept. Isostatic 300-900 MPa pressure preserves food (Patterson, 2005). Flavor, color, nutrition, and chemical contents are unaffected. Protein molecules unfold only when high pressure changes non-covalent connections. Proteins, carbohydrates, polysaccharides, and nucleic acid can be altered by high pressure, but amino acids, vitamins, and taste compounds cannot (Rastogi et al., 2007). HPP processing is typically done at room temperature, 400–600 MPa. Many HPP uses need lengthier processes to deactivate pathogens and decay bacteria, but pressure treatments alone cannot ensure food safety.

High pressure at low, ambient, and high temperatures generates unique process and product possibilities (Sevenich et al., 2013). mpHHP inactivated microorganisms in fruit juice, dairy products, whole egg liquid, beef, and fish better than single pulsed HHP (Buzrul, 2015). mpHHP inactivated E. coli O157:H7 and Salmonella enteritidis in hamburger meat and chicken fillets better than standard HHP (Buzrul, 2015). mpHHP inactivates food enzymes and prolongs shelf life (Buzrul, 2015). Sensory and nutritional content is unaffected by flavor and scent (Naik et al., 2013). HPP of milk is gentler than standard shelf-life-extending procedures since B1 and B6 vitamin levels did not alter, according to Sierra et al., (2000). Huppertz et al., (2005) discovered that warm high-pressure-treated milk produced more cheese curd. High-pressure blanched frozen potatoes used 40% less oil for frying than thermally blanched ones (Knorr et al., 1999). HPT decreased potato frying oil consumption by compressing and decompressing moisture.

Ohmic Heating (OH)

Ohmic heating, or Joule heating, electro-heating, direct electrical resistance heating, or electroconductive heating, is a novel terminal processing technology that uniformly warms food using electricity. Electrical conductivity is critical to ohmic burner design. Unlike microwave and RF heating, energy flows through food. (Wongsa-Ngasri and Sastry 2016). Ohmic heating is microwave heating despite the wavelength difference (Yin et al., 2018). Conductive material warms fast when electricity is transmitted between two conductors (Ramaswamy et al., 2016). OH, which distributes temperature homogeneously inside the material, is beneficial for high-viscosity or particle meals. Pasteurises and sterilizes food (Vicente et al., 2016).

Cell membrane electroporation enhances extraction and decreases gelatinization temperature and enthalpy under OH. It cooks fast and retains color and nutrition. It affects water absorption, solubility, heat, and pasting. Ohmic-heated pre-gelatinized starches reduce processing energy. Ohmic techniques heat fruit slices in syrups, stews, sauces, soups, and heat-sensitive liquids (Wongsa-Ngasri & Sastry, 2016). Quality, safe, and valuable ohmic-heated food. Defrosting, preservation, microbial inactivation, pasteurization, sterilization, extraction, dehydration, evaporation, fermentation, peeling, bleaching, extraction, packaging, value-added process, and heating food at serving temperature (Fowler & Park, 2015). Ohmic heating reduces protein and milk processing. Pereira et al. (2007, 2008) showed no deleterious effects on goat milk flavor from high-temperature short-time (HTST) ohmic pasteurization or conventional treatment. OH at 72°C for 15 sec with 220 Volts sterilizes milk poorly (Sagar & Suresh Kumar, 2010).

Ozone Processing

Lightning or high-energy UV radiation creates natural ozone in the atmosphere (Jakob & Hansen, 2005). Ozone-based food preservation improves food safety without compromising taste or value (Guzel-Seydimet al., 2004). By adding an oxygen free radical to molecular oxygen, ozone (O3) is created. In 1982, ozone was GRAS for bottled water purification. Since 2001, the FDA has licensed their food for antibacterial use. Ozone has sterilized drinking water and eliminated unpleasant odors and organic/inorganic impurities for decades (Muthukumarappan et al., 2000). In 1997, the USFDA declared ozone GRAS for food processing (Graham, 1997). Ozone is promising in food (Khadre & Yousef, 2001). Ozone, a powerful oxidizer, kills microorganisms without leaving residue in food. Aqueous ozone may disinfect food-producing plant water and sanitize fruit and vegetable surfaces (Muthukumarappan et al., 2000; Khadre & Yousef, 2001; Garcia et al., 2003). Short-term exposure to low-concentration ozone inactivates microorganisms (Kim et al., 1999). Due to its rapid decomposition, ozone in food leaves no trace (Guzel-Seydim et al., 2004).

Restaino et al., (1995) found that pure water gram-negative bacteria are marginally more susceptible to ozone than gram-positive. Khadre and Yousef (2001) found ozone-resistant gram-positive bacteria. Lee and Deniniger (1999) found that gram-positive bacteria predominate in ozonated drinking water. Bacterial spores survive ozone better than vegetative cells (Kim et al., 1999). Ozone is used in drinking water purification, aseptic packaging sterilization, fresh produce disinfection, and cold storage food preservation. Gas or aqueous ozone generators are inexpensive (Balasubramaniam et al., 2014). Ozone is a safe, effective disinfectant in food and beverage goods and equipment (Sopher et al., 2002). CD transfers a dry, dust-free, oil-free, oxygen-containing gas between two electrodes to create ozone. Dielectric and grounded electrodes are used. Diatomic oxygen is split, and free radicals create ozone. Ozone-generated contactors disinfect water in ozone-based systems. Bubble diffuser chambers and turbine-agitator reactors are significant contractors for ozone treatment (Bablon et al., 1991). Food treatments include hygiene, controlling spoilage, overripening, and decontaminating (Baratharaj, 2009). Ozone processing is used in food storage, washing fruits and vegetables, disinfecting stainless steel surfaces in Ultra high temperature (UHT) plants, dairy biofilms on stainless steel surfaces, Clean in place (CIP) systems, insect management, and beverage manufacturing air quality. Ozone processing is eco-friendly, germicidal, and requires no chemical storage.

Plasma Technique

Plasma is the word used in physics and chemistry to describe an ionized gas. Atmospheric Plasma is the fourth state of matter and the dominant state in the universe, consisting of positively and negatively charged ions, activated neutral species, and free electrons. Ionizing gas energy creates Plasma. Electrical, thermal, radioactive (gamma-radiation), optical (UV-light), and electromagnetic radiation can ionize gases to create Plasma (Pankaj & Keener, 2017). Gases regulate plasma performance. The gadget might employ air, nitrogen, argon, helium, or neon. Electricity, microwaves, and lasers drive it. Ionized plasma-generated gases provide conditions without thermal equivalent (Pankaj et al., 2014). Cold Plasma, a new non-thermal food processing method, uses active, cold-reactive gases below 40°C to clean package surfaces and inactivate contaminating bacteria in food, animals, fruit, and vegetables. Electricity and carrier gas—air, nitrogen, oxygen, or helium—sanitize this adaptable system. The main processes are UV light and reactive chemical compounds of cold plasma ionization. Plasma circulates items (e.g., ultraviolet light decontamination) without shadows to treat all parts of a substance.

Many cold plasma devices are created under air pressure or in low-pressure treatment rooms. Labeling jam jars, printing on glass bottles, or sealing liquid products requires reliable, low-cost material processing. In composite product production, ambient pressure plasma pretreatment allows the handling delicate coating materials. To preserve high-gloss plastic-coated surfaces, rests in the bonding surface region should be considered when packaging is handled quickly, and adhesive bonding is needed. Glass jar labels are pretreated with ambient press plasma. Cold Plasma decontaminates food packaging and processing. This also dry disinfects meat, fish, milk powder, and sprouted seeds (Korachi et al., 2015; Lee et al., 2015). Cold plasmas have improved the functional characteristics of banana starch, rice starch, zein, pea protein isolates, wheat rice, and Basmati rice through surface and molecular modification (Ezeh et al., 2018). CP increases swelling efficiency and lowers the cooking temperature, viscosity pasting, water solubility, and food grain retention. Lipid oxidation may be promoted and acceptance diminished by adding cold plasma oxygen to fat-rich diets (Gavahian et al., 2018). Cold Plasma is used in package production to prevent postal package contamination. Industrial production may use in-package testing (Misra et al., 2014). Indirect cold plasma treatment of RTE meat in PE bags reduced Listeria innocua numbers (Rød et al., 2012).

E.coli, salmonella, *L.monocytogenes*, yeast, and mold species on meat surfaces may be removed by CP treatments on beef, pig, and chicken (Rød et al., 2012; Misra & Jo, 2017). Milk processing with cold plasma may preserve color, pH, flavor, and nutritional content. Alkaline phosphatase enzymes and contaminated bacteria were inactivated in seconds (Coutinho et al., 2018). Cold plasma therapy tried berries, cherries, apples, melons, Kiwi, etc. CP treatments on F&V surfaces affected food component pH and acidity. Plasma species react with surface wetness. During storage, treated items increase in texture (firmness) and color (Kovačević et al., 2016). Plasma treatment reduced Salmonella and *E.coli* on apple, melon, and mango surfaces (Tappi et al., 2016).

Ultrasound

The potential of ultrasound advancements in food processing is still being researched. As defined by Yusaf and Al-Juboori (2014), ultrasound consists of mechanical vibrations propagating through solids, liquids, and gases at a frequency greater than 20 kilohertz (kHz). “Ultrasonics” or “Power Ultrasound” refers to sound frequencies between 20 and 100 kilohertz. Many manufacturing processes benefit from the use of ultrasound. Ultrasonic technology is rising to streamline production, improve consistency, and bolster product safety (Knorr et al., 2011). Ultrasonic frequencies have several practical applications in the food processing industry. Mighty low-frequency waves may be found between 20 and 100 kHz. The physicochemical characteristics of food, such as its acidity, hardness, sugar content, ripeness, etc., may be evaluated using high-frequency ultrasound, also known as a power ultrasound. Because of the friction, shear, and temperature change it causes in the medium through which it propagates, low-frequency ultrasound can be used to inactivate microorganisms in food (Piyasena et al., 2003) and to alter the physical and chemical properties of food (Soria & Villamiel, 2010). Free radicals and chemical reactions at frequencies between 100 kHz and 1 MHz are at the heart of sonochemistry. The frequency range of ultrasound applications is typically 20 kHz to 500 MHz (Yusaf & Al Juboori, 2014). Sinusoidal waves at low and medium frequencies propagate sound. When elastic mediums are vibrated, acoustic cavitation occurs. Pre- and post-harvest, ultrasonic technology controls the quality of vegetables and fruits, cheese, agricultural cooking oil, bread, cereal powder, bulk, emulsified fats, food gels, and vented and frozen feed. The pace, amount, and type of protein aggregation in honey and any added adulterants may be determined.

Ultrasound creates acoustic pressure (AP) in a fluid, which differs from hydrostatic pressure. Ultrasound penetrates profoundly and produces thermal and nonthermal effects through alternating current (AC) via a piezoelectric crystal. Its versatility, knowledge, and research discoveries make it a pioneering technology. The system reduces processing time, saves resources, and improves food shelf life and uniformity (Chemat et al., 2011). Ultrasound improves flavor, homogeneity, and energy efficiency over heat treatment. Generators, transducers, and application frameworks make ultrasonic waves. Based on ultrasound sensitivity, the food sector uses two ultrasound technologies. Low-intensity ultrasound needs less than 1 W / cm2 and does not change the object's physical or chemical characteristics. Low-intensity, non-destructive ultrasound is most commonly used to analyze food composition, structure, and physical condition. The flow rate can also be monitored. High-intensity applications employ power levels from 10 to 1000 W/cm2, which can cause physical damage or chemical reactions like oxidation. High-intensity ultrasound may kill germs and enzymes, produce emulsions, tenderize meat, increase drying, extract, and filter. Ultrasonic atomization requires a fluid feeder via a vibrating horn and ultrasonic nebulizers, which act at higher frequencies and generate a "fountain" structure in a thin liquid film between ultrasonic nozzles that create an aerosol.

Ultrasonic nozzles are more significant than ultrasonic nebulizer droplets. Centrifugal and spinning nozzles are currently employed in industrial spray dryers. However, they do not use ultrasonic nebulizers, which give high speeds with limited particle size and benefit from compact spray dryers (Ashokkumar et al., 2010). Rastogi (2010) stated that immersion in a high-osmotic solution is utilized to dehydrated meals. Fernandes et al., (2008) assessed melon osmotic dehydration after ultrasonic pretreatment at air pressure. Tiwari et al., (2009) found that ultrasonic therapy preserves orange juice vitamin C better than heat treatment. Mahdi et al., (2007) demonstrated that ultrasound might make 0.5-mm ultrasonic emulsions via microfluidization. Commercial dairy production uses ultrasonic cutting instruments primarily. Ultrasound is non-toxic, energy-efficient, and environmentally beneficial. Food and science employ ultrasonics to analyze food composition (fruit, vegetables, and dairy products) to discover international contamination in canned and dairy goods.

Microfluidics Technologies

Whitesides (2006) defines *microfluidics* as systems that process or alter nanoliters to attoliters of fluids. Microfluidics manipulates tiny fluid volumes using hundreds of micrometer-sized channels. Microfluidics has a cheap cost, rapid reaction and analysis time, robustness, minimal emissions, and high throughput (Whitesides, 2006; Chen et al., 2013). Miniaturization lets devices employ minimal chemicals and integrate multiple analytical procedures (Mairhofer et al., 2009). Microfluidic technology includes physics, chemistry, biochemistry, micro/nanotechnology, engineering, and biotechnology. Electrical kinetics, electro-hydrodynamics, and thermal efficiency in microfluidic devices solve critical technological difficulties. Microfluidic instruments' laminar flow, huge surface-to-volume ratios, and micrometer-scale surface tension and capillary effects make handling and analyzing complicated samples more efficient. Microfluidics is used in the agriculture and food industries for food safety, processing, and animal sciences (Neethirajan et al., 2011). Its simplicity, reliability, flexibility, accuracy, convenience, pace, and robustness make it one of the most exciting diagnostic technologies (Mao & Huang, 2012). Microfluidics aids traditional animal food preparation and tracking. High-data analysis is complex in embryogenesis-based biotechnology investigations. Microfluidic instruments efficiently monitor fecundation and embryogenesis and handle microorganisms (Hamon et al., 2013).

Modified Atmospheres Technology

Modified-atmosphere Packaging (MAP) preserves perishable foods in a modified air composition to slow physiological and microbiological deterioration. MAP is a complicated way to change gaseous composition in active or passive products. Fresh and least processed respiring feed items require a different MAP than non-respiring goods. To decolorize beef brown or prevent microbiological spread in cheese and confectioneries, non-breathing food is stored in oxygen-free modified atmospheres (MA). High-gas-barrier films or laminates prevent the package from exchanging gases (especially O_2), which would create a less ideal environment. Respiring products maintain consistency by remaining metabolically active after processing. An appropriate packaging method must supply O_2 to allow the environment to include a simple metabolism to prolong the shelf life of respiring items via MAP. Various commodities (type and maturity) have various metabolic needs, which rely on storage temperature and processing (cutting, grinding, slicing). The product's intensive respiratory activities need the adjustment of the gas environment within the container during storage and expert knowledge about each product's packaging arrangement to maximize quality shelf life.

Modified atmospheres technology improves food shelf life following packing and processing. Most MAPs include flexible or semi-flexible frames, bags with and without perforations, and variable thicknesses (Mangaraj et al., 2009). The most common polymer films for MAP are synthetic polyolefins, low-density polyethylene (LDPE), linear low-density polyethylene (LLDPE), high-density polyethylene (HDPE), polypropylene (PP), polyvinyl chloride (PVC), polyesters, polyethylene terephthalate (PET), polyvinylidene chloride (PVDC), ethylene-vinyl alcohol (EVOH), polyamide (Nylon), PVOH, and ethylene vinyl acetate (Sandhya, 2010; Soltani et al., 2015). This method uses a gas-air combination inside Packaging to maintain product uniformity and extend shelf life. Technology has helped food companies extend shelf life and keep prices down.

MAP improves the product environment to prevent natural deterioration and microbiological spoiling (Caleb et al., 2013). In a MAP system, the dynamic interactions between metabolic and biochemical processes of the substance being packaged on one side and Packaging on the other cause the concentration of gases in the packaging headspace to vary. O2 from the environment enters the package to replace product use, while the other gases are expelled via the package system (Sandhya, 2010; Selcuk & Erkan, 2015; Castellanos et al., 2016). Combining these two procedures ensures that a steady volume of gases enters the packing headspace and maintains equilibrium levels to protect the product (Caleb et al., 2013). After packaging fruit or vegetables, the headspace O_2 concentration decreases, and the CO_2 concentration rises before the constant values reach equilibrium. In low O_2 and high CO_2 circumstances, plant metabolism and microbial activity reduce product shelf life. Storage temperatures increase product shelf life, which slows process growth (Sandhya, 2010). Low O_2, high CO_2, and low temperatures significantly reduce product-related and other potentially harmful bacteria. In these unfavorable conditions for microbial growth (low O_2, high CO2, and low temperature), the product, natural microflora, and active and potentially pathogenic microbes compete to inhibit the latter's production (Bourke & O'Beirne, 2004). MAP's produce storage is nonsterile. Fruits and vegetables have a complex microbiota of bacteria, molds, and yeasts that ruin food but are safe. However, product packaging can alter microorganisms' microenvironment, disrupting this healthy equilibrium.

The respiration rate of the material and the flow of gases into the packing medium determine MAP without affecting the initial gas composition (Caleb et al., 2012). To maintain ideal gas concentration throughout time, passive MAP should be established within the package using natural respiration and film permeability (Charles et al., 2003). Although active MAP is a rapid gas replacement technique, gas scavenging via absorber establishes a desirable gas mixture in a package (Kader & Watkins, 2000; Charles et al., 2003). This approach relies on it. O_2, CO_2, and ethylene scavengers must be utilized in processed food (Philips, 1996; Sandhya, 2010). CO_2 absorbers can avoid harmful CO_2 levels (Kader & Watkins, 2000). MAP preserved broccoli, cauliflower, carrots, and garlic. Packaging affects broccoli and cauliflower consistency (Artés et al., 2001). Micro or macro-perforated polyethylene and polypropylene bags were utilized to preserve these goods by allowing respiration and transpiration and achieving appropriate equilibrium gas levels (Serrano et al., 2006). MAP preserves cabbage, cucumber, onion, spinach, and tomatoes (Plestenjak et al., 2008). In both cases, refrigeration temperatures were employed, although the packing mechanism may have a preservation impact at ambient temperatures, albeit slightly lower (Caleb et al., 2013; El-Ramady et al., 2015). MAP methods preserve apples, bananas, blueberries, feijoa, grapes, guava, mandarin, medlar, papaya, pear, pomegranate, and strawberry (Selcuk and Erkan 2015; Castellanos et al., 2016). In both cases, the product's shelf life can be doubled.

Membrane Emulsification Technology

An "emulsion" is a structure formed by spreading each of the two tiny droplets (usually oil and water) of one immiscible liquid into the other (McClements et al., 2007). Emulsions are distributed systems of insoluble liquids like water and oil. Food processing requires emulsification. Food firms use emulsion science's fundamental ideals to improve quality and productivity. High-pressure homogenizers, rotor/stator ultrasonic devices, stirred vessels, colloid mills, and toothed disc mills are used to make emulsions. These machines use high shear stress in the dispersion zone to distort and disrupt big droplets. Thus, high energy inputs are necessary, and shear-sensitive compounds like proteins and starches may not work.

Membrane emulsification has received more attention in recent years. Applying pressure over a membrane pushes the scattered process into the continuous phase. Water-in-oil (W/O) and oil-in-water (O/W) emulsions are categorized by their continuous and dispersed phases. In most food emulsions, the "disperse phase" produces droplets, and the "continuous phase" surrounds droplets and the interface (McClements et al., 2007). Membrane emulsification processes emulsions at 104–106 J m-3, compared to 106–108 J m-3 for mechanical methods. Membrane emulsification produces 5 O/W, W/O, and W/O/W.

Coarse, micro, and nanoemulsions are classified by droplet size and stability (Komaiko & McClements, 2016). Traditional or macroemulsions are coarse. Thermodynamically metastable particles are > 200 nm. Destabilizing influences progressively degrade them. Traditional emulsions are optically turbid because the droplet size is near the wavelength, dispersing light and making it opaque. Thermodynamically stable microemulsion droplets are <100 nm in diameter. However, even little changes in composition and temperature might affect their stability. Microemulsions develop spontaneously due to their lower free energy than phase-separated components. They are optically transparent because their particle size is less than the wavelength, and light spreads poorly (Anton & Vandamme, 2011). Nanoemulsion droplets are typically 200–100 nm (Saifullah et al., 2016). Phase separation over time makes nanoemulsions thermodynamically metastable, like stable emulsions. The lessened attractive force between microscopic droplets prevents gravitational separation and droplet aggregation, making nanoemulsions kinetically stable (McClements & Rao, 2011). Nanoemulsions are unaffected by temperature and pH, unlike thermodynamically stable microemulsions. Their preparation requires fewer surfactants. Nanoemulsion droplet size impacts its rheology, stability, and optical characteristics. Nanoemulsions are better for diverse purposes than microemulsions.

Nanoemulsions are sustainable food processing and packaging. Nanoemulsions encapsulate antioxidants and nutraceuticals. Active ingredients. It controls food flavor components (McClements & Rao, 2011; Ines et al., 2015). Nanoemulsion enhances bioactive substance solubility, gastrointestinal release, and cell absorption (Chen et al., 2006; McClements & Rao, 2011). To extend the shelf life of fruits, vegetables, dairy products, confectionaries, and meat, nanoemulsion-based edible nanocoatings with antimicrobials, antibrowning agents, antioxidants, coloring components, enzymes, and flavoring agents can be applied. Nanoemulsion coatings minimize food humidity loss, oxidation, and moisture and gas exchange (Rojas-Graü et al., 2009; Salvia-Trujillo et al., 2015; Donsi, 2018). The "disperse phase" produces droplets, the "continuous phase" that surrounds droplets, and the interface make up most food emulsions (McClements et al., 2007). Non-polar and polar molecules remain on the interface in oil and water, whereas amphiphile molecules prefer to position. Many natural and processed foods contain emulsions or are emulsified during processing (McClements, 2007). Milk, cream, butter, margarine, juice, soup, cakes, pastries, mayonnaise, cream liqueur, coffee creamers, sauce, and ice cream are emulsion-type foods (Dickinson & Stainsby, 1982; Charcosset, 2009). However, some meat products, such as bread dough, require emulsions, which might be complex (Dalgleish, 2004). Emulsion-based foods vary in color, fragrance, shape, taste, shelf life, and physicochemical and organoleptic qualities. Strawberry yogurt is a pink viscoelastic gel, milk is white, and margarine is semi-solid yellow.

Robotics Technologies

Food companies may use them. Robots can process, store, palletize, and serve food. Thus, food robotics has grown tremendously in recent years. Thus, robot kinematics, logistics, cleanliness, economic performance, human-robot interactions, health and security, and service and maintenance are crucial. Robots seed, water spray, harvest, chop, prepare, and package food (Sun et al., 2016). Robots are used in meat processing and bakery product identification. Robots clean, count, fill, and organize bottles on a conveyor belt in the beverage sector (Saravacos & Kostaropoulos, 2016). Advanced vision systems with many HD cameras identify and analyze flaws (Saldaña et al., 2013). Robots enhance efficiency by 25% for food businesses. Food markets execute at different speeds (Gebbers and Adamchuk, 2010).

Several food firms and packaging machinery manufacturers use robots in dairy, poultry, bread, confection, snack, and beverage activities. Robots handle unwrapped things and package them. Food manufacturing can benefit from robot control. Food printers use technological skills in fine dining to accelerate culinary industry adaption. This enables a high-value, low-volume food processing operation with work tools to regulate solid food material production on various scales. The research's relevance and market potential have yet to be determined. Technical advances and implementations will be controlled to assess how contemporary technology meets customer expectations and changes lifestyles.

Food printing technology may allow home users to customize food forms, colors, and flavors. Custom-shaped chocolate and personalized full-color images in enhanced nutritious food templates may be generated (Causer, 2009). In addition to existing food selections, a personalized diet has been shown to improve health and body size (Watzke & German, 2011). Food printing ensures diet monitoring and healthy, tasty food. Food additives must be customized for varied manufacturer settings, even with confirmed material qualities. We must continue to deliver these highly customized foods to every home.

To comprehend 3D food printing, a mathematical framework including inputs, outputs, and process form is essential. Custom food processing and printer architecture are the primary reasons for such a platform. Process factors include temperature, moisture, food properties, density, thermal and electrical conductivity, viscosity, and permeability. Digitalizing complete culinary systems before mathematically modeling them differs from typical food production. Measurement, computerized database, manual, and theoretical computations collect food property data. Food qualities vary a lot due to formulations, etc. A simulation model can hold property data and structure and estimate food processing results for other properties. Researchers are also developing a 3D object geometry printing process, quantifying data for each process (measuring and combining materials, printing, baking, etc.), and defining protocols for coordinating functions or processes to refine this model for simulation (Figure 2).

Nanotechnologies

Nanotechnology manipulates atoms and molecules under 100 nanometres. Nanotechnology is the creation, characterization, fabrication, and use of nanometer-scale structures, devices, and systems by regulating form and size. Nanotechnology involves processing and using nanomaterials (sizes up to 100 nm) to use their likely physiochemical characteristics. Nanomaterials are designed using "top down" and "bottom up" methods. Nanomaterials can self-assemble latterly (Marlowe, 2005; Morrissey, 2006).

Nanostructured meals are designed to taste, look, and feel good (Cientifica Report, 2006). Nanotechnology extends food shelf life and reduces microbe-induced food loss (Pradhan et al., 2015). Nanocarriers transfer food additives without altering their shape. In some cell lines, only submicron nanoparticles can be absorbed; hence particle size directly affects bioactive component dispersion. Nanopolymers replace conventional food packaging materials. Nanosensors confirm infections, mycotoxins, and microorganisms (Bratovčić et al., 2015). The food business uses nanotechnology. Nanotechnology can enhance food

Figure 2.

packaging uniformity and prevent lumps: Food texture gelator: Nano additions and neutraceuticals for food nutrition: Nanocapsulation and nanocarriers for food scent, flavor, and ingredients.

Nanotechnology can improve food packaging by boosting plastic mechanical strength or creating new materials with specialized features. Active packaging nanoparticles are antibacterial (Figure 3). Polymer/inorganic nanocomposites can increase mechanical, barrier, thermal, and flammable characteristics by spontaneously improving polymer structure with ultra-small inorganic particles. Nano-filler produces equivalent performance at lower concentrations (1-5% weight) as silica (50%). These systems decouple, enhance, and evolve material characteristics independently. However, new ideas and technical approaches are used for nanomaterials to generate bioactive chemicals and micronutrients for food and nutraceuticals. Nanoemulsions, liposomes, micelles, biopolymer complexes, and cubosomes increased

Figure 3.

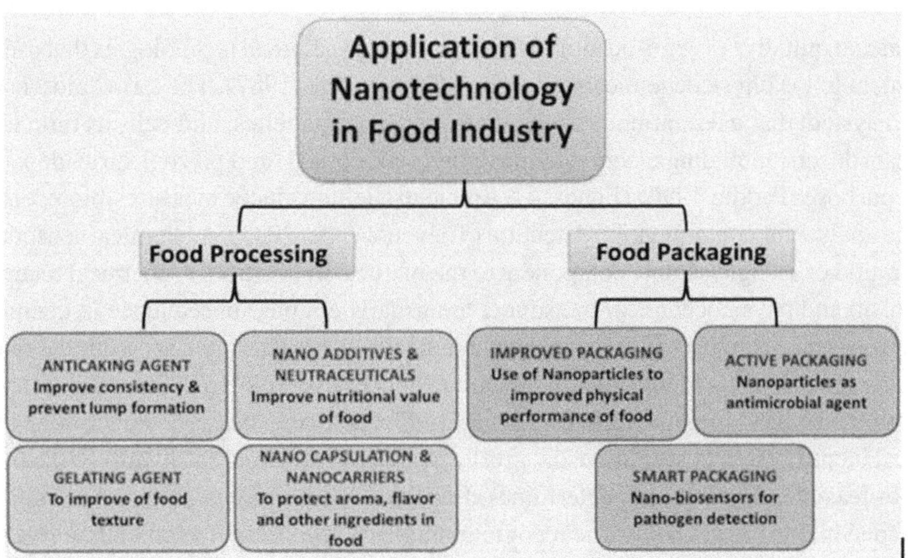

bioactive molecule storage, dispersion, food penetration, and flavor masking. Nanomaterials encapsulate and produce active food additives better than traditional agents. Enzymes can improve food processing by utilizing nanotechnology. Enzymes are given to meals to hydrolyze anti-nutritional substances and make minerals and vitamins easier to biodegrade. Due to their substantial surface-to-volume ratios, nanomaterials can make these enzymes highly active, long-lived, and cost-effective.

Encapsulation Technique

Encapsulation is a method for incorporating active substances into a carrier medium and may improve the food delivery of bioactive compounds and living cells. Encapsulating protective layers must be organic, biodegradable, and able to separate internal and exterior phases. Food encapsulation often uses polysaccharides. Encapsulating proteins and lipids is common. Encapsulation isolates an active ingredient (core) by covering it with a wall material called an encapsulating agent. This prevents significant changes to the active component or shields consumer-undesirable sensory qualities. When stimuli like pH or heat are added, the encapsulated material's protection breaks down, and the active product is transported to the exact target location or in ideal circumstances (Suave et al., 2006). Encapsulation suppresses odors and tastes, monitors active ingredients' interactions with their food matrix, regulates their release, ensures their supply at the desired pace and time, and protects them from moisture and heat (Ubbink & Kruger, 2006).

Food manufacturers have utilized encapsulation for over 75 years. Encapsulation covers minute particles, including acidulants, leavening agents, organic fats, minerals, vitamins, enzymes, essential oils, and culinary flavors. Bioactive substances and whole foods like raisins, almonds, and pastries can be microencapsulated or macro-coated. Bio-nanotechnological controlled releases include nano encapsulation. In 'pure' beverages, nano-capsulated particles do not cause turbidity like other microencapsulated additives. Nanoencapsulation reduces functional elements to 30 nm and encapsulates them in diets/ compounds. Once the primary material is finalized, the procedure is mainly automated.

Biosensors

Biosensors are quantitative or semi-quantitative instrumental analytical technologies that use a biological sensing element in the physicochemical transducer (Turner et al., 1987). The term "biosensor" refers to an analytical system that uses antibodies, enzymes, tissues, organelles, and cells as bioreceptors and a variety of transducers, including electrochemical, thermal, optical, and piezoelectric, depending on the biosensor's purpose (Paddle 1996) (Figure 4). An electrode transducer measures bioreceptor response electrons and analyte for contamination detection (Thévenot et al., 2001). A chemical sensor can measure the concentration of a single sample component to the whole composition. A chemical receptor recognition mechanism and physicochemical transducer are usually coupled in sequence in chemical sensors. The reconnaissance system uses biosensors and chemical sensors that interface with the optoelectronic system (Turner et al., 1987)—identifying chemical substances using electrical, thermal, or optical signals from separation enzymes, immunosystems, tissues, organelles, or cells.

Biological identification gives biosensors greater sensitivity and binding affinity to the target analyte, making them feasible. This binding determines if an interaction creates an electrical signal captured and amplified (Sharma et al., 2017). The recognition agent's specificity to the analyte produces signals primarily from these exact interactions, independent of matrix complexity (Idili et al., 2014). Food re-

Figure 4.

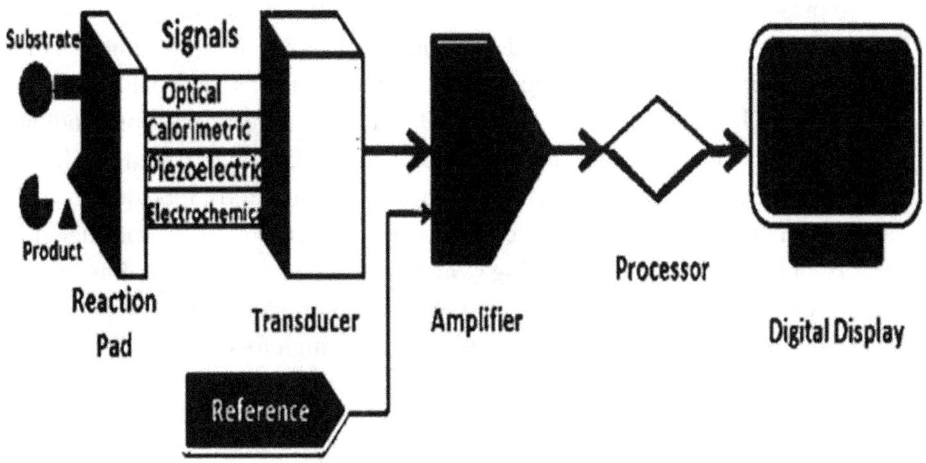

searchers and companies are creating quick, responsive, effective, and cost-effective analytical methods to detect foodborne infections and contaminants (Ibrišimović et al., 2015). Biosensors analyze food well. Logistics, time savings, real-time control, and, most importantly, removing highly qualified employees make them better than chromatography, spectrophotometry, etc. (Mello & Kubota, 2002). Biosensors employ bioreceptors and transducers to convert bioreceptor reactions into analytical signals. Biological binding processes can involve enzyme/substrate, antigen/antibody, receptor/ligand, chemical, and nucleic acid hybridization (Thakur & Ragavan, 2013). However, current advances in protein technology offer incentives to change protein structure, ligand binding, etc., to increase target selectivity and sensitivity.

Bio-sensors and other sensing improvements for quickly detecting microbial activity, biochemical activity, and other food-degradation responses have led to quality and safety advances. New bio- or nanomaterials can detect food quality and safety. Rapid, non-contact, non-destructive, and highly specialized micro and nano biosensing technologies are available. Micro and nanosensors for food microbiological and biochemical changes are being studied using various transduction routes. The bioreceptor or transducer classifies biosensors (Gaudin, 2017). Biosensors depend on a transducer (Figure 4) that detects physical changes during reactions. Calorimetric biosensors generate heat through reaction, while electrochemical biosensors modify the electrical or electronic output.

Biosensors analyze food quality, authenticity, and health (Rotariu et al., 2016). Food health monitoring detects chemicals, antibiotics, allergies, bacterial toxins, and pathogenic microorganisms. A similar methodology examines or validates a food product's nutritional content (Ribeiro et al., 2014). Authenticity studies verify food origin and manufacture, revealing adulteration or fraud (Inbaraj & Chen, 2015). Pathogenic Escherichia and Salmonella, aflatoxins, and pesticides like organophosphates and carbamates can only be identified post-production (Castillo et al., 2015). Biosensors, which monitor food from raw material to shelf products, can overcome this restriction and improve food safety and outbreak prevention (Jayas, 2017). RFID is a recent agri-food business breakthrough (Kumari et al., 2015). Optimizing knowledge reliably, improving process quality and dependability, reducing personnel costs and operational expenditures, and eliminating product defects, RFID works with biosensors, electronic surveillance, and other intelligent devices.

Intelligent food packaging commercialization requires current and effective consumer quality and safety practices to preserve packaging and reduce food loss (Vanderroost et al., 2014). Sensing indicators and RFID tags with sensors are new goods in fancy packaging. Sensing indicators measure color strength, color fluctuations, and dye diffusion for instant consistency, semi-quantitative, and the visual state of packaged goods. Gases (water vapor, ethanol, O_2, and CO2), freshness, time, and thermochromic (temperature-dependent color) indicators are used. Touch-activated, cold-activated, and high-temperature thermochromic inks are used in innovative packaging to notify customers of food quality and consistency. RFID, like bar codes, magnetic inks, biometrics, and voice recognition, is an automated identification (Auto ID) category of intelligent food packaging (Vanderroost et al., 2014). Food health and protection biosensing systems detect nutrients, pathogens, and toxins. Nanomaterials with biosensing and antibacterial capabilities and intelligent and robotic technologies with high food packing capacity are potential food packaging advances.

CONCLUSION

Food manufacturing is vital to developed and developing nations. Due to rising demand for authentic local products, food waste from large supply chains, consumer demand for food sustainability, and concerns about the long-term resilience of complex food systems, traditional business models that centralize and mass-produce food products are being challenged. Distributed food manufacturing has several promising technologies. Meal safety and shelf stability must be prioritized to meet customer needs for wholesome meals. Radiofrequency heating (RF), microwave heating (MH), High Hydrostatic Pressure (HHP), Ultrasound (US), Cold Plasma (CP), Ohmic Heat (OH), Plasma Technique (PT), Ozone processing (OP), Robotics technologies (RT), 3D printing (3DP), Microfluidics technologies (MT), Extrusion technology (ET), Biosensor Technology (BT), Nanotechnology (NT), Modified Atmospheres technology (MAT), etc. are innovative, flexible, and efficient food processing methods.

This study topic has focused on nanometer-scale structures in the food sector in the last few years. Nanobiotechnology shrinks and becomes more sensitive to its instruments and materials. Food processing and health are its main uses. Nanomaterials have also been shown to protect food against moisture, lipids, pollutants, flavors, and odors during processing.

3D food printing may generate customized chocolates or uniform nibbles. These implementations still have simple internal components and repetitive textures. Printing materials, platform architecture, printing methods, and food manufacture must be researched comprehensively. Cuisine printers will be an ecological invention with an immersive, open Web-based user interface that picks fresh ingredients and cooks favorite cuisine on demand. Innovative technology is anticipated to grow in the food business. Most aspects of new technology will increase product consistency and diversity and dramatically enhance existing and accepted technology. It is crucial that customers can choose items from these new technologies and understand their pros and cons. Food processors and merchants might improve food health using modern technology, benefiting customers. Personalization and customization are the future. Improve food health and uniformity first. Innovation is the food industry's goal.

REFERENCES

Abbasi, S., & Azari, S. (2009). Novel microwave-freeze drying of onion slice. *International Journal of Food Science & Technology*, *44*(5), 947–979. doi:10.1111/j.1365-2621.2008.01774.x

Alfaro, L., Siramard, S., Chouljenko, A., & Sathivel, S. (2018). Effects of liquid nitrogen pretreatment on the osmotic dehydration and quality of cryogenically frozen blueberries (*Vaccinium angustifoliumAit.*). *Food Bioscience*, *22*, 165–169. doi:10.1016/j.fbio.2018.02.006

Ammendolia, M. G., Iosi, F., De Berardis, B., Guccione, G., Superti, F., Conte, M. P., & Longhi, C. (2014). Listeria monocytogenes behaviour in presence of non-UV-irradiated titanium dioxide nanoparticles. *PLoS One*, *9*(1), e84986. doi:10.1371/journal.pone.0084986 PMID:24416327

Anton, N., & Vandamme, T. F. (2011). Nano-emulsions and micro-emulsions: Clarifications of the critical differences. *Pharmaceutical Research*, *28*(5), 978–985. doi:10.100711095-010-0309-1 PMID:21057856

Artés, F., Vallejo, F., & Martinez, J. (2001). Quality of broccoli as influenced by film wrapping during shipment. *European Food Research and Technology*, *213*(6), 480–483. doi:10.1007002170100390

Ashok, Y., & Satya, S. (2014). Osmotic dehydration of fruits and vegetables: A review. *Journal of Food Science and Technology*, *51*(9), 1654–1673. doi:10.100713197-012-0659-2 PMID:25190823

Ashokkumar, M., Haskaracharya, R., Entish, S., Lee, J., Almer, M., & Bogdan, Z. (2010). The ultrasonic processing of dairy products: An overview. *EDP Sciences*, *90*(2-3), 147–168. doi:10.1051/dst/2009044

Bablon, G., Bellamy, W. D., Bourbigot, M. M., Daniel, F. B., Dore, M., Erb, F., Gordon, G., Langlais, B., Laplanche, A., Legube, B., Martin, G., Masschelein, W. J., Pacey, G., Reckhow, D. A., & Ventresque, C. (1991). Fundamental aspects. In G. Langlais, D. A. Reckhow, & D. R. Brink (Eds.), *Ozone in water treatment: Application and engineering* (pp. 11–132). Lewis Publishers.

Bai, Y. X., Yang, G. J., Hu, Y. C., & Qu, M. (2012). Physical and sensory properties of electrohydrodynamic (EHD) dried scallop muscle. *Journal of Aquatic Food Product Technology*, *21*(3), 238–247. doi :10.1080/10498850.2011.590271

Balasubramaniam., Ramaswamy, R., Rodriguez, R. I., Mustafa, V. and Yousef, A. (2014). Ozone Technology Fact Sheet for Food Processors. pp 1-3.

Banerjee, R., & Verma, A. (2015). Minimally Processed Meat and Fish Products. In M. W. Siddiqui & M. S. Rahman (Eds.), *Minimally Processed Foods* (pp. 193–250). Springer International Publishing. doi:10.1007/978-3-319-10677-9_10

Bansal, V., Siddiqui, M., & Rahman, M. (2015). Minimally Processed Foods: Overview. In M. W. Siddiqui & M. S. Rahman (Eds.), *Minimally Processed Foods* (pp. 1–15). Springer International Publishing. doi:10.1007/978-3-319-10677-9_1

Baratharaj, V. (2009). *Use of Ozone in Food Processing and Cold Storage.* OTSIL. www.otsil.net

Barkworth, H. (1938). Taints and off-flavors of milk. *Dairy Inds*, *3*, 367.

Bedane, T. F., Chen, L., Marra, F., & Wang, S. (2017). Experimental study of radio frequency (RF) thawing of foods with movement on conveyor belt. *Journal of Food Engineering*, *201*, 17–25. doi:10.1016/j.jfoodeng.2017.01.010

Bhandari, B., Datta, N., & Howes, T. (1997). Problems Associated with Spray Drying Of Sugar-Rich Foods. *Drying Technology*, *15*(2), 671–684. doi:10.1080/07373939708917253

Bhatia, S., Stevanovic, J. T., & Ratti, C. (2020). Freeze-Drying of Plant-Based Foods. *Foods*, *9*(1), 87. doi:10.3390/foods9010087 PMID:31941082

Bintsis, T., Litopoulou-Tzanetaki, E., & Robinson, R. (2000). Existing and potential applications of the ultraviolet light in the food industry- A critical review. *Journal of the Science of Food and Agriculture*, *80*(6), 637–645. doi:10.1002/(SICI)1097-0010(20000501)80:6<637::AID-JSFA603>3.0.CO;2-1 PMID:29345786

Bourke, P., & O'Beirne, D. (2004). Effects of packaging type, gas atmosphere and storage temperature on survival and growth of Listeria spp. in shredded dry coleslaw and its components. *International Journal of Food Science & Technology*, *39*(5), 509–523. doi:10.1111/j.1365-2621.2004.00811.x

Bratovčić, A., Odobašić, A., Ćatić, S., & Šestan, I. (2015). Application of polymer nanocomposite materials in food packaging. *Croatian Journal of Food Science and Technology*, *7*(2), 86–94. doi:10.17508/CJFST.2015.7.2.06

Buzrul, S. (2015). Multi-pulsed high hydrostatic pressure treatment of foods. *Food*, *4*(4), 173–183. doi:10.3390/foods4020173 PMID:28231197

Caleb, O., Mahajan, P., Al-Said, F. A.-J., & Opara, U. L. (2013). Modified Atmosphere Packaging Technology of Fresh and Fresh-cut Produce and the Microbial Consequences—A Review. *Food and Bioprocess Technology*, *6*(2), 303–329. doi:10.100711947-012-0932-4 PMID:32215166

Caleb, O. J., Opara, U. L., & Witthuhn, C. R. (2012). Modified atmosphere packaging of pomegranate fruit and arils: A review. *Food and Bioprocess Technology*, *5*(1), 15–30. doi:10.100711947-011-0525-7

Castellanos, D. A., Cerisuelo, J. P., Hernández-Muñoz, P., Herrera, A. O., & Gavara, R. (2016). Modelling the evolution of O2 and CO2 concentrations in MAP of a fresh product: Application to tomato. *Journal of Food Engineering*, *168*, 84–95. doi:10.1016/j.jfoodeng.2015.07.019

Castillo, G., Spinella, K., Poturnayová, A., Šnejdárková, M., Mosiello, L., & Hianik, T. (2015). Detection of aflatoxin b1 by aptamer-based biosensor using pamam dendrimers as immobilization platform. *Food Control*, *2015*, 529–18. doi:10.1016/j.foodcont.2014.12.008

Causer, C. (2009). They've got a golden ticket. *Potentials IEEE*, *28*(4), 42–44. doi:10.1109/MPOT.2009.933608

Chaplin, M. (2004). *What are biosensors?* London south bank university. www.lsbu.ac.uk/biology/enztech/biosensors.html

Charcosset, C. (2009). Preparation of emulsions and particles by membrane emulsification for the food processing industry. *Journal of Food Engineering*, *92*(3), 241–249. doi:10.1016/j.jfoodeng.2008.11.017

Charles, F., Sanchez, J., & Gontard, N. (2003). Active modified atmosphere packaging of fresh fruits and vegetables: Modeling with tomatoes and oxygen absorber. *Journal of Food Science*, *68*(5), 1736–1742. doi:10.1111/j.1365-2621.2003.tb12321.x

Chemat, F., Zill-e-Huma, & Khan, M. K. (2011). Applications of ultrasound in food technology: Processing, preservation, and extraction. *Ultrasonics Sonochemistry*, *18*(4), 813–835. doi:10.1016/j.ultsonch.2010.11.023 PMID:21216174

Chen, J., Chen, D., Xie, Y., Yuan, T., & Chen, X. (2013). Progress of Microfluidics for Biology and Medicine. *Nano-Micro Letters*, *5*(1), 66–80. doi:10.1007/BF03354852

Chen, L., Remondetto, G. E., & Subirade, M. (2006). Food protein-based materials as nutraceutical delivery systems. *Trends in Food Science & Technology*, *17*(5), 272–283. doi:10.1016/j.tifs.2005.12.011

Chen, L., Wei, X., Irmak, S., Chaves, B. D., & Subbiah, J. (2019). Inactivation of Salmonella enterica and Enterococcus faecium NRRL B-2354 in cumin seeds by radiofrequency heating. *Food Control*, *103*, 59–69. doi:10.1016/j.foodcont.2019.04.004

Cheng, Y., Liu, L., Zhao, G., Shen, C., Yan, H., Guan, J., & Yang, K. (2015). The effects of modified atmosphere packaging on core browning and the expression patterns of PPO and PAL genes in 'Yali' pears during cold storage. *Lebensmittel-Wissenschaft + Technologie*, *60*(2), 1243–1248. doi:10.1016/j.lwt.2014.09.005

Compton, M., Willis, S., Rezaie, B., & Humes, K. (2018). Food processing industry energy and water consumption in the Pacific northwest. *Innovative Food Science & Emerging Technologies*, *47*, 371–383. doi:10.1016/j.ifset.2018.04.001

Coutinho, N. M., Silveira, M. R., Rocha, R. S., Moraes, J., Ferreira, M. V. S., Pimentel, T. C., Freitas, M. Q., Silva, M. C., Raices, R. S. L., Ranadheera, C. S., Borges, F. O., Mathias, S. P., Fernandes, F. A. N., Rodrigues, S., & Cruz, A. G. (2018). Cold plasma processing of milk and dairy products. *Trends in Food Science & Technology*, *74*, 56–68. doi:10.1016/j.tifs.2018.02.008

Dalgleish, D. G. (2004). Food emulsions: their structures and properties. In Journal of Food Emulsions, 4th Edn, eds. Friberg SE, Larsson K, Sjöblom, 1-44. New York: Marcel Dekker.

Datta, A. K., & Anantheswaran, R. C. (2000). *Handbook of microwave technology for food applications*. Marcel Dekker Inc.

Dickinson, E., & Stainsby, G. (1982). *Colloids in foods*. Elsevier Applied Science Publishers.

Donsi, F. (2018). In S. M. Jafari & D. J. McClements (Eds.), *"Applications of nanoemulsions in foods," in Nanoemulsions: Formulation, Applications, and Characterization* (pp. 349–377). Academic Press.

Duan, X., Zhang, M., Mujumdar, A. S., & Wang, R. (2010). Trends in microwave-assisted freeze drying of foods. *Drying Technology*, *28*(4), 444–453. doi:10.1080/07373931003609666

El-Ramady, H. R., Domokos-Szabolcsy, É., Abdalla, N. A., Taha, H. S., & Fári, M. (2015). Postharvest management of fruits and vegetables storage. *Sustainable Agriculture Research*, *15*, 65–152.

Escriche, I., Chiralt, A., Moreno, J., & Serra, J. A. (2000). Influence of blanching-osmotic dehydration treatments on volatile fraction of strawberries. *Journal of Food Science*, *65*(7), 107. doi:10.1111/j.1365-2621.2000.tb10247.x

Ezeh, O., Yusoff, M. M., & Niranjan, K. (2018). Nonthermal processing technologies for fabrication of microstructures to enhance food quality and stability. In *Food Microstructure and Its Relationship with Quality and Stability* (pp. 239–274). Woodhead Publishing. doi:10.1016/B978-0-08-100764-8.00012-5

Falade, K. O., & Igbeka, J. C. (2007). Osmotic dehydration of tropical fruits and vegetables. *Food Reviews International*, *23*(4), 373–405. doi:10.1080/87559120701593814

Fernandes, F. A. N., Gallao, M. I., & Rodrigues, S. (2008). Effect of osmotic dehydration and ultrasound pretreatment on cell structure: Melon dehydration. *Lebensmittel-Wissenschaft + Technologie*, *41*(4), 604–610. doi:10.1016/j.lwt.2007.05.007

Fito, P., Chiralt, A., Betoret, N., Gras, M., Chafer, M., Martnez-Monzo, J., Andres, A., & Vidal, D. (2001). Vacuum impregnation and osmotic dehydration in matrix engineering applications in functional fresh food development. *Journal of Food Engineering*, *49*(2-3), 175–183. doi:10.1016/S0260-8774(00)00220-X

Flink, J. M. (1979). Dehydrated carrot slices: influence of osmotic concentration on drying behaviour and product quality. In Food Process Engineering, Eds. Linko, P., Malkki, Y. Olkku J. and Larinkari, J., 412-418. Applied Science Publishers: London.

Fowler, M. R., & Park, J. W. (2015). Effect of salmon plasma protein on Pacific whiting surimi gelation under various ohmic heating conditions. *Lebensmittel-Wissenschaft + Technologie*, *61*(2), 309–315. doi:10.1016/j.lwt.2014.12.049

Garcia, A., Mount, J. R., & Davidson, P. M. (2003). Ozone and chlorine treatment of minimally processed lettuce. *Journal of Food Science*, *68*(9), 2747–2751. doi:10.1111/j.1365-2621.2003.tb05799.x

Gaudin, V. (2017). Advances in biosensor development for the screening of antibiotic residues in food products of animal origin - A comprehensive review. *Biosensors & Bioelectronics*, *90*, 363–377. doi:10.1016/j.bios.2016.12.005 PMID:27940240

Gavahian, M., Chu, Y. H., Khaneghah, A. M., Barba, F. J., & Misra, N. N. (2018). A critical analysis of the cold plasma induced lipid oxidation in foods. *Trends in Food Science & Technology*, *77*, 32–41. doi:10.1016/j.tifs.2018.04.009

Gebbers, R., & Adamchuk, V. I. (2010). Precision agriculture and food security. *Science*, *327*(5967), 828–831. doi:10.1126cience.1183899 PMID:20150492

Giri, S. K., & Prasad, S. (2007). Drying kinetics and rehydration characteristics of microwave vacuum and convective hot-air dried mushrooms. *Journal of Food Engineering*, *78*(2), 512–521. doi:10.1016/j.jfoodeng.2005.10.021

Graham, D. M. (1997). Use of ozone for food processing. *Food Technology*, *6*(51), 72–75.

Gupta, M., & Wong, W. L. E. (2007). Microwaves and metals. Singapore: John Wiley & Sons. doi:10.1002/9780470822746

Guzel-Seydim, Z. B., Greene, A. K., & Seydim, A. C. (2004). Use of ozone in the food industry. *Lebensm-Wiss-UTechnologie*, *37*(4), 453–460. doi:10.1016/j.lwt.2003.10.014

Hamon, M. O., Oyarzabal, A., & Hong, J. W. (2013). Nanoliter/picoloter scale fluidic systems for food safety. In B. Park & M. Appell (Eds.), *Advances in applied nanotechnology for agriculture* (pp. 145–165). ACS. doi:10.1021/bk-2013-1143.ch008

Harnkarnsujarit, N., & Charoenrein, S. (2011). Influence of collapsed structure on stability of ß carotene in freeze-dried mangoes. *Food Reviews International*, *44*(10), 3188–3194. doi:10.1016/j.foodres.2011.08.008

Hassan, A. B., Mahmood, N. S., Elmamoun, K., Adiamo, Q., Isam, A., & Ahmed, M. (2017). Effects of gamma irradiation on the protein characteristics and functional properties of sesame (*Sesamum indicum* L.) seeds. *Radiation Physics and Chemistry*, *2017*, 144.

Hawkes, J., & Flink, J. M. (1978). Osmotic concentration of fruit slices prior to freeze dehydration. *Journal of Food Processing and Preservation*, *2*(4), 265–284. doi:10.1111/j.1745-4549.1978.tb00562.x

He, S., & Shi, X. (2021). Microbial Food Safety in China: Past, Present, and Future. *Foodborne Pathogens and Disease*, *18*(8), 510–518. doi:10.1089/fpd.2021.0009 PMID:34242111

Herman, L. E., Martínez-Sánchez, C. E., Pacheco-Angulo, H., Carmona-García, R., Ruiz-Espinosa, H., & Ruiz-López, I. I. (2013). Mass transfer modeling of equilibrium and dynamic periods during osmotic dehydration of radish in NaCl solutions. *Food and Bioproducts Processing*, *9*(3), 216–224. doi:10.1016/j.fbp.2012.10.001

Hitkova, H., Stoyanova, A., Ivanova, N., Sredkova, M., Popova, V., Iordanova, R., & Bachvarova-Nedelcheva, A. (2012). Study of antibacterial activity of nonhydrolytic synthesized TiO_2 against E. coli, P. aeruginosa and S. aureus. *Journal of Optoelectronics and Biomedical Materials*, *4*(1), 9–17.

Huang, H. W., Hsu, C. P., Yang, B. B., & Wang, C. Y. (2013). Potential utility of high-pressure processing to address the risk of food allergen concerns. *Comprehensive Reviews in Food Science and Food Safety*, *13*(1), 78–90. doi:10.1111/1541-4337.12045 PMID:33412689

Huang, H. W., Hsu, C. P., Yang, B. B., & Wang, C. Y. (2014). Potential utility of high-pressure processing to address the risk of food allergen concerns. *Comprehensive Reviews in Food Science and Food Safety*, *13*(1), 78–90. doi:10.1111/1541-4337.12045 PMID:33412689

Huang, Y. W., & Toledo, R. (1982). Effect of high doses of high and low intensity UV irradiation on surface microbiological counts and storage life of fish. *Journal of Food Science*, *47*(5), 1667–1669. doi:10.1111/j.1365-2621.1982.tb05007.x

Huang, Z., Marra, F., & Wang, S. (2016). A novel strategy for improving radio frequency heating uniformity of dry food products using computational modeling. *Innovative Food Science & Emerging Technologies*, *34*, 100–111. doi:10.1016/j.ifset.2016.01.005

Huppertz, T., Hinz, K., Zobrist, M. R., Uniacke, T., Kelly, A. L., & Fox, P. F. (2005). Effects of high pressure treatment on the rennet coagulation and cheese-making properties of heated milk. *Innovative Food Science & Emerging Technologies*, *6*(3), 279–285. doi:10.1016/j.ifset.2005.03.005

Ibrišimović, N., Ibrišimović, M., Kesić, A., & Pittner, F. (2015). Microbial biosensor: A new trend in the detection of bacterial contamination. *Monatshefte für Chemie*, *146*(8), 1363–1370. doi:10.100700706-015-1451-6

Idili, A., Amodio, A., Vidonis, M., Feinberg-Somerson, J., Castronovo, M., & Ricci, F. (2014). Folding-upon-binding and signal-on electrochemical DNA sensor with high affinity and specificity. *Analytical Chemistry*, *86*(18), 9013–9019. doi:10.1021/ac501418g PMID:24947124

Inbaraj, B. S., & Chen, B. H. (2016). Nanomaterial-based sensors for detection of foodborne bacterial pathogens and toxins as well as pork adulteration in meat products. *Yao Wu Shi Pin Fen Xi*, *24*(1), 15–28. PMID:28911398

Ines, G. R. M., Juliana, M. C., Araceli, O. M., Laura, S. T., & Olga, M. B. (2015). Long-term stability of food-grade nanoemulsions from high methoxyl pectin containing essential oils. *Food Hydrocolloids*, *52*, 438–446.

Islam, M. N., & Flink, J. N. (1982). Dehydration of potato II. Osmotic concentration and its effect on air drying behavior. *Journal of Food Technology*, *17*(3), 387–403. doi:10.1111/j.1365-2621.1982.tb00194.x

Jakob, S. J., & Hansen, F. (2005). New Chemical and Biochemical Hurdles. In D. Sun (Ed.), *Emerging technologies for Food Technology* (pp. 387–418). Elsivier Ltd.

Jayas, D. S. (2017). The role of sensors and bio-imaging in monitoring food quality. *Resource Magazine*, *2017*, 12.

Kader, A. A., & Watkins, C. B. (2000). Modified atmosphere packaging-toward 2000 and beyond. *HortTechnology*, *10*(3), 483–486. doi:10.21273/HORTTECH.10.3.483

Karam, M. C., Petit, J., Zimmer, D., Djantou, E. B., & Scher, J. (2016). Effects of drying and grinding in production of fruit and vegetable powders: A review. *Journal of Food Engineering*, *188*, 32–49. doi:10.1016/j.jfoodeng.2016.05.001

Kasuriya S. and Atong D. (2004). Rapid Drying of Ceramic and Efficient Food Processing with a Continuous Microwave Belt Furnace, 27-29.

Khadre, M. A., & Yousef, A. E. (2001). Decontamination of a multilaminated aseptic food packaging material and stainless steel by ozone. *Journal of Food Safety*, *21*(1), 1–13. doi:10.1111/j.1745-4565.2001.tb00304.x

Kim, B., Yun, H., Jung, S., Jung, Y., Jung, H., Choe, W., & Jo, C. (2011). Effect of atmospheric pressure plasma on inactivation of pathogens inoculated onto bacon using two different gas compositions. *Food Microbiology*, *28*(1), 9–13. doi:10.1016/j.fm.2010.07.022 PMID:21056769

Kim, J. G., Yousef, A. E., & Dave, S. (1999). Application of ozone for enhancing the microbiological safety and quality of foods: A review. *Journal of Food Protection*, *62*(9), 1071–1087. doi:10.4315/0362-028X-62.9.1071 PMID:10492485

Knorr, D., Eshtiaghi, M. N., & Rastogi, N. K. (1999). Accelerated mass transfer during osmotic dehydration of high intensity electrical field pulse pretreated carrots. *Journal of Food Science*, *64*(6), 1020–1023. doi:10.1111/j.1365-2621.1999.tb12272.x

Knorr, D., Froehling, A., Jaeger, H., Reineke, K., Schlueter, O., & Schoessler, K. (2011). Emerging technologies in food processing. *Annual Review of Food Science and Technology*, *2*(1), 203–235. doi:10.1146/annurev.food.102308.124129 PMID:22129381

Kodandaram, R. D., Waghray, K., & Sathyanarayana, S. V. (2020). Infrared Heating - A New Green Technology for Process Intensification in Drying of Purslane Leaves to Reduce the Thermal Losses. In *International Conference on Emerging Trends in Engineering*. Springer Nature: Switzerland. 10.1007/978-3-030-24314-2_56

Komaiko, J. S., & McClements, D. J. (2016). Formation of food-grade nanoemulsions using low-energy preparation methods: A review of available methods. *Comprehens Rev Food Sci Safety*, *15*(2), 331–352. doi:10.1111/1541-4337.12189 PMID:33371595

Korachi, M., Ozen, F., Aslan, N., Vannini, L., Maria, E. G., Davide, G., & Fatma, Y. E. (2015). Biochemical changes to milk following treatment by a novel, cold atmospheric plasma system. *International Dairy Journal*, *42*, 64–69. doi:10.1016/j.idairyj.2014.10.006

Kovačević, D. B., Putnik, P., Dragović-Uzelac, V., Pedisić, S., Jambrak, A. R., & Herceg, Z. (2016). Effects of cold atmospheric gas phase plasma on anthocyanins and color in pomegranate juice. *Food Chemistry*, *190*, 317–323. doi:10.1016/j.foodchem.2015.05.099 PMID:26212976

Kowalska, H. (2006). Kinetykaosmotycznegoodwadnianiadyni. *Żywność. Nauka. Technologia. Jakość*, *2*(47), 135–142.

Krishnamurthy, K., Khurana, H. K., Soojin, J., Irudayaraj, J., & Demirci, A. (2008). Infrared heating in food processing: An overview. *Comprehensive Reviews in Food Science and Food Safety*, *7*(1), 2–13. doi:10.1111/j.1541-4337.2007.00024.x

Kumari, L., Narsaiah, K., Grewal, M. K., & Anurag, R. K. (2015). Application of RFID in agri-food sector. *Trends in Food Science & Technology*, *43*(2), 144–161. doi:10.1016/j.tifs.2015.02.005

Lagnika, C., Huang, J., Jiang, N., Li, D., Liu, C., Song, J., Wei, Q., & Zhang, M. (2018). Ultrasound-assisted osmotic process on quality of microwave vacuum drying sweet potato. *Drying Technology*, *36*(11), 1367–1379. doi:10.1080/07373937.2017.1402786

Lee, H., Kim, J. E., Chung, M. S., & Min, S. C. (2015). Cold plasma treatment for the microbiological safety of cabbage, lettuce, and dried figs. *Food Microbiology*, *51*, 74–80. doi:10.1016/j.fm.2015.05.004 PMID:26187830

Lee, J. Y., & Deniniger, R. A. (1999). Survival of bacteria after ozonation. *Ozone Science and Engineering*, *22*(1), 65–75. doi:10.1080/01919510008547229

Leonard, A., Blacher, S., Nimmol, C., & Devahastin, S. (2008). Effect of far-infrared radiation assisted drying on microstructure of banana slices: An illustrative use of X-ray microtomography in microstructural evaluation of a food product. *Journal of Food Engineering*, *85*(1), 154–162. doi:10.1016/j.jfoodeng.2007.07.017

Lewicki, P. P., & Lenart, A. (2015). Osmotic dehydration of fruits and vegetables. In A. S. Mujumdar (Ed.), *Handbook of industrial drying* (4th ed., pp. 661–674). CRC Press.

Liu, W., & Lanier, T. C. (2016). Rapid (microwave) heating rate effects on texture, fat/water holding, and microstructure of cooked comminuted meat batters. *Food Research International*, *81*, 108–113. doi:10.1016/j.foodres.2016.01.005

Loaharanu, P., & Ahmad, M. (1991). Advantages and disadvantages of the use of irradiation for Food Preservation. *Journal of Agricultural & Environmental Ethics*, *4*(1), 14–30. doi:10.1007/BF02229144

Mahapatra, A. K., Muthukumarappan, K., & Julson, I. J. (2007). Applications of ozone, bacteriosins and irradiation in Food processing: A review. *Critical Reviews in Food Science and Technology*, *45*(6), 447–461. doi:10.1080/10408390591034454

Mahdi, J. S., Yinghe, H., & Bhandari, B. (2007). Production of sub-micron emulsions by ultrasound and microfluidization techniques. *Journal of Food Engineering*, *82*(4), 478–488. doi:10.1016/j.jfoodeng.2007.03.007

Mairhofer, J., Roppert, K., & Ertl, P. (2009). Microfluidic systems for pathogen sensing: A review. *Sensors (Basel)*, *9*(6), 4804–4823. doi:10.339090604804 PMID:22408555

Maloney, N., & Harrison, M. (2016). Advanced heating technologies for food processing A2. In C. E. Leadley (Ed.), *Innovation and Future Trends in Food Manufacturing and Supply Chain Technologies* (pp. 203–256). Woodhead Publishing Series in Food Science, Technology and Nutrition. doi:10.1016/B978-1-78242-447-5.00008-3

Mangaraj, S., & Goswami, T. (2009). Modified atmosphere packaging of fruits and vegetables for extending shelf-life: A review. *Fresh Produce*, *3*(1), 1–31.

Mao, X., & Huang, T. J. (2012). Microfluidic diagnostics for the developing world. *Lab on a Chip*, *12*(8), 1412–1416. doi:10.1039/c2lc90022j PMID:22406768

Marlowe, D. E. (2005). Nanotechnology and the U.S. Food and Drug Administration. *ASTM Standardization News*, *33*, 29–31. PMID:16237872

Marra, F., Zhang, L., & Lyng, J. (2009). Radio frequency treatment of foods: Review of recent advances. *Journal of Food Engineering*, *91*(4), 497–508. doi:10.1016/j.jfoodeng.2008.10.015

McClements, D. J. (2019). *Future Foods: How Modern Science Is Transforming the way We Eat*. Switerzerland, Springer Scientific. doi:10.1007/978-3-030-12995-8

McClements, D. J., Decker, E. A., & Weiss, J. (2007). Emulsion-based delivery systems for lipophilic bioactive components. *Journal of Food Science*, *72*(8), R109–R124. doi:10.1111/j.1750-3841.2007.00507.x PMID:17995616

McClements, D. J., & Gunasekaran, S. (1997). Ultrasonic characterization of foods and drinks: Principles, methods, and applications. *Critical Reviews in Food Science and Nutrition*, *37*(1), 1–46. doi:10.1080/10408399709527766 PMID:9067087

McClements, D. J., & Rao, J. (2011). Food-grade nanoemulsions: Formulation, fabrication, properties, performance, biological fate, and potential toxicity. *Critical Reviews in Food Science and Nutrition*, *51*(4), 285–330. doi:10.1080/10408398.2011.559558 PMID:21432697

Mello, L. D., & Kubota, L. T. (2002). Review of the use of biosensors as analytical tools in the food and drink industries. *Food Chemistry, 77*(2), 237–256. doi:10.1016/S0308-8146(02)00104-8

Misra, N. N., & Jo, C. (2017). Applications of cold plasma technology for microbiological safety in meat industry. *Trends in Food Science & Technology, 64*, 74–86. doi:10.1016/j.tifs.2017.04.005

Misra, N. N., Patil, S., Moiseev, T., Bourke, P., Mosnier, J. P., Keener, K. M., & Cullen, P. J. (2014). In-package atmospheric pressure cold plasma treatment of strawberries. *Journal of Food Engineering, 125*, 131–138. doi:10.1016/j.jfoodeng.2013.10.023

Morrissey, S. (2006). Nanotechnology in food and agriculture. *Chemical and Engineering News, 84*, 31–31.

Moyer, J. C., & Stotz, E. (1947). The blanching of vegetables by electronics. *Food Technology, 1*, 252–257.

Muthukumarappan, K., Halaweish, F., & Naidu, A. S. (2000). Ozone. In A. S. Naidu (Ed.), *Natural food antimicrobial systems* (pp. 783–800). CRC Press. doi:10.1201/9781420039368.ch29

Naik, L., Sharma, R., Rajput, Y. S., & Manju, G. (2013). Application of High Pressure Processing Technology for Dairy Food Preservation - Future Perspective: A Review. *J. Anim Prod Adv, 3*(8), 232–241. doi:10.5455/japa.20120512104313

Nathakaranakule, A., Jaiboon, P., & Soponronnarit, S. (2010). Farinfrared radiation assisted drying of longan fruit. *Journal of Food Engineering, 100*(4), 662–668. doi:10.1016/j.jfoodeng.2010.05.016

Neethirajan, S., Kobayashi, I., Nakajima, M., Wu, D., Nandagopal, S., & Lin, F. (2011). Microfluidics for food, agriculture and biosystems industries. *Lab on a Chip, 11*(9), 1574–1586. doi:10.1039/c0lc00230e PMID:21431239

Nic, M., Jirat, J., & Kosata, B. (2006). IUPAC compendium of chemical terminology, the "Gold Book" 2. Blackwell Scientific Publications: Oxford: Oliveira, M.E.C. & Franca, A.S. (2002). Microwave heating of foodstuff. *Journal of Food Engineering, 53*, 347–359.

Ozkoc, S. O., Sumnu, G., Sahin, S., & Turabi, E. (2009). Investigation of physicochemical properties of breads baked in microwave and infrared-microwave combination ovens during storage. *European Food Research and Technology, 228*(6), 883–893. doi:10.100700217-008-1001-0

Paddle, B. M. (1996). Biosensors for chemical and biological agents of defence interest. *Biosensors & Bioelectronics, 11*(11), 1079–1113. doi:10.1016/0956-5663(96)82333-5 PMID:8828162

Pandiselvam, R., Manikantan, M. R., Subhashree, N., Mathew, A. C., Balasubramanian, D., Shameena Beegum, P. P., Ramesh, S. V., Niral, V., Ranjini, T. N., & Hebbar, K. B. (2019). Correlation and principal component analysis of physical properties of tender coconut (Cocos nucifera L.) in relation to the development of trimming machine. *Journal of Food Process Engineering, 42*(6), e13217. doi:10.1111/jfpe.13217

Pankaj, S. K., Bueno-Ferrer, C., Misra, N. N., Milosavljević, V., O'Donnell, C. P., Bourke, P., Keener, K. M., & Cullen, P. J. (2014). Applications of cold plasma technology in food packaging. *Trends in Food Science & Technology, 35*(1), 5–17. doi:10.1016/j.tifs.2013.10.009

Pankaj, S. K., & Keener, K. M. (2017). Cold plasma applications in food packaging. In *Reference Module in Food Science*. Elsevier. doi:10.1016/B978-0-08-100596-5.21417-0

Patterson, M. F. (2005). Microbiology of pressure treated foods. *Journal of Applied Microbiology*, *98*(6), 1400–1409. doi:10.1111/j.1365-2672.2005.02564.x PMID:15916652

Pereira, R. N., Martins, J., Mateus, C., Teixeira, J. A., & Vicente, A. A. (2007). Death kinetics of in goat milk and in cloudberry jam treated by ohmic heating. *Chemical Papers*, *61*(2), 121–126. doi:10.247811696-007-0008-5

Pereira, R. N., Martins, R. C., & Vicente, A. A. (2008). Goat milk free fatty acid characterization during conventional and ohmic heating pasteurization. *Journal of Dairy Science*, *91*(8), 2925–2937. doi:10.3168/jds.2007-0873 PMID:18650269

Philips, C. A. (1996). Review: Modified atmosphere packaging and its effects on the microbiological quality and safety of produce. *International Journal of Food Science & Technology*, *31*(6), 463–479. doi:10.1046/j.1365-2621.1996.00369.x

Pingen, S., Sudhaus, N., Becker, A., Krischek, C., & Klein, G. (2016). High pressure as an alternative processing step for ham production. *Meat Science*, *118*, 22–27. doi:10.1016/j.meatsci.2016.03.014 PMID:27019422

Piyasena, P., Mohareb, E., & McKellar, R. C. (2003). Inactivation of microbes using ultrasound: A review. *International Journal of Food Microbiology*, *87*(3), 207–216. doi:10.1016/S0168-1605(03)00075-8 PMID:14527793

Plestenjak, A., Pozrl, T., Hribar, J., Unuk, T., & Vidrih, R. (2008). Regulation of metabolic changes in shredded cabbage by modified atmosphere packaging. *Food Technology and Biotechnology*, *46*(4), 427–433.

Pradhan, N., Singh, S., Ojha, N., Srivastava, A., Barla, A., & Rai, V. (2015). Facets of nanotechnology as seen in food processing, packaging, and preservation industry. *BioMed Research International*, *2015*, 365672. doi:10.1155/2015/365672 PMID:26613082

Ramaswamy, H. S., Marcotte, M., Sastry, S. K., & Abdelrahim, K. (2016). Ohmic heating. In food processing, Eds. Abdelrahim K, pp. 93–102. CRC Press: Boca Raton, USA.

Ramesh, R., Nayak, B., Amirbahman, A., Tripp, C. P., & Mukhopadhyay, S. (2016). Application of ultraviolet light assisted titanium dioxide photocatalysis for food safety: A review. *Innovative Food Science & Emerging Technologies*, *38*, 105–115. doi:10.1016/j.ifset.2016.09.015

Rastogi, N. (2013). Introduction. In *Recent Developments in High Pressure Processing of Foods* (pp. 1–7). Springer. doi:10.1007/978-1-4614-7055-7_1

Rastogi, N. K. (2010). Opportunities and challenges in non-thermal processing of foods. In Innovation in Food Engineering, 3–58. New Techniques and Products, CRC Press.

Rastogi, N. K. (2012). Recent trends and developments in infrared heating in food processing. *Critical Reviews in Food Science and Nutrition*, *52*(9), 737–760. doi:10.1080/10408398.2010.508138 PMID:22698266

Rastogi, N. K. (2015). Infrared heating of foods and its combination with electron beam processing. In *Electron Beam Pasteurization and Complementary Food Processing Technologies* (pp. 61–82). Woodhead Publishing Series in Food Science, Technology and Nutrition. doi:10.1533/9781782421085.2.61

Rastogi, N. K., Raghavarao, K. S., Balasubramaniam, M. S., Niranjan, V. M., & Knorr, D. (2007). Opportunities and challenges in high pressure processing of foods. *Critical Reviews in Food Science and Nutrition*, *47*(1), 69–112. doi:10.1080/10408390600626420 PMID:17364696

Rastogi, N. K., Raghavarao, K. S. M. S., Niranjan, K., & Knorr, D. (2002). Recent developments in osmotic dehydration: Methods to enhance mass transfer. *Trends in Food Science & Technology*, *2*(13), 48–59. doi:10.1016/S0924-2244(02)00032-8

Report, C. (2006). *Nanotechnologies in the Food Industry.* Cientifica. http://www.cientifica.com/www/details.php?id47

Restaino, L., Frampton, E. W., Hemphil, J. B., & Palnikar, P. (1995). Efficacy of ozonated water against various food-related microorganisms. *Applied and Environmental Microbiology*, *61*(9), 3471–3475. doi:10.1128/aem.61.9.3471-3475.1995 PMID:7574656

Ribeiro, F. W. P., Barroso, M. F., Morais, S., Viswanathan, S., de Lima-Neto, P., Correia, A. N., Oliveira, M. B. P. P., & Delerue-Matos, C. (2014). Simple laccase-based biosensor for formetanate hydrochloride quantification in fruits. *Bioelectrochemistry (Amsterdam, Netherlands)*, *95*, 957–14. doi:10.1016/j.bioelechem.2013.09.005 PMID:24161938

Rød, S. K., Hansen, F., Leipold, F., & Knøchel, S. (2012). Cold atmospheric pressure plasma treatment of ready-to-eat meat: Inactivation of Listeria innocua and changes in product quality. *Food Microbiology*, *30*(1), 233–238. doi:10.1016/j.fm.2011.12.018 PMID:22265306

Rojas-Graü, M. A., Soliva-Fortuny, R., & Martín-Belloso, O. (2009). Edible coatings to incorporate active ingredients to fresh-cut fruits: A review. *Trends in Food Science & Technology*, *20*(10), 438–447. doi:10.1016/j.tifs.2009.05.002

Rotariu, L., Lagarde, F., Jaffrezic-Renault, N., & Bala, C. (2016). Electrochemical biosensors for fast detection of food contaminants – trends and perspective. *Trends in Analytical Chemistry*, *2016*, 7980–7987. doi:10.1016/j.trac.2015.12.017

Sagar, V. R., & Suresh Kumar, P. (2010). Recent advances in drying and dehydration of fruits and vegetables: A review. *Journal of Food Science and Technology*, *47*(1), 15–26. doi:10.100713197-010-0010-8 PMID:23572596

Saifullah, M., Ahsan, A., & Shishir, M. R. I. (2016). Production, stability and application of micro- and nanoemulsion in food production and the food processing industry. In A. M. Grumezescu (Ed.), *Nanotechnology in the Agri-Food Industry, Emulsions* (pp. 405–442). Academic Press. doi:10.1016/B978-0-12-804306-6.00012-X

Salazar-Gonzalez, C., San Martin-Gonzalez, M. F., Lopez-Malo, A., & Sosa-Morales, M. E. (2012). Recent studies related to microwave processing of fluid foods. *Food and Bioprocess Technology*, *5*(1), 31–46. doi:10.100711947-011-0639-y

Saldaña, E., Siche, R., Luján, M., & Quevedo, R. (2013). Review: Computer vision applied to the inspection and quality control of fruits and vegetables. *Brazilian Journal of Food Technology*, *16*(4), 254–272. doi:10.1590/S1981-67232013005000031

Salvia-Trujillo, L., Rojas-Graü, M. A., Soliva-Fortuny, R., & Martín-Belloso, O. (2015). Use of antimicrobial nanoemulsions as edible coatings: Impact on safety and quality attributes of fresh-cut Fuji apples. *Postharvest Biology and Technology*, *105*, 8–16. doi:10.1016/j.postharvbio.2015.03.009

Sandhya, K. V. (2010). Modified atmosphere packaging of fresh produce: Current status and future needs. *Lebensmittel-Wissenschaft + Technologie*, *43*(3), 381–392. doi:10.1016/j.lwt.2009.05.018

Saravacos, G., & Kostaropoulos, A. E. (2016). *Equipment for novel food processes handbook of food processing equipment*. Springer. doi:10.1007/978-3-319-25020-5

Sastry, S. (2008). Safety of Foods Processed Using Four Alternative Processing Technologies. *National Integrated Food Safety Initiative* USDA, 1-2.

Sastry, S. K., Datta, A. K., & Worobo, R. W. (2008). Ultraviolet light. *Journal of Food Science*, (Supplement), 90–92.

Sau, S., Datta, P., Sarkar, T., & Sarkar, S. (2018). Impact of low doses of gamma irradiation on off- season guava at ambient storage conditions. *International Journal of Current Microbiology and Applied Sciences*, *1*(1), 295–307. doi:10.20546/ijcmas.2018.701.032

Selcuk, N., & Erkan, M. (2015). The effects of modified and palliflex controlled atmosphere storage on postharvest quality and composition of 'Istanbul' medlar fruit. *Postharvest Biology and Technology*, *99*, 9–19. doi:10.1016/j.postharvbio.2014.07.004

Serrano, M., Martinez-Romero, D., Guillen, F., Castillo, S., & Valero, D. (2006). Maintenance of broccoli quality and functional properties during cold storage as affected by modified atmosphere packaging. *Postharvest Biology and Technology*, *39*(1), 61–68. doi:10.1016/j.postharvbio.2005.08.004

Sevenich, R., Bark, F., Crews, C., Anderson, W., Pye, C., Riddellova, K., Hradecky, J., Moravcova, E., Reineke, K., & Knorr, D. (2013). Effect of high pressure thermal sterilization on the formation of food processing contaminants. *Innovative Food Science & Emerging Technologies*, *20*, 42–50. doi:10.1016/j.ifset.2013.07.006

Sharma, A., Goud, K. Y., Hayat, A., Bhand, S., & Marty, J. L. (2017). Recent advances in electrochemical-based sensing platforms for aflatoxins detection. *Chemosensors (Basel, Switzerland)*, *5*(1), 1. doi:10.3390/chemosensors5010001

Sharma, G. P., & Prasad, S. (2001). Drying of garlic cloves by microwave–hot air combination. *Journal of Food Engineering*, *50*(2), 99–105. doi:10.1016/S0260-8774(00)00200-4

Shi, J., Pan, Z., McHugh, T. H., Wood, D., Hirschberg, E., & Olson, D. (2008). Drying and quality characteristics of fresh and sugarinfused blueberries dried with infrared radiation heating. *Lebensmittel-Wissenschaft + Technologie*, *41*(10), 1962–1972. doi:10.1016/j.lwt.2008.01.003

Sierra, I., Vidal-Valverde, C., & Lopez-Fandino, R. (2000). Effect of high pressure on the Vitamin B1 and B6 content in milk. *Milchwissenschaft. Milk Science International*, *55*, 365–367.

Singh, T., Shukla, S., Kumar, P., Wahla, V., Bajpai, V. K., & Rather, I. A. (2017). Application of nanotechnology in food science: Perception and overview. *Frontiers in Microbiology*, *8*, 1501. doi:10.3389/fmicb.2017.01501 PMID:28824605

Soltani, M., Alimardani, R., Mobli, H., & Mohtasebi, S. (2015). Modified atmosphere packaging: A progressive technology for shelf-life extension of fruits and vegetables. *Journal of Applied Packaging Research*, *7*, 2.

Sopher, C. D., Graham, D. M., Rice, R. G., & Strasser, J. H. (2002). Studies on the use of ozone in production agriculture and food processing. In *Proceedings of the International Ozone Association, Pan American Group, Raleigh-Durham,* NC, USA, 18–22 May, 1–15.

Soria, A. C., & Villamiel, M. (2010). Effect of ultrasound on the technological properties and bioactivity of food: A review. *Trends in Food Science & Technology*, *21*(7), 323–331. doi:10.1016/j.tifs.2010.04.003

Steinhart, H. (2006). Novel foods and novel processing techniques as threats and challenges to a hypersensitive world. *Allergy Matters: New Approaches to Allergy Prevention and Management.*, *10*, 65–77. doi:10.1007/1-4020-3897-6_8

Suave, J., Dall'agnol, E. C., Pezzin, A. P. T., Silva, D. A. K., Meier, M. M., & Soldi, V. (2006). Microencapsulação: Inovaçãoemdiferentesáreas. *RevistaSaúde e Ambiente*, *7*, 12–20.

Sun, J., Peng, Z., Zhou, W., Fuh, J. Y. H., Hong, G. S., & Chiu, A. (2015). A review on 3D printing for customized food fabrication. *Procedia Manufacturing*, *1*, 308–319. doi:10.1016/j.promfg.2015.09.057

Talens, P., Martínez-Navarrete, N., Fito, P., & Chiralt, A. (2002). Changes in optical and mechanical properties during osmodehydrofreezing of kiwi fruit. *Innovative Food Science & Emerging Technologies*, *3*(2), 191–199. doi:10.1016/S1466-8564(02)00027-9

Tappi, S., Gozzi, G., Vannini, L., Berardinelli, A., Romani, S., Ragni, L., & Rocculi, P. (2016). Cold plasma treatment of fresh cut melon stabilization. *Innovative Food Science & Emerging Technologies*, *33*, 225–233. doi:10.1016/j.ifset.2015.12.022

Thakur, M. S., & Ragavan, K. V. (2013). Biosensors in food processing. *Journal of Food Science and Technology*, *50*(4), 625–641. doi:10.100713197-012-0783-z PMID:24425965

Thévenot, D. R., Toth, K., Durst, R. A., & Wilson, G. S. (2001). Electrochemical biosensors: Recommended definitions and classification1international union of pure and applied chemistry: Physical chemistry division, commission i.7 (biophysical chemistry); analytical chemistry division, commission v.5 (electroanalytical chemistry)1. *Biosensors & Bioelectronics*, *16*(1), 121–131. doi:10.1016/S0956-5663(01)00115-4 PMID:11261847

Tiwari, B. K., Donnell, C. P. O., Muthukumarappan, K., & Cullen, P. J. (2009). Ascorbic acid degradation kinetics of sonicated orange juice during storage and comparison with thermally pasteurized juice. *Lebensmittel-Wissenschaft + Technologie*, *42*(3), 700–704. doi:10.1016/j.lwt.2008.10.009

Tiwari, R. B. (2005). Application of osmo-air dehydration for processing of tropical fruits in rural areas. *Indian Food Ind*, *24*(6), 62–69.

Tribst, A. A. L., de Castro Leite Júnior, B. R., de Oliveira, M. M., & Cristianini, M. (2016). High pressure processing of cocoyam, Peruvian carrot and sweet potato: Effect on oxidative enzymes and impact in the tuber color. *Innovative Food Science & Emerging Technologies*, *34*, 302–309. doi:10.1016/j.ifset.2016.02.010

Turner, A. P. F., Karube, I., & Wilson, G. S. (1987). *Biosensors, fundamentals and applications*. Oxford University Press.

Ubbink, J., & Kruger, J. (2006). Physical approaches for the delivery of active ingredients in foods. *Trends in Food Science & Technology*, *17*(5), 244–254. doi:10.1016/j.tifs.2006.01.007

Uyar, R., Bedane, T. F., Erdogdu, F., Palazoglu, T. K., Farag, K. W., & Marra, F. (2015). Radio-frequency thawing of food products—A computational study. *Journal of Food Engineering*, *146*, 163–171. doi:10.1016/j.jfoodeng.2014.08.018

Vanderroost, M., Ragaert, P., Devlieghere, F., & De Meulenaer, B. (2014). Intelligent food packaging: The next generation. *Trends in Food Science & Technology*, *39*(1), 47–62. doi:10.1016/j.tifs.2014.06.009

Vicente, A. A., Pereira, R. N., Penna, T. C. V., & Knirsch, M. (2016). Electricity effects on microorganisms and enzymes. In H. S. Ramaswamy, M. Marcotte, S. Sastry, & K. Abdelrahim (Eds.), *Ohmic heating in food processing* (pp. 93–102). CRC Press.

Wang, Y., Li, X., Sun, G., Dong, L., & Pan, Z. (2014). A comparison of dynamic mechanical properties of processing-tomato peel as affected by hot lye and infrared radiation heating for peeling. *Journal of Food Engineering*, *126*, 27–34. doi:10.1016/j.jfoodeng.2013.10.032

Watzke, H., & German, J. (2011). Personalizing foods. In An integrated approach to new food product development, Ed. Howard, R., Moskowitz, I. Sam, S., Tim, S., Boca, R., 133-173. CRC Press: US.

Whitesides, G. M. (2006). The origins and the future of microfluidics. *Nature*, *442*(7101), 368–373. doi:10.1038/nature05058 PMID:16871203

Wongsa-Ngasri, P., & Sastry, S. K. (2016). Tomato peeling by ohmic heating: Effects of lye-salt combinations and post-treatments on weight loss, peeling quality and firmness. *Innovative Food Science & Emerging Technologies*, *34*, 148–153. doi:10.1016/j.ifset.2016.01.013

Wunderlich, S., & Martinez, N. M. (2018). Conserving natural resources through food loss reduction: Production and consumption stages of the food supply chain. *International Soil and Water Conservation Research*, *6*(4), 331–339. doi:10.1016/j.iswcr.2018.06.002

Yin, Z., Hoffmann, M., & Jiang, S. (2018). Sludge disinfection using electrical thermal treatment: The role of ohmic heating. *The Science of the Total Environment*, *615*, 262–271. doi:10.1016/j.scitotenv.2017.09.175 PMID:28972902

Yusaf, T., & Al-Juboori, R. A. (2014). Alternative methods of microorganism disruption for agricultural applications. *Applied Energy*, *114*, 909–923. doi:10.1016/j.apenergy.2013.08.085

Zhang, M., Tang, J., Mujumdar, A. S., & Wang, S. (2006). Trends in microwave-related drying of fruits and vegetables. *Trends in Food Science & Technology*, *17*(10), 524–534. doi:10.1016/j.tifs.2006.04.011

Chapter 2
Technology's Role in Sustainability:
How the Gastronomy Is Becoming More Eco-Friendly

Yeliz Demir
https://orcid.org/0000-0003-0184-6828
Isparta University of Applied Sciences, Turkey

Serkan Bertan
https://orcid.org/0000-0002-9424-5011
Pamukkale University, Turkey

ABSTRACT

Sustainable gastronomy is a concept associated with economic, social, and environmental factors. In this book chapter, the social dimension of sustainable gastronomy is examined. In this context, supporting sustainable local agricultural activities, labeling and standardizing local food products, supporting local producers, encouraging consumers to buy local products, preserving and maintaining local food culture, and eliminating or reducing food waste can be considered sustainable gastronomy's main social objectives. For sustainable gastronomy activities to be carried out successfully, all stakeholders (producers, consumers, employees, and businesses) must be involved in the process. For this reason, transparency and information are among the most important concepts in sustainable gastronomy process. It has been determined that websites and social media tools can be effective in achieving the social goals of sustainable gastronomy activities implemented today. The main purpose of these websites is to increase the awareness of the people who visit these sites by providing information about sustainable gastronomy. According to the purposes of creating the websites, these platforms contain information on supporting producers, supporting local women, influencing demand, encouraging sustainable food practices, online training, exams, and various certification processes about sustainable gastronomy. It is seen that various contents related to sustainable gastronomy are shared on the YouTube platform. It has been determined that information about sustainable gastronomy-related activities is shared throughout these posts and that experiences, local practices, and country gastronomy are promoted within the scope of sustainability. It is seen that there are posts within the scope of sustainable gastronomy activities on Instagram and Twitter, and the purpose of these posts is to inform about sustainable gastronomy activities through short posts, videos, and photos.

DOI: 10.4018/978-1-6684-9094-5.ch002

INTRODUCTION

The concept of gastronomy is related to the nutrition process for individuals to survive and to food production, preparation, cooking, and consumption. Additionally, the World Tourism Organization states that the concept of gastronomy reflects the culture, heritage, traditions, and sense of community of different peoples. From a broad perspective, gastronomy encompasses both tangible and intangible cultural values. In this respect, it should be considered that resources are used sustainably in the production of cuisine products within the scope of gastronomy and that people and other living things are not adversely affected by this process. The relevant literature shows that "green" practices stand out within the scope of sustainable gastronomy. Within the scope of these practices, it is seen that besides the protection of the environment (Green Restaurant Association, 2023a), the "Green Cuisine" label application initiated in Slovenia focuses on local gastronomic values as well as the impact of gastronomy on the environment (Slovenia, 2023a; Slovenia, 2023b). In this context, the book section's main purpose is to determine how technology is used to achieve the social goals of sustainable gastronomy activities.

Sustainable Gastronomy

The United Nations (2023) states that the UN General Assembly has designated June 18 as the International Day of Sustainable Gastronomy. The concepts of economic, social, and environmental sustainability are included in the scope of the sustainable food system (Food and Agriculture Organization of the United Nations (FAO), 2018). FAO (2018) explains that a sustainable food system provides food security and nutrition for all without compromising economic, social, and environmental fundamentals.

Transparency is one of the important subjects in sustainable gastronomy activities. The United Nations (2023) states that sustainable gastronomy means cuisine that considers where the ingredients come from, how the food is grown, and how these foods reach the markets and eventually the plates. FAO (2019) states that food labels determined by global standards provide consumers with information about the nutritional properties of food, its contents, storage method, expiry date, origin, and originality of food.

Michelin Guide (2023d) states that gastronomy can be defined as the cooking style of a particular region and is a cultural expression of the natural and cultural diversity of the world. In other words, gastronomy generally refers to local food and cuisine (FAO, 2020). It is seen that the sustainable agricultural heritages of the countries are trying to be protected with global systems (FAO, 2021a). FAO (2021a) notes that the Shizuoka region was recognized as Globally Important Agricultural Heritage Systems (GIAHS) in 2018 due to the traditional agricultural methods maintained on Wasabi farms.

Another aim of sustainable gastronomy is to increase the use of local food products and in this way to support local producers and farmers. A cooking competition has been organized in Oman since 2013 by FAO and local partners. Professional chef assistants participating in the competition in December 2021 were asked to prepare gourmet meals using only local products. This event raised awareness that chef assistants can cook creatively and deliciously with local products instead of imported products. In addition, farmers' markets were designed to showcase the food products used in this competition, thus establishing a link between producers and potential buyers in the accommodation sector (FAO, 2022). FAO (2021b) states that Geographical Indication (GI) labels given to traditionally produced products will help the consumers associate product characteristics such as taste or quality with the Geographical Indication status. It is stated that this tagging system could result in higher incomes for rural households. Within the scope of sustainable gastronomy activities, while trying to protect and reflect the culinary culture specific to the regions, it also provides an economic benefit for the local people. For example, with the support of the FAO, the Government of Canada, and Lebanon's Ministry of Agriculture, rural women in Deir al-Ahmar have had wheat grown for them so that they can make a popular "Freekeh" dish from green wheat (an ancient dish in Lebanon). It is seen that women in the region turn into entrepreneurs in this way, where they are trained in harvesting, processing, and packaging (FAO, 2023) Additionally, within the scope of sustainable gastronomy, it is seen that recommendations such as purchasing food products from local producers, feeding according to the season, reducing meat consumption, and avoiding food waste are given (UNtoday, 2023; Michelin, 2023d).

Eco-Friendly in Sustainable Gastronomy

Environmentally friendly and eco-friendly concepts appear to be prominent in the relevant literature. Ecotsy (2023) emphasizes that the concept of environmentally friendly is an umbrella concept that expresses goods and services that cause little or no harm to the environment and that it can mean using renewable energy sources, avoiding the use of dangerous substances, and reducing waste within the scope of environmentally friendly practices. Hossain (2022) states that the concept of eco-friendly refers to goods and materials that are environmentally friendly and do not harm the environment and that this product supports green living and green production practices that use fewer or different types of resources. Ecotsy (2023) emphasizes that the concept of eco-friendly refers to products and activities designed with a particular focus on sustainability and that these products often have features such as being biodegradable, made from recycled materials, or reusable.

Another name for the concept of "environmentally friendly" is "green" (Campbell et al., 2015). Within the scope of "green" practices in the field of gastronomy, the international non-profit "The Green Restaurant Association (GRA) was established in 1990 (Green Restaurant Association, 2023a). Energy, Water, Reusable Items, Disposables, Food and Beverage, Chemical and Pollution Prevention, Grease Collector, Linens, Pest Control, Recycler, and Composter are GRA's product and service certification standards (Green Restaurant Association, 2023b). On the GRA website, which evaluates the concept of "being green" with a holistic perspective, it is seen that there are certificate programs, online training, and exam programs related to becoming a green restaurant, becoming a green vendor, becoming a green consumer, and becoming a green employee (Green Restaurant Association, 2023c).

In recent years, it has been seen that the concept of "Green Cuisine" has emerged in the field of gastronomy. Regarding this concept, it is seen that the "Slovenian Green Cuisine" label has been applied in Slovenia (Slovenia, 2023a). It is emphasized that if a restaurant or other food provider in Slovenia wishes to participate in this program, they must sign the "Commitment to Sustainable Gastronomy", which consists of seven sustainability principles (Slovenia, 2023b). The commitment is based on serving authentic and seasonal dishes with a local character, Slovenia's abundant and high-quality fresh tap water that does not require packaging, reducing packaging waste, ending the use of single-use plastics, sustainable use of natural resources, and using high hygiene and sanitation standards (Slovenia, 2023a; Slovenia, 2023b;). Another application related to the concept of "green" is the Michelin Guide. Michelin Guide (2021a) as of January 18, 2021, started to give Michelin Green Star to food and beverage businesses. Michelin identified sustainable gastronomy themes as the products and ingredients used (seasonality, locality, production quality, and transparency of this information), the composition of the menu, the ability to raise customer awareness of the chef's philosophy, initiatives to reduce and/or recycle food waste, and management of organizational resources (Michelin Guide, 2021a). It is seen that the concept of eco-gastronomy has emerged in the field of gastronomy. "Eco-gastronomy is a 21st century extension of that appreciation, with an eye on sustainability, social justice and environmental well-being" (Eco-Gastronomy, 2023). In addition, it is stated that environmental concerns are an important factor in the selection, preparation, and marketing of the materials to be used in the preparation of eco-gastronomy meals (Güneş, 2019; Gün & Kılıç, 2022). In addition, Görkem et al. (2016) state that eco-gastronomy focuses on the sustainability of agricultural practices using natural production methods and traditional eating habits.

Technological Eco-Friendly Practices and Examples in Gastronomy

Trying to protect the sustainable agricultural heritage of countries (FAO, 2021a), announcing international special days to increase the awareness of societies about sustainable food production (such as tea, and honey) (FAO, 2021c; Michelin Guide 2023b), increasing the use of local food products (FAO, 2022), training rural women in the region on subjects such as the production, processing, and packaging of special cultural agricultural products within the scope of sustainable gastronomic activities and turning them into potential entrepreneurs (FAO, 2023). It can be thought that these activities represent the social aspect of sustainable gastronomy. Another aim of sustainable gastronomy societies is to raise awareness among consumers about sustainable gastronomy and to involve them in the process. In other words, bringing together supply and demand within the scope of sustainable gastronomy activities is one of the most important subjects. For this purpose, it is important to provide information to consumers about the origin of food products, the method of growing, distribution processes (United Nations, 2023), nutritional properties of food, contents, storage method, and expiry date (FAO, 2019). For example, it takes about 15 tons of water to produce 1 kg of beef. It should be considered that even if this information is stated on the food label of the meat purchased by the consumer, it will cause the consumer to think about the water consumed in meat production (Demir, 2023).

The social purpose of sustainable gastronomy is to spread the sustainability principles, which it considers, to a wide area to inform, raise awareness, and mobilize society on this subject. For this purpose, it is seen that social media tools and websites are used to include societies in sustainable gastronomy activities carried out today. The source, topic, and purposes of the websites related to sustainable gastronomy activities are shown in Table 1.

As can be seen in Table 1, the main purpose of the websites created about sustainable gastronomy is to inform the reader and raise awareness on this subject. In addition, while websites created for sustainable gastronomy differ according to the purpose of establishment, it is seen that the purposes of these platforms include supporting the producer, supporting local women, and influencing demand. In addition, it has been determined that there is information about sustainable gastronomy practices, online education, accreditation, certificate, and donation. Detailed information about the purposes of the websites created within the scope of sustainable gastronomy is provided in Table 1.

The sources and purposes of the posts made within the scope of sustainable gastronomy activities were examined by typing the terms "sustainable gastronomy", "eco-gastronomy" and "environmentally friendly gastronomy" in the search bar of "Instagram", "Twitter" and "YouTube" tools.

It is seen that various contents related to sustainable gastronomy are shared on the YouTube platform. It has been determined that information about sustainable gastronomy-related activities is shared throughout these posts and that experiences, local practices, and country gastronomy are promoted within the scope of sustainability (as can be seen in Table 2).

Table 1. The source, topic, and purposes of the websites created for sustainable gastronomy activities

Source	Topic	Purpose
Eco-Gastronomy (2023)	Eating Is an Agricultural Act'-Wendell Berry	Providing information for awareness
FAO (2019)	No one likes labels… unless it comes to food	Providing information for awareness
FAO (2020)	Calling all foodies: this one's for you!	Providing information for awareness
FAO (2021a)	A spicy agricultural heritage: there's more to wasabi than meets the eye	Providing information for awareness and supporting the manufacturer
FAO (2021b)	We are what we eat: Transforming diets to transform agri-food systems	Providing information for awareness, and influencing demand
FAO (2021c)	A cup of tea… or cha?	Providing information for awareness and supporting manufacturers
FAO (2022)	Local food for healthy locals	Providing information for awareness, influencing demand, and supporting manufacturers
FAO (2023)	Freekeh: The ancient dish firing up rural women's futures in Lebanon	Providing information for awareness, influencing demand, and supporting local women
Green Restaurant Association (2023a)	Mission Green the Restaurant Industry	Providing information for awareness
Green Restaurant Association (2023b)	Product & Service Certification Standards	Providing information for awareness, and apply
Green Restaurant Association (2023c)	Green your restaurant-Become a green vendor-Be a green consumer-Become a green employee	Providing information for awareness, education, accreditation, certificate, and donation
Hivos people unlimited (2023)	SUSTAINABLE DIETS FOR ALL	Providing information for awareness
Slovenia (2023a)	Slovenia Green Cuisine	Providing information for awareness, and apply
Slovenia (2023b)	News from Slovenia-The first International Day of Zero Waste is here	Providing information for awareness, applying, and influencing demand
Intellidigest (2023)	Course Catalog	Providing information for awareness, apply
Michelin Guide (2020)	MICHELIN Debuts New Sustainable Gastronomy Icon	Providing information for awareness
Michelin Guide (2021b)	What is a MICHELIN Green Star?	Providing information for awareness
Michelin Guide (2022)	What is a Michelin Green Star and How Do You Earn One?	Providing information for awareness, and apply
Michelin Guide (2023a)	Sustainability Leads the Way at Le Manoir aux Quat'Saisons, A Belmond Hotel	Providing information for awareness and influencing demand
Michelin Guide (2023b)	World Bee Day – 20th May	Providing information for awareness
Michelin Guide (2023c)	Green Star Focus: Apricity, London	Providing information for influence demand
Michelin Guide (2023e)	Chantelle Nicholson's Green Star Recipe	Providing information for implementation and food
Sustainable Gastronomy Initiatives Map (2023)	Sustainable Gastronomy Initiatives Map	Providing information for awareness and influencing demand
United Nations (2023)	Sustainable Gastronomy Day-18 June	Providing information for awareness
UNtoday (2023)	Sustainable Gastronomy Day – June 18, 2023	Providing information for awareness, conservation, and food

It has been determined that there are posts about sustainable gastronomy, eco-gastronomy, and environmentally friendly gastronomy on Instagram and Twitter. It has been determined that Instagram and Twitter share the tag's sustainable gastronomy, green gastronomy, eco gastronomy, and environmentally friendly gastronomy. It can be said that these shares are made in the form of short-term videos and photographs and that these shares are to draw the attention of the audience to the subject of sustainability in gastronomy in a short time and to give various news on this subject.

Table 2. The subject and purposes of YouTube shares created about sustainable gastronomy activities

Source	Topic	Purpose
Barcelona Visio (2009)	"Slow Food". A boost to ecogastronomy	Providing information
Bestdestination (2022)	Macao Gastronomy City and Sustainable Gastronomy Day	Country gastronomy
Breakthrough Institute (2017)	Sustainable Gastronomy	Providing information
bitividi (2020)	Sustainable Gastronomy Day	Providing information
CGTN (2019)	What is sustainable gastronomy?	Providing information
CGTN America (2019)	Nature is the key to a happy kitchen for an eco-friendly chef in Ecuador	Sharing experience
David Szanto (2016)	Stories of Eco-Gastronomy: Torino	Country gastronomy
David Szanto (2017)	Stories of Eco-Gastronomy: Montreal	Country gastronomy
EcoCircuitos Panama (2017).	Eco-Story Telling: Panama's Gastronomy with Patricia Miranda	Sharing experience
Future Food Institute (2020)	Food is a Conversation: Sustainable Gastronomy Day - Japanese Session	Providing information
Galiano Conservancy Association (2021)	John Volpe on EcoGastronomy I Galiano Conservancy Association	Providing information
Gastro Tourism Academy (2021)	Eco-Gastronomy: Minjangan Grass	local applications
Gastro Tourism Academy (2023)	Climate Eaters Webinar-Climate Action for Food Sustainability: Eco Gastronomy and Greenpreneurs	Providing information
Michelin Guide Asia (2020)	MICHELIN Introduces New Sustainable Gastronomy Icon	Providing information
Nico Dingemans (2023)	Sustainable Gastronomy Day-18.06.2023-From Farm to Fork Dinner in Amman	Sharing experience
Nespresso (2021)	Sustainable Gastronomy with Adam Handling I Edition 28	Sharing experience
New York Botanical Garden (2019)	Creating a More Eco-friendly and Sustainable Dining Experience	Sharing experience
Prof. M. THANGA DARWIN (2019)	Sustainable Gastronomy Day 18 JUNE 2023	Providing information
Shelbizleee (2022)	what a "climatarian" actually eats in a day I most eco-friendly food	Sharing experience
The Atlantic (2020)	The Struggle of a Zero-Waste Restaurant	Sharing experience
UNH Paul College of Business and Economics (2020)	Inside the Classroom: Ecogastronomy	Sharing experience
US Embassy BSB (2022a)	Sustainable Gastronomy Ep. 1	Providing information
US Embassy BSB (2022b)	Sustainable Gastronomy Ep. 2	Providing information
US Embassy BSB (2022c)	Sustainable Gastronomy Ep. 3	Providing information

CONCLUSION

This book determines how sustainable gastronomy activities use technology to achieve their social goals. First, it should be considered that the sustainable gastronomy activities carried out should reach people to achieve their social goals. It has been determined that sustainable gastronomy activities carried out for this purpose benefit from websites and social media tools to reach more people. After the examinations, it is seen that it is common to create a website about sustainable gastronomy activities. The main purpose of the websites created within the scope of sustainable gastronomy activities is to increase the awareness of people who access the websites by providing information about sustainable gastronomy.

According to the purposes of creating the websites, it is seen that these platforms contain information about supporting the producer, supporting local women, influencing demand, encouraging sustainable food practices, online training on sustainable gastronomy, exams, and various certification and accreditation processes. It is seen that various contents related to sustainable gastronomy are shared on the YouTube platform. It has been determined that information about sustainable gastronomy-related activities is shared throughout these posts and that the gastronomy of countries, experiences, and local practices is promoted within the scope of sustainability.

Finally, it has been determined that there are short videos and photographs about sustainable gastronomy on Instagram and Twitter. It can be said that these shares are made in the form of short-term videos and photographs and that these shares are to draw the attention of the audience to the subject of sustainability in gastronomy in a short time and to give various news on this subject.

REFERENCES

America, C. G. T. N. (2019). Nature is the key to a happy kitchen for an eco-friendly chef in Ecuador. Retrieved from https://www.youtube.com/watch?v=O0OT4UYyxkE Access Date: 26.06.2023.

Barcelona Visio. (2009). "Slow Food". A boost to ecogastronomy. Retrieved from https://www.youtube.com/watch?v=QWVgcd_rqVk Access Date: 10.06.2023.

Bestdestination (2022). Macao Gastronomy City and Sustainable Gastronomy Day. Retrieved from https://www.youtube.com/watch?v=YxZ55B4XtdM Access Date: 10.06.2023.

Bitividi (2020). Sustainable Gastronomy Day. Retrieved from https://www.youtube.com/watch?v=6cGxObtNaBU Access Date: 22.06.2023.

Breakthrough Institute. (2017). Sustainable Gastronomy. Retrieved from https://www.youtube.com/watch?v=s6dkVmsCyUw Access Date: 15.06.2023.

Campbell, B., Khachatryan, H., Behe, B., Dennis, J., & Hall, C. (2015). Consumer perceptions of eco-friendly and sustainable terms. *Agricultural and Resource Economics Review*, *44*(1), 21–34. doi:10.1017/S1068280500004603

CGTN. (2019). What is sustainable gastronomy? Retrieved from https://www.youtube.com/watch?v=isGKvrauay4 Access Date: 11.06.2023.

David Szanto. (2016). Stories of Eco-Gastronomy: Torino. Retrieved from https://www.youtube.com/watch?v=d43gt6xw4IQ Access Date: 13.06.2023.

David Szanto. (2017) Stories of Eco-Gastronomy: Montreal. Retrieved from https://www.youtube.com/watch?v=hNlyaTHxI1o Access Date: 13.06.2023.

Demir, Y. (2023). Sığır eti üretiminde su ayak izi durumu (Evaluation of water footprint in beef production). *Aydın Gastronomy*, *7*(1), 161–171. doi:10.17932/IAU.GASTRONOMY.2017.016/gastronomy_v07i10011

Eco-Gastronomy. (2023) Eating Is an Agricultural Act'-Wendell Berry. Retrieved from https://eco-gastronomy.com/ Access Date: 12.05.2023.

EcoCircuitos Panama. (2017). Eco-Story Telling: Panama's Gastronomy with Patricia Miranda. Retrieved from https://www.youtube.com/watch?v=O2fL2uugKHc Access Date: 17.06.2023.

Ecotsy (2023). Being Environmentally Friendly vs Eco Friendly: Which is Better? Retrieved from https://ecotsy.com/environmentally-friendly-vs-eco-friendly/ Access Date: 6.06.2023.

FAO. (2019). No one likes labels… unless it comes to food. Retrieved from https://www.fao.org/fao-stories/article/en/c/1177378/ Access Date: 25.06.2023.

FAO. (2020). Calling all foodies: this one's for you. Retrieved from https://www.fao.org/fao-stories/article/en/c/1198076/ Access Date: 10.06.2023.

FAO. (2021a). A spicy agricultural heritage: there's more to wasabi than meets the eye. Retrieved from https://www.fao.org/fao-stories/article/en/c/1391113/ Access Date: 17.06.2023.

FAO. (2021b). We are what we eat: Transforming diets to transform agri-food systems. Retrieved from https://www.fao.org/fao-stories/article/en/c/1393055/ Access Date: 19.06.2023.

FAO. (2021c). A cup of tea… or cha? Retrieved from https://www.fao.org/fao-stories/article/en/c/1401152/ Access Date: 10.06.2023.

FAO. (2022). Local food for healthy locals. Retrieved from https://www.fao.org/fao-stories/article/en/c/1481073/ Access Date: 12.06.2023.

FAO. (2023). Freekeh: The ancient dish firing up rural women's futures in Lebanon. Retrieved from https://www.fao.org/fao-stories/article/en/c/1634616/ Access Date: 21.06.2023.

Food and Agriculture Organization of the United Nations (FAO). (2018). Sustainable food systems-Concept and framework. Retrieved from https://www.fao.org/3/ca2079en/CA2079EN.pdf Access Date: 12.05.2023.

Future Food Institute. (2020). Food is a Conversation: Sustainable Gastronomy Day - Japanese Session. https://www.youtube.com/watch?v=gr7U5BsSeBM Access Date: 19.06.2023.

Galiano Conservancy Association. (2021). John Volpe on EcoGastronomy | Galiano Conservancy Association. Retrieved from https://www.youtube.com/watch?v=lqeFMwzxql4 Access Date: 14.06.2023.

Gastro Tourism Academy. (2021). Eco-Gastronomy: Minjangan Grass. https://www.youtube.com/watch?v=FjISfzcr7Vw Access Date: 10.06.2023.

Gastro Tourism Academy. (2023). Climate Eaters Webinar- Climate Action for Food Sustainability: Eco Gastronomy and Greenpreneurs. Retrieved from https://www.youtube.com/watch?v=Az1NvVpeNaM Access Date: 17.06.2023.

Görkem, O., Bayram, M., Bertan, S., & Bayram, Ü. (2016). Approach of the Young to Eco-Gastronomy: The Case of Turkish Cuisine. Global Issues and Trends in Tourism, 602.

Green Restaurant Association. (2023a). Greening Restaurants Since 1990. Retrieved from https://www.dinegreen.com/about Access Date: 21.06.2023.

Green Restaurant Association. (2023b). Product & Service Certification Standards. Retrieved from https://www.dinegreen.com/product-certification-standards Access Date: 21.06.2023.

Green Restaurant Association. (2023c). Green your restaurant-Become a green vendor-Be a green consumer-Become a green employee. Retrieved from https://www.dinegreen.com/ Access Date: 21.06.2023.

Gün, S., & Kılıç, G. D. (2022). Eko Gastronomi ve Sürdürülebilir Gastronomi (Eco Gastronomy and Sustainable Gastronomy). *Journal of Tourism and Gastronomy Studies*, *10*(2), 999–1015.

Güneş, S. G. (2019). Eco-gastronomy, Tourism and Sustainability: The Rise of Sustainable Restaurants in the World. *Erasmus International Academic Research Symposium on Educational and Social Sciences*, 5-6 April 2019, Izmir, (pp.67-84).

Hivos people unlimited (2023). Sustaınable Dıets For All. Retrieved from https://hivos.org/program/sustainable-diets-4-all/ Access Date: 23.06.2023.

Hossain, M. M. (2022). Consumers Buying Intention towards Eco-Friendly Products: A Perspectives of Coronavirus Outbreak. *International Journal of Humanities and Social Science Review*, *8*(1), 11–24.

Intellidigest (2023). Course Catalog. Retrieved from https://intellidigest.com/courses/ Access Date: 20.06.2023.

Michelin Guide. (2020). MICHELIN Debuts New Sustainable Gastronomy Icon. Retrieved from https://Guide.michelin.com/sg/en/article/news-and-views/michelin-new-sustainable-gastronomy-icon Access Date: 20.06.2023.

Michelin Guide. (2021a). The MICHELIN Guide and sustainable gastronomy. Retrieved from https://mb.cision.com/Public/55/3413982/85252c3e3c870f10.pdf Access Date: 10.06.2023.

Michelin Guide. (2021b). What is a MICHELIN Green Star? Retrieved from https://Guide.michelin.com/kr/en/article/features/what-is-a-michelin-green-star-kr Access Date: 19.06.2023.

Michelin Guide. (2022). What is a Michelin Green Star and How Do You Earn One? Retrieved from https://Guide.michelin.com/ie/en/article/features/what-is-the-michelin-green-star-and-how-do-you-earn-one Access Date: 10.06.2023.

Michelin Guide. (2023a). Sustainability Leads the Way at Le Manoir aux Quat'Saisons, A Belmond Hotel. https://Guide.michelin.com/gb/en/article/sustainable-gastronomy/sustainability-leads-the-way-at-le-manoir-aux-quat-saisons-a-belmond-hotel Access Date: 15.06.2023.

Michelin Guide. (2023b). World Bee Day–20th May. https://Guide.michelin.com/ie/en/article/sustainable-gastronomy/world-bee-day Access Date: 20.06.2023.

Michelin Guide. (2023c). Green Star Focus: Apricity, London. Retrieved from https://Guide.michelin.com/ie/en/article/sustainable-gastronomy/green-star-focus-apricity-london Access Date: 20.06.2023.

Michelin Guide. (2023d). Sustainable Gastronomy Day–18th June. Retrieved from https://Guide.michelin.com/gb/en/article/features/sustainable-gastronomy-day---18th-june Access Date:20.06.2023.

Michelin Guide. (2023e). Chantelle Nicholson's Green Star Recipe. Retrieved from https://Guide.michelin.com/ie/en/article/dining-in/chantelle-nicholson-s-green-star-recipe Access Date: 20.06.2023.

Michelin Guide Asia. (2020). MICHELIN Introduces New Sustainable Gastronomy Icon. Retrieved from https://www.youtube.com/watch?v=BzsQ4Uouyvs Access Date: 13.06.2023.

Nespresso (2021). Sustainable Gastronomy with Adam Handling | Edition 28. Retrieved from https://www.youtube.com/watch?v=qzewMPs8jZ0 Access Date: 10.06.2023.

New York Botanical Garden. (2019). Creating a More Eco-friendly and Sustainable Dining Experience. Retrieved from https://www.youtube.com/watch?v=FBA_QNq2MTI Access Date: 10.06.2023.

Nico Dingemans. (2023). Sustainable Gastronomy Day-18.06.2023-From Farm to Fork Dinner in Amman. Retrieved from https://www.youtube.com/watch?v=IsiXsILPqNo Access Date: 10.06.2023.

Prof. M. THANGA DARWIN (2019). Sustainable Gastronomy Day 18 JUNE 2023. Retrieved from https://www.youtube.com/watch?v=6xGczlEcqMI Access Date: 10.06.2023.

Shelbizleee (2022). What a "climatarian" actually eats in a day|most eco-friendly food. Retrieved from https://www.youtube.com/watch?v=KY8WDiGqAEI Access Date: 11.06.2023.

Slovenia. (2023a). A sustainability label to delight and Guide all foodies who care about the environment. Retrieved from https://www.tasteslovenia.si/en/taste-slovenia/european-region-of-gastronomy-2021/slovenia-green-cuisine/ Access Date: 20.06.2023.

Slovenia. (2023b). News from Slovenia-The first International Day of Zero Waste is here. Retrieved from https://www.slovenia.info/en/press-centre/news-of-the-tourism-press-agency/23006-the-first-international-day-of-zero-waste-is-here Access Date: 20.06.2023.

Sustainable Gastronomy Initiatives Map. (2023). Retrieved from https://www.google.com/maps/d/u/0/viewer?mid=10_pkm4ERhUZrsZDLQDeL6icHGgesnxVf&ll=-3.81666561775622e-14%2C0&z=1 Access Date: 21.06.2023.

The Atlantic (2020). The Struggle of a Zero-Waste Restaurant. Retrieved from https://www.youtube.com/watch?v=Gj38m08QLoE Access Date: 13.06.2023.

UNH Paul College of Business and Economics. (2020). Inside the Classroom: Ecogastronomy. Retrieved from https://www.youtube.com/watch?v=QuKyDeljE8Q Access Date: 13.06.2023.

United Nations. (2023) Sustainable Gastronomy Day 18 June. Retrieved from https://www.un.org/en/observances/sustainable-gastronomy-day#:~:text=Sustainable%20gastronomy%2C%20therefore%2C%20means%20cuisine,and%20eventually%20to%20our%20plates. Access Date: 15.06.2023.

UNtoday. (2023). Sustainable Gastronomy Day June 18, 2023. Retrieved from https://nationaltoday.com/sustainable-gastronomy-day/ Access Date: 20.06.2023.

US Embassy BSB. (2022a). Sustainable Gastronomy Ep. 1. Retrieved from https://www.youtube.com/watch?v=j1BBpBznugU Access Date: 17.06.2023.

US Embassy BSB. (2022b). Sustainable Gastronomy Ep.2. Retrieved from https://www.youtube.com/watch?v=JRzgeRCw7Pk&list=PLjWN0y_JL94pQWTXt-Qq_EYeIN-dNKJCo&index=2 Access Date: 17.06.2023.

US Embassy BSB. (2022c). Sustainable Gastronomy Ep.3. Retrieved from https://www.youtube.com/watch?v=jy1iUOv4L0w&list=PLjWN0y_JL94pQWTXt-Qq_EYeIN-dNKJCo&index=3 Access Date: 17.06.2023.

World Tourism Organization (UNWTO). (2023). Gastronomy. Retrieved from https://www.unwto.org/gastronomy Access Date: 12.04.2023.

Chapter 3
Breakthrough of Cutting-Edge Technologies Into the Culinary World:
Trends and Challenges

Ananiyasri Raghavan
https://orcid.org/0009-0003-8850-4288
SASTRA University, India

Chidambaram Vijayabanu
https://orcid.org/0000-0002-0125-4534
SASTRA University, India

ABSTRACT

Technology has undoubtedly revamped the food industry into its current glory. The food industry, through a complex network of diverse businesses, has managed to maintain its momentum and massive domination over other industries by adapting and modifying technology to pique the interests of its customers. The object of the chapter is trace the impact of various technologies on the food industry including the internet, social media, and the recently adopted technologies namely Artificial Intelligence (AI) and Augmented Reality (AR). The current chapter aims to equip the targeted audience with global food and food service trends, challenges, and continued innovation due to competition. The chapter will further deal with recent technological inventions and new trends and discuss the long-term impact technology will have on the future of the food industry. The chapter will facilitate the learning of trends and innovations in the food industry and provide insight to hoteliers, chefs, entrepreneurs, engineers, software technicians, data analysts, economists, and other learned audiences.

1. INTRODUCTION

Technology has changed how the food industry operates, from using coal and wood logs for cooking food to using induction stoves and microwave ovens. Food intake, which was only for sustenance and survival, has also become a pleasurable experience. Technology has undoubtedly revamped the

DOI: 10.4018/978-1-6684-9094-5.ch003

food industry into its current glory. One of the highest-earning industries, the food service industry utilizes technological innovation to yield and process nearly 64% of the world's food products. The food industry, through a complex network of diverse businesses, has managed to maintain its momentum and massive dominance over other industries by adapting and modifying technology to pique the interests of its customers. Since its early stages, technology has significantly changed whichever field it applies to. However, it has revamped the entirety of the food industry. The invention of new technological appliances has attracted consumers who adapt them to their daily lives. The launch of television and media broadened the horizons of the culinary world, and the discoveries were limitless. Social media has influenced every individual's lifestyle and food choices. High-end machines and devices catering to large-scale food production have overtaken the food industry. A paradigm shift has occurred in the food service industry due to the popularisation of applications, software systems, and services. Artificial intelligence has made the functioning of this industry very efficient. Augmented reality is used by various outlets to enhance dining experiences. These inventions have led to breakthrough discoveries, resulting in a monumental transformation in the functioning and working of the gigantic food industry.

The scope and object of the act are to trace the impact of various technologies on the food industry, including the Internet, social media, and the recently adopted technologies of Artificial Intelligence (A.I.) and Augmented Reality (A.R.). The current chapter aims to equip the targeted audience with global food and food service trends, challenges, and continued innovation due to competition. The chapter will further deal with recent technological inventions and new trends and discuss the long-term impact technology will have on the future of the food industry. The chapter will facilitate the learning of trends and innovations in the food industry and provide insight to hoteliers, chefs, entrepreneurs, engineers, software technicians, data analysts, economists, and other learned audiences.

2. TRANSFORMATION OF THE CULINARY WORLD BY TECHNOLOGY

The culinary world has always believed in the sanctity of making things by hand. It has always been believed that freshly made food attracts everyone, irrespective of cuisine. However, times have changed as people always run against the clock, work for more money and fame, and seek more worldly pleasures. This population does not care about making fresh food cooked with fresh ingredients every day and would rather spend cooking time watching Netflix or catching up on their sleep. Researchers and innovators understood the change in perspectives and acted upon it to increase their businesses. The current generation has welcomed the introduction of ready-made meals and the availability of ingredients and pantries in every store and supermarket. This practice, which started due to the necessity and increase of famine, has become an industry catering to people's needs and opening employment opportunities for many of the population. The food business adopted a few food processing approaches to meet consumer demand and produce food instantly. As time progressed, each sector associated with the food industry began implementing novel methods to boost efficiency and reduce waste. (Freeman, 1987). The food industry was modernized by employing cutting-edge equipment for agriculture and food processing, which was supplanted by smart machines and production lines. (Mahalik & Nambiar, 2010).

2.1 The Kraft Revolution

The use of technology in the food industry started during the Green Revolution, wherein technology helped produce crops and products at a larger scale to meet the ever-growing demand for food (Rollin et al., 2011). The introduction of similar technologies inspired the idea of producing groceries, condiments, and similar cooking ingredients in large-scale quantities, which proved to be a saviour for the U.S. when it was attacked by famine in the early 20th century. One such product manufactured then was the Kraft Dinner, which created the market for ready-made dinners. A salesperson came up with the idea of selling pasta noodles like macaroni and cheese as pairs during the Great Depression, so he started rubber-banding grated cheese to boxes of pasta. Kraft released this product in the United States and Canada in 1937. The product's success was strongly influenced by the time of its release. The product, considered a full supper for families, acquired a substantial market during World War II due to the rationing of milk and dairy products, an increase in the consumption of vegetarian meals, and the number of women working outside the home. When many Canadian houses did not have refrigerators, its ten-month shelf life was appealing. Customers might also get two food ration stamps for every one they receive.

Following the example set by Kraft, the mass food production industry has changed drastically. Today's technology can guarantee the availability of a wide range of goods, from fresh produce and fruits to carefully vetted meat, partially cooked food, and meals requiring minimal preparation. Other similar products are also found in almost every store or supermarket. Even products such as wine, alcohol, etc., which needed ageing and time to be made, are now being produced and sold in bulk. The introduction of heavy machinery and large-scale appliances started the mass production and sale of food products. These machines can manufacture tonnes of products every day. The range includes moulds, mixers, ovens, freezers, blenders, coolers, and packagers and covers the entire process of manufacturing food products and produce.

2.2 Ovens to OTG

Despite changing how food manufacturing works, technology has made making food much more efficient and effortless. For example, in the pre-technological discovery era, people had to collect wood, clay, or coal to build ovens and utensils that would not sustain themselves for a long time. Baking was a laborious process, and hence bakeries were a hot business. However, creating one device, the OTG, oven, toaster, and grill, has made the process less time-consuming and attracted many consumers. The creation of stovetops, microwaves, ovens, and other technological appliances has revolutionized the kitchen forever. Appliances like mixers, juicers, blenders, and kitchen aids have made the cooking process effortless. Devices like refrigerators and freezers have led to food storage for a long time without getting spoiled, thus increasing the food product's longevity and shelf life and reducing the time spent in the kitchen daily.

2.3 Robots and Their Multi-Functionality

The introduction of robots into the food industry has increased their systematic functioning. Robots cover various functions throughout the process due to their multi-tasking capabilities. The concept of robotics is now applied to milking cows and crop cultivation, including crop imaging (McKinley, 2014) which has shown how robotics can help boost the development of the agriculture industry. As just another technological device, robots are taking over the world with their efficiency and automation. Robots in the food industry can check inventory, deliver produce, receive orders from customers, and perform various functions. These cutting-edge machines can benefit the food business by helping employees and replacing risky practices. For example, "robot butchers" help by handling intricate cuts of meat in factories and minimizing the risk undertaken by employees performing the same task. Companies can meet their demands much faster as robots speed up the packaging and production processes. A report states, "The rise of robotics in the food industry is a tangible example of food tech. The number of robots in the European food industry is well over 30,000, while the number of robots per 10,000 employees rose from 62 in 2013 to 84 in 2017. Although Germany is the largest market, robot density is relatively highest in Sweden, Denmark, the Netherlands, and Italy." Robots have also increased productivity and minimized costs. When there is a great demand for products, and the requirement is to be met within a short period with minimal capital, robotic devices help create a more efficient environment to meet the need of the hour.

2.4 Dynamic Drones

The food industry is highly dependent on agricultural production and the functioning of the agriculture industry. The weather, quality, and fertility of the soil and land are all important factors in agriculture. Without technology, workers must physically and manually inspect the crops to ensure they are healthy. The agricultural industry's profitability has changed as a result of precision agriculture. It is termed so, as this mode of agriculture uses satellite imagery and GPS tracking systems to monitor the yield of crops, weather patterns, water table level, soil condition, and other such necessary information to improve farming standards. Not only can they see all that is happening across the fields, but the farmers and other involved parties can test the crops' health and soil fertility by analyzing the results.

One of the most popular ways to practice precision technology is by using drones. They can quickly locate and detect damaged crops, taking urgent action to fix the problem. The industry may evaluate and ensure that the ideal conditions, such as seed selection, crop monitoring, watering, and temperature monitoring, can be enhanced by employing a computerized system, which will result in better quality in the products produced by the food sector (Kumar et al., 2021). Using such technologies helps workers in the agriculture industry increase efficiency in their work and reduce time spent manually monitoring the fields.

2.5 Environment-Friendly Production and Packaging

One of the recent positive trends worldwide is to "go green" and take actions that protect Mother Nature. Technology helps ensure that the packaging materials used during production and containers used to store food can be reused later or are biodegradable. The latest technologies have ensured that the food industry does not affect the environment and can perform at its best while being sustainable. The eco-friendly packing alternatives made possible by the rise in technology include recyclable and edible packaging. These modes of packaging reduce waste to a minimal amount, and the nanoparticles used in the packaging material also help retain the product's food standards by killing harmful bacteria. In a recent transition, PepsiCo has committed to a sustainability goal by using 100% recycled plastic beverage bottles for its Pepsi Brand and ensuring they are reusable. In a similar initiative to reduce the amount of plastic ending up in oceans and polluting the environment, top beverage producers, namely Coca-Cola, PepsiCo, and Bisleri, have started fixing a repurchase value on all plastic bottles. The buyback amount is printed on the bottles, and customers can return empty plastic bottles to receive payment.

Technology has also made it possible to maintain sustainability by creating apps and websites that ensure food or products are not wasted. Users can browse these websites to find places to donate unused food products. Food banks can be easily located via these systems. One such app, Copia, works similarly by connecting shops and restaurants with a surplus of available food with numerous non-profit organizations that need food to feed people in need.

2.6 Tales of the Glory of 3D Printing

3D printing is the process of creating three-dimensional objects through a layering method with the help of computer programs. This procedure layers materials such as plastics, composites, or biomaterials to produce merchandise that ranges in size, form, stiffness, and colour. The utilization of 3D printing has now become crucial to meeting the demands of a certain market. It recently introduced soft foods, so people who cannot consume hard foods can eat them. Along with the Dutch business TNO 2016, one of Italy's most well-known producers of pasta shells, Barilla, created the first 3D printer for fresh pasta. Using it, they have been able to produce high-quality pasta recipes as well as novel forms. The business has declared that at least 15 3D-printed pasta shapes are available on its website. The Silicon Valley startup BeeHex recently printed fresh pizzas using 3D printing technology.

NASA provided funding for this project to spare astronauts from having to endure the monotony of gulping down freeze-dried, pre-packaged "space cuisine" daily, weekly, and monthly. The goal of this creation was to enable astronauts to enjoy excellent meals while in space. This innovation-enabling technology encourages sustainability and is highly eco-aware of the environment.

2.7 Sensors and Their Significant Role

Smart sensors help transfer aspects of digital data and offer priceless insights into the industrial process. Sensors play a multi-faceted role in the food industry. They count inventory, monitor the production process, detect any defects in the system, and indicate any harmful contamination, making the manager's work more efficient. Sensors monitor the food and beverage conditions while dispersing the goods to multiple locations. Sensors can monitor if the products are only delivered to regular distributors. Workers receive information via a desktop or mobile app if an issue arises. Sensors also play a role in daily food activities. Sensors are used in soda, coffee, and beverage vending machines and eliminate the need for an employee to monitor the process. Sensors indicate the temperature in refrigerators and, in the case of supermarkets, also indicate if a product has been stolen.

We see that these breakthrough technologies have transformed the functioning of the culinary world, for what started as a harsh and manpower-intensive industry has now fully transformed into a machine-made industry with billions of turnovers, irrespective of minimal capital investment.

3. FUTURE RETROSPECTION OF RESTAURANTS

The launch of television and media broadened the horizons of the culinary world, and the discoveries were limitless. People across the globe could watch and learn about the cuisines of different cultures and countries. Travel shows target the inquisitiveness of people to try international dishes. Restaurants took advantage of people's curiosity and started introducing the concept now popularly known as "multi-cuisine." International dishes are sold in the localities of customers, and they get to rejoice in the experience of a world tour without moving anywhere. Restaurants became the new talk of the town, where people only had to pay for a luxurious culinary experience. Restaurants were costly, so going to restaurants was seen as a status symbol. Restaurants make people feel welcomed and respected as servers open the doors, wait at your table, take your food order, deliver it to your table, and treat you with the highest care and diligence. However, over time, the growing economy enabled more people to visit restaurants, and thus the craze about restaurants slowly died down and became just another business.

However, introducing technology has been a bane to the hotel industry. While introducing exciting technological appliances in restaurants has kindled people's curiosity, technology has also created new experiences of delivering food to people's doorsteps, thus feeding the fantasy of people who do not want to dress up, travel, and eat at restaurants. The following trends have changed the functioning of restaurants:

3.1 Online Delivery Services

Technology has introduced various apps via which people can call the restaurant and order food, which will be delivered to their doorstep quickly. Since many restaurants found it difficult to run their own home delivery, services such as Uber, Tomato, and Swiggy took advantage and opened a new market for themselves. These services established their websites and applications, and people could order food from their desktops and mobile phones. They collaborated with various food outlets and restaurants, and now people can order food from different restaurants through these areas, and the employees will deliver the food from these services. The introduction of online ordering has created employment opportunities for a huge number of people who are not educated and are from average or poor financial backgrounds.

People can also order online and get the food from the restaurant or food outlet to eat elsewhere, as sitting and eating at the outlet is no longer necessary. The introduction of drive-throughs has revolutionized take-out food service. People can go to drive-throughs, give their order, collect it in a few minutes, and leave. This concept has attracted the interest of the current working generation. They pick food from drive-throughs as it is efficient, reasonably priced, and saves them the hassle of cooking. Popular food outlets like McDonald's, Starbucks, and Taco Bell thrive due to their fast and efficient drive-thru systems.

3.2 Digital Payment Options

Paying via phone, debit cards, online portals, and Q.R. codes have become the norm. It used to take a long time to wait in the queue, order the food, and pay the cashier. Introducing services such as Google Pay, Apple Pay, etc., is efficient as one need not wait in a long queue. They can order via the app, pay via online portals, and get the food delivered to their doorstep or pick it up in the drive-thru on the way to work or home. These digital payment modes became popular during the COVID-19 pandemic, as contactless interactions were the need of the hour. It removes contact with others and speeds up the process of buying food.

3.3 Restaurant Robots

Robots in restaurants are the talk of the town. Restaurants have started including robots on their staff. These robots wait on tables, take orders, bring food to the table, and assist in every way a waiter would. Robotics in the kitchen preparation process has many benefits. To begin with, robots can produce large quantities of food efficiently without taking breaks.

Robots became the need of the hour when countries were facing labour shortages due to pandemics and other factors. These robots follow commands and save a lot of waiting time. Some robots also perform major cooking functions such as mixing dough, scoring meats, etc. These robots are also cost-effective, reducing the need to employ waitstaff. Companies such as Servi produce programmed robots that act as restaurant waitstaff.

3.4 The Ghost Kitchen

The name implies that they are invisible. However, the concept entails that these food outlets do not have a physical storefront. It means there is no place to sit, order food, eat, and go back. In these outlets, only the cooking and packaging processes happen. These kitchens rely on online food delivery services such as Uber Eats, where people can order food online. The delivery boy takes the meal from the ghost kitchen and gives it to the customer. The mere introduction of online delivery apps has revolutionized the restaurant system. Ghost kitchens operate out of areas with lower rents because there is no need for a coveted downtown or a high-traffic site, making the functioning cost-effective with increased turnover despite a minimum investment. The idea of the ghost kitchen is developing daily. The restaurant chain Chick-fil-A recently embraced the idea of a ghost kitchen. Little Blue Menu, a ghost kitchen concept expanded from Nashville to Atlanta, was only recently launched. Customers may also place orders from their three newest virtual restaurants, Outfox Wings, Garden Day, and Flock & Farm, along with their standard Chick-fil-A menu.

3.5 Asian Discoveries

Though very odd, introducing conveyer belts in restaurants led to a cost-effective, time-conscious, and efficient food delivery system within the restaurants. Many restaurants in South Asia have introduced conveyor belts, where the food is delivered to every table by the device without any manual assistance. Sushi conveyor belts are a trend of their own. Asia is known for its hard work and dedication, and its vending machines best exemplify Japan's fast-paced way of life. These mini-stores typically sell coffee and other beverages so that customers can buy what they need on the spot and get to work on time. However, these vending machines have become top-notch for selling fresh, hot food. These hot meals range from soups to French fries to ramen and have every kind of meal that would fill the entire day.

Many convenience stores in Asia offer food products and appliances to cook the food there, sit, eat, and then leave the store. People can buy ready-made noodles, cook them in the microwave, eat them in the store, and then leave. They are also reasonably priced, making it easier for employees to have fresh food without worrying about finances. Before the introduction of robots, restaurants in Asia used centralized communication systems where customers could order food from the speakers at their table directly to the kitchen, and the food would arrive at their table. These restaurants also use walkie-talkies to interact with the wait staff. This method is cost-effective and efficient, as there is no need for a server to take orders.

3.6 Self-Service Kiosks

Self-service kiosks are among the latest technologies commonly used in the restaurant industry throughout various nations. Previous studies have shown that customers can reduce order errors because fast food kiosks do not require vocal communication (Lee, 2020). McDonald's has been testing them since 2003 and was one of the first businesses to have a kiosk. The fast-food restaurant declared that self-service kiosks would be introduced at all its U.S. sites in 2020. Fast-food establishments have quickly embraced the technology due to the fierce competition the sector faces and how it gives them a competitive edge by displacing human service workers. (Seo, 2020). These kiosks display the money and the customization options, accept payment, provide food receipts, and deliver the food. More restaurants focusing on takeout and fast service will likely start, including self-service kiosks. These devices are easy to maintain as they are programmed, take minimal cost, and create a unique customer experience. Introverts especially appreciate these machines, as they do not have to interact with people to order food.

3.7 Facial Recognition Technology

Facial recognition technology has expanded its usage from identification processes such as passport verification in airports to remembering frequent customers in restaurants. Several restaurants are testing this technology for self-service kiosks and reordering purposes. Self-service kiosks that use facial recognition technology can identify consumers and their prior orders to make restocking quick and simple. Making ordering quick and simple can enhance and personalize the customer's experience and boost client retention. Since their preferences have already been preset, the consumer would feel special, and the food would arrive swiftly.

3.8 Self-Ordering Tablets

After various predictions and anticipation, the serverless restaurant experience has started functioning. At restaurants with modern technology, self-ordering tablets have proven that there is no need for human servers. These innovations have been incorporated by companies like Panera Bread and McDonald's in numerous well-known locations. Clients do not need to interact with staff to place their food orders. Thanks to this AI-based technology, customers may browse the meal menu and select their favourite products, making the experience speedy and distinctive.

3.9 Bluetooth Temperature Sensors

Restaurant owners are always pressured to ensure that the hygiene and quality standards mandated are maintained properly. Bluetooth temperature sensors have been introduced to make their lives simpler and less stressful. It is used to maintain the same degree of quality in both their food and equipment. With fixed probes and sensors, restaurants may measure the temperature of their critical assets manually or automatically. They can maintain the temperatures of refrigerators, freezers, ovens, stoves, fryers, and just about any other cooking device. These sensors can also indicate any contamination, hazard, or danger, thus reducing the burden on the shoulders of the managers and employees.

3.10 Handheld Entertainment

Restaurant tablets offer more than simply a speedy ordering and payment option. Many businesses, including Buzz Time, offer iPads with fun games and trivia to eateries. Also, it enables customers to participate in contests held across the country twenty-five hours a day, seven days a week. A live poker or trivia tournament might be planned using these tablets in the restaurant. With this technology, restaurants may draw in a huge number of clients. They can incorporate events that pique their customers' interests by comprehending the area's regional interests.

Technology impacts the restaurant sector more than just providing tools for eateries, just like it does in every other industry. Intelligent use of technology in restaurants will determine whether these discoveries impact the industry positively or negatively.

4. SOCIAL MEDIA AND ITS SIGNIFICANCE IN THE FOOD INDUSTRY

Social Media has broken every communication barrier that previously existed due to differences in geographical locations, religions, ethnicities, and religions. Now, anyone can know what is happening in every corner of the world. We use free social media sites online while choosing our menu, shopping, and cooking (UDA, 2014). Social media significantly influences the industry and how food and cuisines are valued and celebrated; therefore, this outbreak concentrates more on the humanistic side of the food sector, where customers and consumers have complete authority.

4.1 Social Sharing: The Unwritten Rule

To be considered worthwhile, every new product used, restaurant visited, and food item ordered must be shared in social forums. The desire of chefs to create aesthetically beautiful cuisine is expanding as society becomes more aware of how much people like taking pictures of the food they have ordered in restaurants. (Spence et al., 2016). People write evaluations on TripAdvisor and Google, but it has been taken a step further since these images are now posted in open forums. Restaurant owners and managers must pay close attention to how their business is run and ensure all patrons are happy with the level of service. If the meals are gorgeous, social media filters and phone cameras have been adjusted to make them even more enticing when shared online. On the one hand, the restaurants benefit from free advertising since customers will come in because of the cuisine. The chefs are now under pressure to ensure the dinner tastes delicious and is delivered flawlessly. So, people will quickly discover if the guest is unsatisfied with the food. The desire to share dining experiences on social media includes a desire to gain a following or feel validated about their purchase by a community of people with similar interests. When exposed to enticing food imagery, visual processing, and reward-related brain regions people become more attentive and neutrally active. (Sina et al., 2022). According to some research, perceived standards about real consumption influenced by social media may affect how much low- and high-energy foods and drinks are consumed. (Simeone & Scarpato, 2020). Food is still used today, as it has been for a long time, to build support networks. Surprisingly, individuals seldom gather around the dinner table to connect over food; instead, they share their dishes on social networking sites to engage in discussion and share common interests.

4.2 Online Reviews and Their Credibility

Social media sites are regarded as trusted review resources. Social media is what consumers go to when wanting to discover restaurants people are focused on, particularly testimonials from former clients. The cost and time for data analysis have been reduced thanks to internet apps (Hudson & Gilbert, 2006). The restaurant sector has undergone a tremendous transformation due to the authority that digital recommendations give consumers. Customers are more important to a restaurant's success than efficiency or service standards. For instance, a study by Restaurant Insights revealed that 68% of American diners visited an eatery they had never been to before based solely on excellent online evaluations. Restaurants must upgrade their customer experience, sanitation, and food quality to create their niche, especially in a society where any negative outcome might turn prospective consumers away. To prevent social criticism, restaurants embracing new requirements must closely monitor their activities. Customers will criticize the world through online reviews and social media if an employee is wearing a mask improperly or not wearing one, if a table seems dirty, if the capacity of guests is exceeded in person, or if gloves are not changed regularly enough. Reviews nowadays are accessible everywhere beyond TripAdvisor and Google. Other social networking platforms are striving to capture more users by establishing food review platforms. Facebook, for instance, offers a food review option in its graphical interface. Without visiting the website, prospective customers can find reviews regarding the restaurant they want to see on its Facebook page. These reviews also do not stay on that platform anymore because they are spread everywhere and made viral on every social platform.

4.3 Garnering Loyalty

To maintain a strategic advantage in a particular market, a brand must address customer loyalty. With a larger range of choices available to customers nowadays, this is particularly true for restaurants. Restaurants must maintain a high level of brand engagement and distinguish their businesses by utilizing rewards programs. One may establish powerful customer loyalty methods and loyalty programs with social media platforms to encourage customer retention. To roll out new offers, discount clientele, identify regular customers with scorecards, and foster a rapport, restaurants can establish their sites or profiles on social media.

4.4 Social Media Analytics

The food culture and the services are the primary elements that attract loyal clients to return to a restaurant. Customers like to eat at well-known establishments where they are greeted with a warm welcome. This is not a simple sensation since everyone has an entirely distinct concept of or desire for what they want to feel or how they like to engage with the people at a restaurant. Some diners might enjoy the outgoing bartender narrating anecdotes, while others might appreciate the secluded corner booth. The more businesses grasp how consumers behave and connect with their brand or restaurant, the more they will enhance consumer engagement and build a loyal fan base. Much of the data obtained by mobile and social media apps is about users, which companies and outlets might employ to learn more about them and their interactions with their franchises. Every time a person tags a restaurant and tends to make an online order or utilizes their mobile app, data about that individual is discovered, which includes details about who they are, their preferences for menu items, the days during the week that they visit, how they feel about a specific restaurant and everything in between. Mass establishments engaged with social media and even replied to interactions and comments on their pages (Facebook, Instagram, Twitter, etc.). Major outlets such as Starbucks and Krispy Kreme are very active on social media to connect with their audience. They follow current events and give discounts to garner more customers. For example, Domino's, the pizza outlet, gives huge discounts during the Indian Premier League season to attract the loyalty of their customers.

4.5 Promotion of Dietary Lifestyles

The reach of social media worldwide has led to a growing culture of consuming healthy food. People from all over the world have started sharing their journeys to a healthy lifestyle. People talk about their struggles with various disorders, such as juvenile diabetes, anorexia, and thyroid disease, and the diets they follow to maintain their health. People post healthy alternatives to junk food, and professionals from various fields, such as doctors, nutritionists, dieticians, biologists, food analysts, and other similar professionals, share the secrets of various foods and freely talk about various health disorders to educate people and spread awareness. Social media, though just another technology, has influenced a mass population to change their food lifestyle. Many influencers also promote products they feel are healthier alternatives to already-existing products. People prefer consuming visually appealing products that have become viral on all social media platforms or are endorsed by celebrities. This influencing culture plays a major role in the customer's product-consuming choices.

A study that "sought to investigate the acute influence of socially approved social media posts on participants' eating behaviour" was carried out by researchers from the Faculty of Health and Life Sciences at Aston University. One hundred sixty-nine adult women between 18 and 65 (mean age: 21) participated. These participants were instructed to view a mock Instagram account with pictures and videos of various meals, such as fruits, vegetables, cookies, cakes, and non-food imagery, with the postings denoted by the number of "Likes" as either high or low. Following this workout, participants accessed a buffet that included cookies and grapes. Ultimately, the researchers discovered that individuals ate more grapes than those who did not know when they saw highly liked photos of better food alternatives.

One of the study's authors, Doctoral student Lily Hawkins, stated that the study's findings "suggest that not only exposure to healthy food images on social media, but those that are also heavily endorsed with 'likes,' may nudge people to choose to eat more healthy foods rather than less nutritious foods." One explanation for this would be that when people think that others "like" and consume fruit and vegetables, they are motivated to alter their behaviour to fit in with what they perceive to be the norm.

The University of Birmingham conducted similar research on how social media, particularly Facebook, affects our eating habits. According to researchers, the study's central thesis was that individuals naturally want to participate in whatever popular alternatives their social media followers have discovered because they want to feel like they are part of a community.

Over 80% of customers desire to cut back on their sugar intake. Snack and beverage options that are healthier are popular on many platforms. This means that many people are looking for healthier foods and beverages. The study examined whether Facebook users' eating choices and behaviours were predicted by the type of food consumed. It investigated whether social media posts about food and eating habits influenced people's eating behaviours and preferences for various foods. The study found that perceived eating behaviour standards strongly predicted the consumption of fruits and vegetables. Perceived societal standards regarding what people should and should not consume are another factor that predicts the consumption of foods and beverages with added sugar.

The researchers concluded their study by saying, "Our results demonstrate that perceived norms regarding actual intake and norms associated with approval may influence consumption of low- and high-energy-dense foods and beverages differentially." "If you believe that people who use social media regularly are knowledgeable about eating behaviours that are conducive to health, this might sound like great news. Having a network that frequently promotes fad diets, spreads misleading information, or inhibits the use of nutrient-dense foods like fruits and vegetables may also suggest that such a network negatively impacts eating habits. This study was the first to investigate whether marketing healthy foods on the social media site Instagram by a well-known actual fit influencer led to a more favourable attitude towards the product and a higher buy intention than promoting healthy foods by an anonymous fictional fit influencer. We found that, in line with our expectations, viewers exposed to the well-known real-life fit influencer had a more positive attitude towards the healthy food product and a higher buy intention than those exposed to the fake fit influencer due to higher parasocial interaction.

Internet celebrities (i.e., social influencers) are consequently believed to have a bigger impact on consumer socialization since they are trusted more than traditional celebrity endorsers because they are perceived as peers and laypeople. Fixation time was longer when peers shared bad items but longer when famous people uploaded good foods. Nonetheless, participants were better able to remember and identify bad food brands than those healthy when they came from celebrities and firms rather than their peers, especially among older teens. (Murphy et al., 2020)

Social media has also spread various food lifestyles, such as veganism, where people do not eat any sort of animal product, including dairy products, and various styles of gluten-free and nut-free diets. Throughout the last 50 years, many food types have been in demand. They include some unusual types, such as functional foods, which have proven essential to leading a healthy lifestyle (Baxter, 2003). A brand-new market for healthy goods has emerged due to the advent of varied food lifestyles. As an illustration, consider the variety of milk options available, which include full-fat, non-fat, non-dairy, soy, and almond milk, to name just a few. Consumers are heavily influenced by these healthy lifestyles, leading to increased consumption of products such as vegan butter, vegan cheese, and other products catering to specific groups of people, making them feel unique and validated.

4.6 Diversity Through Social Media

Even though reality television stars and food programming continue to have cultural clout, many of their offerings are in direct competition with the vast and expanding amount of online food content, which ranges from the "serious amateurs" and "ordinary experts" of YouTube food advice channels to the microcelebrities and media influencers of foodie lifestyle blogs and the high-end aesthetics of Instagram (Lewis & Phillipov, 2018).

Social media allows people worldwide, from different countries and continents, to share their food culture and culinary heritage. Regional boundaries no longer narrow the cuisine, as people can now know the kinds of dishes cooked in each country and culture. While people used to have to visit restaurants to try multi-cuisine dishes, that is not the case anymore. More people from different regions take to social media to let people know about their delicacies and the intricacies behind making them. Social media has made huge barriers of geography and heritage seem invisible, thanks to its spread and connectivity worldwide. The spread of multi-national cuisine via social media has helped increase the market for food production. Due to the recognition of international dishes, the production of seasonings and raw produce catering to the ingredients of different cuisines is present in all supermarkets. Social media has influenced food, so Biryani can be cooked in Switzerland and Pasta in Mexico.

Thus, we see how social media has had a tremendous influence on consumers, irrespective of their age, and is ultimately deciding the profitability of the food industry.

5. ARTIFICIAL INTELLIGENCE AND ITS IMPACT ON THE FOOD INDUSTRY

The replication of human intelligence functions by machines, particularly computer systems, is known as artificial intelligence. Artificial intelligence, a recent technology, has shifted the food industry to a different paradigm. Though the definition may make the entire technology seem minimal, its functions are endless. The use of artificial intelligence in various stages of production in the food industry, from agriculture to storing, producing, manufacturing, distributing, and selling, shows the adaptability of this technology to various functions of the mass-working industry.

5.1 AI in the Factories

Artificial intelligence controls and regulates the processes of manufacturing and production in factories. Eliminating the need for monitoring, A.I. controls the entirety of the process using different programming languages, software, and sensors. It systematizes the process, ensures conformity and regular functionality, and eliminates danger. It is also programmed to sense any defect or fault in the system. The introduction of A.I. into this system has made the work of factories uniform and efficient.

5.2 AI in Restaurants

As discussed above concerning the concerning role of technology in the food industry, A.I. majorly governs the systematic running of restaurants. From taking orders via voice kiosks or digital kiosks, which transition from processing straightforward programmed commands to handling complex, multi-part questions, resulting in more lifelike conversational interactions, to allowing payments in the same kiosk with the help of digital transaction platforms, it has reduced the hassle of scrolling through the menu, waiting for the staff to order, or standing in a queue to order and pay, making the entire process of ordering food effortless and unique. Unaddressed phone calls can be detrimental to the restaurant industry. Many restaurants are implementing technology to prevent them from missing calls from consumers. A.I. systems can now take messages, make reservations, add people to a waitlist, and even respond to frequently asked queries. A.I. can assist in menu planning by analyzing the available inventory and products. Watson, an A.I. developed by International Business Machines Corporation (IBM), became the first A.I. chef in 2016 by generating creative new meals based solely on the ingredients. (Pinel, 2015).

The introduction of A.I. via food service apps such as Uber Eats and Zomato has revolutionized the food industry in many ways. With the help of these apps, customers can sit in the comfort of their homes, choose the restaurant from which they want to order food, customize it to their preferences, pay via digital payment apps, track the processing of their order, estimate the time of delivery, and get the food delivered to their home. Restaurants can delegate monotonous tasks such as telling the menu, collecting orders, and collecting payments to A.I. software and employ staff only for cooking the food and other necessary operations, thus reducing labour costs and increasing productivity. A.I. can be used to detect the age and preferences of customers and recommend dishes according to their tastes. A.I. can also record customers' voices to remember frequent customers and have their orders ready. For the benefit of customers, A.I. can assist restraints in understanding data. A.I. can identify trends in consumer behaviour, offering business owners information on replicating winning promotions and removing old incentives for deals, coupons, and discounts that did not work. To ensure a better client experience, it can roll out limited-time incentives for devoted customers. It can also monitor its reception and performance.

5.3 AI in Stores and Supermarkets

A.I. can be used to forecast inventory and demand and help decide the price of certain items and stock them accordingly in the storeroom to avoid product wastage and cut unnecessary costs. It can also track manufacturing and expiration dates and differentiate between old and new stocks. With the help of dedicated algorithms, A.I. can track fraudulent transactions. This feature can be useful, especially in huge supermarkets where losses are incurred from internal employee theft. A.I. can help analyze employees' interactions with customers and help score and improve track performance while evaluating employee records during promotions and appraisals. This fosters a positive, competitive online environment that supports the growth of brands.

5.4 AI and its Assistance to Customers

A.I. can assist businesses in adjusting to and mastering an increasingly dynamic market environment by finding and learning from patterns in large volumes of data from a variety of different sources, including past transactions, weather forecasts, social media trends, shopping patterns, online viewing history, facial expression analysis, and seasonal shopping patterns (Chui, 2017). Over 40% of online users prefer voice search to their cell phones when looking for nearby eateries, and 27% of all internet users now use voice search. Home appliances have become more common as voice assistants such as Amazon Alexa, Google Home, and Siri are now equipped to place orders in online delivery service apps, make reservations in restaurants, pre-order food from the menu while on the go, and multi-task, to mention a few. They have grown in complexity and scope, as have the tasks they can assist users with.

These functionalities of A.I. are just the start of the history of the revolution of A.I. and its impact on a particular niche, such as the food industry. The scope of A.I. is endless, and while the thought is exhilarating, the repercussions of A.I. taking over all sorts of jobs and functions are yet unknown.

6. AUGMENTED REALITY: THE MAGIC MAGNET

An interactive experience known as augmented reality involves adding artificially produced visuals, audio, and other stimuli to a real-world environment. It can give users a more realistic, immersive experience than they often have, adding to its distinctiveness. Augmented reality is the icing on the cake of the new food industry's glitz. Stakeholders in the food sector are being compelled to reevaluate traditional methods of managing food chains and adapt to the most recent technological advances. Food companies have a genuine interest in investigating augmented reality as technology opens numerous potentials in supply chains (Rejeb et al., 2021).

While its scope might seem very limited in the food industry, the way Augmented Reality has made its way into this mammoth industry is shocking. This technology is now enhancing many aspects of the industry.

6.1 The Magical Transformation of the Restaurant Experience

With the crumbling economy and growth of delivery service apps, restaurants find it difficult to attract customers to enjoy the restaurant experience. However, this technology has piqued the curiosity of customers of every age group, from retired people to small kids, to tech enthusiasts to moms who would like a break from their screaming children. Although in the past, restaurants had to use colouring books for kids and arcade games for adults to keep them entertained, A.R. allows eateries to create that "wow" factor.

While waiting for their pizzas to be made, patrons of the Italian restaurant Pizza Express can use an augmented reality (A.R.) feature built into the business's smartphone to play a 60-second football game. Similar to how Robinsons Fruit Shot, a well-known soft drink brand in the U.K., leveraged this technology to change kids' colouring placemats at restaurants across the country into fun and engaging activities, seeing it as a fantastic opportunity to market their product. Even take-out restaurants are being impacted by this technology, which is currently the topic of conversation. The American fast-food chain Pizza Hut collaborated with Pac-Man to produce a limited-edition pizza box that brought the classic game to life, sparked customers' childhood memories and exploited nostalgia to increase sales. Users can scan the pizza box's Q.R. code to access the game and relive their childhood nostalgia. According to the statement on the packaging, augmented reality packaging "creates an engaging brand experience and empowers people to directly connect with a brand via uber-creative visual and audio content." In the food and beverage business, it can also be used to make products more appealing (Konopelko, 2019).

Le Petit Chef, an augmented reality dining experience that has delighted numerous customers in several restaurants worldwide, did away with smartphones and instead used 3D projection mapping technology to give the impression that a thumb-sized chef was preparing their dish right before them. The promotions' creators said they "transported the customers from a snowy landscape to a tropical island, all while remaining seated at their table."

To engage guests and make them eager for the eating experience, the menus themselves were modified using A.R. The food on augmented reality menus comes to life, piquing clients' interest and possibly increasing their likelihood of placing an order. In contrast to what a traditional menu would offer, A.R. menus let customers view every aspect of a dish, including the feel of the components used, portion sizes, and other details like ingredients, calorie counts, etc. Due to the huge decrease in the need to ask to wait for staff questions about dishes, these A.R. menus provide guests with sufficient information about the meal options, enabling them to place orders with confidence and minimizing touch. Also, a digital payment system has been integrated with the A.R. menu in addition to menu, enabling contact-free payment right from the customer's table and improving both convenience and security throughout the entire process.

In the U.K., a Mexican restaurant chain called Wahaca was the first to provide an A.R. menu. It updated its physical menus in collaboration with Poplar Studio by adding a Q.R. code that would launch an augmented reality (A.R.) experience when scanned with a smartphone. Customers can view further information, such as photographs and descriptions of dishes, as well as details on ingredients, allergens, and sustainability practices, after being met by the Wahaca co-founder, who instructs them on navigating the menu. Due to the experience's reliance on Web AR technology, patrons will have to physically visit the restaurant to view the menu rather than download an app.

In 2018, Domino unveiled a system akin to this one called "New Pizza Chef." Consumers can design their pizza and experience it in augmented reality before placing an order. Customers can select their crust, cheese, and swirls via Domino's mobile app before choosing virtual toppings. These choices also activated animated characters who could be seen "making" the pizza. The AR menu was connected to a digital payment system that enabled contactless payments from the client's table, making the entire experience more convenient and secure. The intriguing possibility was that the pizza the consumer received would perfectly resemble the one they designed.

6.2 Product Launches and A.R.

Customers can be excited and drawn to products launched in collaboration with suppliers and restaurants when restaurants debut A.R. products. While physically waiting in a restaurant for their friends to arrive or for their meals to be prepared, customers can learn about new products inventively and virtually experience diverse environments with the help of A.R. product releases.

Jack Daniels offered one of the first innovative product launches to promote its partnership with the UK-based Slug & Lettuce restaurant chain. The experience, which Poplar Studio developed, could be started by simply scanning a Q.R. code with each bottle of Lynchburg Lemonade. After that, clients were virtually taken to the Jack Daniels distillery in Lynchburg, Tennessee, where they could tour the facility, see how the whisky is manufactured, and even don costumes to represent the company's creator, Jasper Newton "Jack" Daniel. This interesting tour piqued the curiosity of numerous clients and raised lemonade sales.

6.3 Social Media Experiences

In social media, augmented reality often uses a filter encouraging users to experiment with different filters while capturing images. Restaurant owners looking for fresh, creative ways to market their goods and provide their clients with memorable experiences chose to create their filters.

Taco Bell's usage of augmented reality (A.R.) as a marketing tool illustrates this, demonstrating the influence augmented reality has over viewers and patrons. In 2016, Taco Bell released a Snapchat lens for Cinco de Mayo that transformed users' faces into enormous taco shells. The filter was well received by Snapchat users, who gave it an average of 24 seconds before sharing images and videos of themselves as tacos with their friends. The filter received a whopping 224 million views in a single day. Importantly, the peculiar filter shattered all prior Snapchat records for views and engagements, making it the most successful ad in Snapchat's history. This event honoured Mexican ancestry and culture and successfully enticed people to the fast-food restaurant as a part of their festivities.

6.4 Employing AR for Employee Training

Because it offers side-by-side education and can overlay further information on what an employee can immediately view, A.R. is also frequently used to train workers in the food service industry. According to a poll, fast food workers currently take an average of 296 hours of training before they are considered ready for the back of the house. In these situations, employees frequently make mistakes that diminish customer satisfaction and affect a restaurant's bottom line. The situation is made worse by the high turnover rate in this sector. A.R. eliminates the need for a trainer to be physically present at the training site, saving human resources and allowing for faster, safer, more efficient, and risk-free staff training. (Jagtap et al., 2021)

A.R. training courses can assist in resolving these problems. For instance, the research claims that a worker in the kitchen preparing hamburgers who is now wearing A.R. glasses would be informed immediately when a client orders from them and may begin making hamburgers by activating the corresponding trigger buttons.

The program would use the chosen trigger to direct the employee to the right ingredient and offer a video tutorial on adding the ingredients and how much to use. The employee might then view another video or rewind the current one to get instructions on each step of the process. The AR program may have a gamification feature to increase employee engagement and productivity. Another study discovered that A.R. helps calculate food portion amounts. The accuracy and consistency of the study's standard serving sizes were improved using A.R.

These aspects, governed by Augmented reality, show that it is attracting consumers and ensuring that nothing stops people from being involved in the food industry, neither the crumbling economy nor the expected monotonous functioning of the world.

7. CHALLENGES

The amount of technology involved in the food industry is immeasurable, and the complexity of tasks being handled by these new technologies makes us realize that technology has become an indispensable part of the food industry. However, though technology has made the entire industry systematized and self-sufficient, we cannot undermine its drawbacks. The employment of technology in even the remotest aspect of the food industry, such as agriculture, seems highly impractical, as farmers and people involved in agriculture will not have the financial stability to invest in such high-range devices. The restraint industry is one of the most employment-friendly industries, which used to generate huge amounts of employment opportunities every year. If the rate at which technology is replacing people in restraint functions continues, it will lead to a huge population of people being unemployed, which would affect the economy's stability. All these technologies need to be understood to put them to their maximum use and considering that people who work in the agriculture sector are mostly people coming from less financially able backgrounds, it would be difficult for them to understand how the equipmets work, how they need to be used and ultimately would lead to inefficient usage. Although social media connects the entire world and makes it appear to be a united society, it is still a medium that allows for free expression. Overusing social media will lead to an even higher spread of negativity and hatred than it currently does. Artificial intelligence is a set of actions programmed by men. However, the growth of A.I. at this stage is threatening people's jobs, as it is equipped enough to replace men easily. A.I. has unlimited power and uncontrollable access to the data of different countries and people across the globe. If not curbed and regulated, its effects on the world would be destructive.

Augmented reality, though a very clever and unique application of technology to ensure exciting experiences, is a result of the endless desires of people. If the needs and wants of people keep on continuing and their expectations keep increasing, no amount of technology can satisfy their need to discover new things. Technology also leads to a huge increase in carbon footprints and generates tonnes of e-waste.

Technology, unlike working people, has warranties and expiration dates. They cannot function forever and would have to be replaced regularly. This would generate huge amounts of non-biodegradable waste in the environment, thus affecting global warming. These drawbacks are minimal in number, as technology is a recent addition to the workings of industries, and thus the repercussions of having employed these technologies are yet to be known.

8. CONCLUSION

The chapter identified the various technologies that have been applied in the food industry in the recent past and analyzed their functions and the impact they have had on the functioning and profitability of the food industry. It states that employing various kinds of technologies, such as Artificial intelligence, augmented reality, etc., has revamped the functioning of the food industry and led to its rapid and steady progress. It notes that technology, though a recent addition, has become an inseparable and indispensable part of the food industry. Through this chapter, a detail analysis has been made with respect to the employing of these cutting edge technologies in different aspects with respect to the manufacturing, consumption and distributing sectors.

Technology has made impossible things happen and has from time to time proved that sky is the limit. Though technologies are creations of human idea, the way they have been created and executed makes one wonder the extent of its growth in the upcoming years and its impact on every aspect and sector in the world. The breakthrough of these revolutionary technologies into the culinary world and food industry is going to change even the remotest of functioning and working of the industry and would ensure that all resources are put to their best use resulting in massive and rapid rise in profits and build the economy of all the regios willing to employ and take advantage the inventions, innovations and creations of mankind.

Technolgy is set out on a mission to ensure that it benefits a larger group of people and is efficient in its functioning. The slow yet steady inclusion of these innovative equipments and inventions in the food industry in various countries in various aspects shows the adaptability and efficiency of these devices in ensuring that it fulfills its purpose. This paper by covering and discussing in details about the impact of evergrowing technology in various sectors and niche of the food industry assists people part of different disciplines in fulfilling their aims and targets with the help of these technologies and introduce them to cutting edge technologies that would increase their efficiency and help them build their brands, businesses and companies. The food industry can tremendously increase its economy and profitability if it cleverly employs all the technology available, systematically monitors its functioning, and evaluates its performance.

REFERENCES

Blaxter, M. (2003). *Health and lifestyles*. Routledge. doi:10.4324/9780203393000

Chui, M. (2017). Artificial intelligence is the next digital frontier. *McKinsey and Company Global Institute, 47*(3.6).

Freeman, C. (1987). *Technology, policy, and economic performance: lessons from Japan*. Burns & Oates.

Hudson, S., & Gilbert, D. (2006). The Internet and small hospitality businesses: B&B marketing in Canada. *Journal of Hospitality & Leisure Marketing, 14*(1), 99–116. doi:10.1300/J150v14n01_06

Jagtap, S., Saxena, P., & Salonitis, K. (2021). Food 4.0: Implementing augmented reality systems in the food industry. *Procedia CIRP, 104*, 1137–1142. doi:10.1016/j.procir.2021.11.191

Konopelko, M. (2019). *Augmented reality packaging in the food & beverages industry.*

Kumar, I., Rawat, J., Mohd, N., & Husain, S. (2021). Opportunities of artificial intelligence and machine learning in the food industry. *Journal of Food Quality, 2021*, 1–10. doi:10.1155/2021/4535567

Lee, S. (2020). A study on the user experience of unmanned payment kiosk systems in fast food restaurants. *Int. J. Smart Bus. Technol, 8*(1), 1–6. doi:10.21742/IJSBT.2020.8.1.01

Lewis, T., & Phillipov, M. (2018). Food/media: Eating, cooking, and provisioning in a digital world. *Communication Research and Practice, 4*(3), 207–211. doi:10.1080/22041451.2018.1482075

Mahalik, N. P., & Nambiar, A. N. (2010). Trends in food packaging and manufacturing systems and technology. *Trends in Food Science & Technology, 21*(3), 117–128. doi:10.1016/j.tifs.2009.12.006

McKinley, J. (2014). With farm robotics, the cows decide when it's milking time. *The New York Times.*

Murphy, G., Corcoran, C., Tatlow-Golden, M., Boyland, E., & Rooney, B. (2020). See, like, share, remember: Adolescents' responses to unhealthy-, healthy-and non-food advertising in social media. *International Journal of Environmental Research and Public Health, 17*(7), 2181. doi:10.3390/ijerph17072181 PMID:32218252

Pinel, F. (2015). What's cooking with Chef Watson? an interview with Lav Varshney and James briscione. *IEEE Pervasive Computing, 14*(4), 58–62. doi:10.1109/MPRV.2015.66

Rejeb, A., Rejeb, K., & Keogh, J. G. (2021). Enablers of augmented reality in the food supply chain: A systematic literature review. *Journal of Foodservice Business Research, 24*(4), 415–444. doi:10.1080/15378020.2020.1859973

Seo, K. H. (2020). A study on the application of kiosk service as the workplace flexibility: The determinants of expanded technology adoption and trust of quick service restaurant customers. *Sustainability, 12*(21), 8790. doi:10.3390u12218790

Simeone, M., & Scarpato, D. (2020). Sustainable consumption: How does social media affect food choices? *Journal of Cleaner Production, 277*, 124036. doi:10.1016/j.jclepro.2020.124036

Sina, E., Boakye, D., Christianson, L., Ahrens, W., & Hebestreit, A. (2022). Social Media and Children's and Adolescents' Diets: A Systematic Review of the Underlying Social and Physiological Mechanisms. *Advances in Nutrition, 13*(3), 913–937. doi:10.1093/advances/nmac018 PMID:35218190

Spence, C., Okajima, K., Cheok, A. D., Petit, O., & Michel, C. (2016). Eating with our eyes: From visual hunger to digital satiation. *Brain and Cognition, 110*, 53–63. doi:10.1016/j.bandc.2015.08.006 PMID:26432045

UDA. (2014). *K.* Social Media, Its Impact on Food Culture in the World.

Chapter 4
The Future of Restaurants: How Technology Is Changing the Way We Dine Out

Sandilyan Ramanujam Pagaldiviti
https://orcid.org/0000-0001-8673-5275
C.V. Raman Global University, India

Birendra Kishore Roy
https://orcid.org/0000-0001-8202-5137
C.V. Raman Global University, India

ABSTRACT

In the present day the restaurant and the food service industry has metamorphosed into such a sophisticated industry where every day some innovative and creative are happening. It's a fact that in the 21st century, and especially after Covid people have become more conscious in choosing any food and any food outlets. Modern technology has come into the picture to provide information and offer solution pertaining to the customers' preferences. This chapter is mainly looks into details mainly of these aspects: use of Artificial Intelligence (AI) and Robotics in food service and precision cooking, the growth of different apps and influencers on social media/internet that tells customer what to eat and where to eat, and the restaurant and the food industry by force had to adopt certain technologies like contact less service and digital menu etc. for survival and sustainability during Covid. The objective is to present both the advantages as well the demerits of technology & come up with recommended guidelines which would help the customer to choose desired food and restaurant preferences.

1. INTRODUCTION

The concept of modern restaurants and traditional restaurants is different. Earlier people were not habituated to visiting any restaurant or food outlets. They were mostly limited in favour of having food at home as they did not have options to eat outside. Compare to today's world people were not having the demand for dining antiquates or aristocracies. Local gatherings were limited; unless there were any

DOI: 10.4018/978-1-6684-9094-5.ch004

social gatherings based on any celebration or occasions people mostly used to have food at home only. But today during any party or celebration people take help from any event manager or caterers who deliver sophisticated services to the guests. If we explore restaurants it would take us early 40s. They travelled around the countries to supply goods on local horses or curt and caravans. At that time people did not have any money for exchanging goods. It was only some medium of exchange of one material to another (Walker, 2021). Exploring foods were limited to the elite class people or the cash cows of society. People did not have a sufficient medium of communication to visit different places to try different foods. However, the situation changes during Raja-Maharajas and Vikings who started to develop their empire in different places. At that time people use to travel with their troops from one place to another mainly for hunting or to explore their Kingdom (Rawson & Shore, 2019). Road facilities, accommodation, and other recreation facilities were developed to facilitate the kings and their families and to provide them with basic facilities like processed food and beverages. In a few places, liquors were served to these elites. Transportation facilities like roadways and river waves also developed which eventually resolved the transportation issues. After the French Revolution food became one of the major commodities to earn money (Lundberg, 1989). During the 50s and 60s food business developed in a better way. Inns became one of the standard accommodations for the local public and visitors. These Inns were mainly located near the countryside where groups of travellers (Goody, 2018) used to visit and have shelter, food, and drinks in exchange for money. Inns were also limited to the elite class people (Lawrence & Ken, 1988) and among those who could expense more (Symons, 2013). After the 2nd world War air transport became cheaper which helped people to travel to different countries. At present every citizen in any country may visit any of the restaurants or hotels to try some new foods in exchange for money. Of course, there is a major number of people who are visiting restaurants or hotels daily or twice a week, but there are also many citizens who are visiting these places individually or with their families at least one day a month and can try any of the foods in exchange of limited amount. Nowadays even people are travelling around the countries in the world to explore different cuisine only because of the availability of multiple international flights and low airfare. Nowadays people are more familiar with visiting different restaurants starting including fine dining, specialty restaurants, QSR, pubs, nightclubs, bars, and many other food outlets where these facilities are available and offer sumptuous meals to their customers (Auty, 1992). At present people are also visiting these places not only to try the food and beverages but also to experience the level of service and ambiance of these places. The concept of visiting a different restaurant or hotel in recent times versus 30 years back was different because at that time people did not have facilities for internet connectivity, Smartphone, and GPS (Breffi & R, 2012). People nowadays have access to different media to choose their food according to their mood daily. People also select and purchase their food items according to their choices from different suppliers/vendors at their fingertips. The usage of the internet, social media, and Smartphone are at its peak and anybody can order any food or any seasonal fruits or vegetable at any particular time of the year (Chen, R, & Si, 2021). For example: if an orange is available during the time of winter in India he or she can also test the same fruit during the time of monsoon because of the facility of improvement for cold storage and packaging system in the food business (Lee, Van Dolen, & Kolk, 2013). Development of air transport facilities helping people to purchase any food within a short duration after ordering. Multiple food bloggers and different food sites nowadays are promoting different restaurants and their menus with the help of the internet daily (Chen, Hsieh, Chang, & Chen, 2015). This food blogging through the internet influences or forces the present generation to decide on food products or outlets. Technology has given the food business and the dine-out concept such a boom that people can decide on a venue for any event by using their Smartphones and the internet (Gupta, 2019). At present every citizen can visit any hotel or restaurant according to their choice or they can order food from the same restaurant with the help of any food delivery app (Cha & Seo, 2020).

In this chapter, we will also discuss how the development of the Restaurant Business, and the role of technology & its impact on the restaurant Business.

2. GROWTH AND DEVELOPMENT OF THE RESTAURANT BUSINESS

"Necessity is the mother of any invention" so the Restaurant Business. As time passes the demand for restaurant operations has also increased. People became more conscious of choosing a menu and an outlet for themselves or their families. This is due to the growing demand (Bendegul, Ahmet, & Anil, 2021) among people and the development of different restaurant concepts around the globe. People became habituated to visiting any of the restaurants in their cities just to try different food and the ambiance of that place. On the other hand, the growths of different outlets are at its peak they want to cover every city with their best facilities and dine-in services. Many of the renowned international and national brands covered the market to get their prospective customer. For example, Kentucky Fried Chicken (KFC) and Dominos Pizza two Quick Service Restaurants (QSRs) which started their business in the USA already spread their business in different countries (Ottenbacher & Harrington, 2009).

Figure 1. Types of accommodation available in India
(Source: (Roday, Biwal, & Joshi, 2009))

Sl. No	Type of Hotel	Sl. No	Type of Hotel
1	International Hotels	13	Tourist Lodges
2	Apartment Hotels	14	Tourist Bungalows
3	Commercial Hotels	15	Government guest houses
4	Floating Hotels	16	Dak Bungalow
5	Capsule Hotels	17	Circuit Houses
6	Transit Hotels	18	Forest Lodges
7	Heritage Hotels	19	Youth Hostels
8	Boutique Hotels	20	Bed and Breakfast establishments
9	Motels	21	Camping sites
10	Resorts	22	Railway Retiring Rooms
11	Timeshare Apartments	23	Dharamshalas
12	Holiday Homes		

Initially, the demands among the people were limited as the facilities were limited in such accommodations. Business travellers or tourists used to stay in inns (Bowie, 2018)or in taverns where the facilities were limited and there were no such luxury facilities compared to present times (Richard, Adams, & Walter, 2012). In general, a local cook, chef, or housekeeper used to stay in those accommodations. They used to prepare food for their guest according to the availability of the ingredients. That time tourists had to consume the food that the chef or the cook was preparing (Ionel, 2016). A fixed menu is generally served at that time. According to the records, restaurant operations and dine-in businesses achieved 'numerous landmarks after the mid-nineteenth century'; when the hotel industry also started to begin their full flagged business.

Table 1. Landmarks in the growth of the Hotel Industry (Bhatnagar, 2002)

Years	Particular Landmark achieved
1650	Pascal opened a cafe in Paris and Coffee House in London
1794	City Hotel (73 rooms) at 115 Broadway, New York. First building specially made for a hotel
1829	Tremont House. Adam and Eve of modem hotel industry was opened at Boston (170 Rooms)
1889	Cesar Ritz introduced luxury hotels such as Savoy. Carlton and Hyde Park hotels in England.
1890-1899	Fredrick Gordon and Sir Blundel Maple launched. Gordon Hotels and Fredrick Hotels Ltd.
Late 19th Century	Famous Waldorf Astoria in New York
18.1.1908	Buffalo Statler by Ellsworth Statler in U.S. Considered to be the birth of commercial hotels.
1927	Stevens Hotel in Chicago, later renamed as Hilton Hotel (3000 rooms)
1950	2 new concepts (a) Motels and (b) International chain operations
Hotel Rossia or Rossiya	Moscow (Remained largest hotel of world for quite some time)
West Inn Stanford	Singapore (Remained tallest building for quite some time)
1930 depression in world economy	Decline in hotel industry
World War Period	Upsurge in hotels

In the 50s and 60s, only elite people can visit the standard restaurant. Therefore, service standard was limited. After the 60s when Hotels and other accommodations started to develop and people were travelling in different cities and countries the requirement for providing food to the mass also increased. People had the intention to order food in their rooms rather than going outside of their accommodation. After World War II when restrictions on international travel were removed and air transport became cost-effective; mainly business travellers became more demanding in receiving personalized services. As employment opportunities were increased people had the option to expense money by visiting hotels and restaurants with their families.

Today different types of food and beverage outlets are there in the market like Fast Food Counter, Speciality Restaurants, Fine-Dining, Restro-BAR, bars, Pubs, Night Club, banquets, Speciality Restaurants, and many others. All of them have their different style of making and serving food to their customer. Some outlets are focusing on serving food at standard dining whereas some of them are more focused on serving their products at a specific ambiance and hospitality. Many restaurants have rotational menus in which they serve their customer according to occasion or time. They generally maintain the proper sequence in their menu, like – Breakfast, Branch, Lunch, Evening Tea, High Tea, Dinner, Supper or as Snacks/Starter, Soup, Main Course, Side Dishes, Options for both veg & non-veg, Rice or Bread, Dessert, Beverages (Hot/Cold), Liquors, etc. Many of them had a desire to have food only in dine-in areas. Therefore, the services like hotel room service or restaurant dine-in service came into the picture. The concept of 'Buffet' service came along with the development of the hotel business into restaurant operations. Customers can consume food by standing at the buffet counter or can be served by any service personnel at the counter.

Earlier it was only a requirement to consume food at any hotel or restaurant but after the 19th century, people became more focused on receiving personalized guest service from these outlets or hotel establishments. History says that more bonafide customers who visited any hotel or restaurant were mainly looking for value for money, service etiquette and authentic ingredients, sequence of service, and hospitality from the waiting staff (Eun-Jung & Loren, 2003). Today people became more health conscious and choose restaurants based on certain qualities like the freshness of the food, hygiene of the place, behaviour of service staff, cost, and distance of the restaurant, special occasions, or diet. Therefore, at present it became more challenging for any business group or restaurant operator to run a food business.

2.1 Challenges Involved in Operating Restaurant Business

Running a food business and delivering fresh products every day to the customer is a challenging task. Therefore, it's become the sole responsibility of any Restaurant Operator to follow the protocols to avoid any critical situation. There are many challenges involved in operating a restaurant business. These are given below:

2.1.1 Fresh Ingredients

Many restaurants fail to provide fresh food to their customer and can get severe guest complaints due to the lack of attention in choosing fresh ingredients in preparing food. It is always advisable to order fresh products over old products. Restaurants must follow the authentic guidelines in storing the ingredients on per First in First out (FIFO) or Last in First out (LIFO) basis. Both raw and cooked foods have to be stored separately to avoid any cross contaminations. Proper temperature is to be maintained in the storage areas to avoid wastage, damage, or pilferages. Maintaining hygiene and sanitation is so important in the food business so providing fresh ingredients to the customer is so important.

2.2.2 Effective Manpower

Having efficient manpower is so important in running a restaurant business. As the customer has a demand for personalized guest service; hotels and restaurants must have trained professionals who can able to interact with the guest effectively, should be there in a position to understand the queries, and fulfilling the same within a given time. The restaurant industry needs trained manpower to prepare food according to the recipe standard and needs trained people to serve the food at guests' tables, and in handling various types of equipment effectively.

2.2.3 Location of the Restaurant

To get prospective customers from the market location plays an important role in the restaurant business. It is advisable to open a restaurant or hotel in a place where the movement of the traffic is huge. Locating a restaurant on the outskirt of the cities or urban areas become challenging for any visitor. The Centre of a city becomes so demanding and competitive that for most business owners it becomes a challenge to expense more money to buy a plot. A competitive market plays a significant role in the success of any restaurant business.

2.2.4 Sustainability

At present restaurant sustainability has become one of the major challenges for any restaurant business. Any business has its impact on the planet in terms of "carbon footprint, shortening their supply chain, food wastage, packaging, water, and energy consumption, recycling, and more." Apart from the environment restaurant sustainability also involved the possible opportunities in providing the best product to the customer to sustain in the competitive market. To improve sustainability restaurants must focus on the following (Burton-Hughes, 2020):

1. Choosing a supplier & cutting down the packaging cost.
2. Having recycled printed or visual display of the menu.
3. The practice of using biodegradables.
4. Effective use of water.
5. Energy conservation and maintenance of equipment.

2.2.5 Value for Money

At present, there are multiple restaurants in any city that are serving similar products every day. But in most cases, it has been observed that people only visit those outlets that are serving better quality products at a competitive price at proper portion. This also increased the faith and brand loyalty among the customer for specific products. However, to do so with the increasing price of daily commodities and necessary raw materials it becomes a challenge to deliver the same product every time to survive in the market.

2.2.6 High Benchmark

Today hospitality, hotel, and restaurant businesses multiple establishments are facilitating their customer with better quality services and facilities; people always have a chance to go for different options with multiple benefits. Expectations of the customer also increased nowadays. Therefore, it is a challenge for this business to meet the standard of their guest and provide them with better facilities and services.

3. THE ROLE OF TECHNOLOGY AND ITS IMPACT ON RESTAURANT BUSINESS

Technology has improved the standard of the service industry which also happened for the Restaurant and Hospitality business. Earlier people used to visit a new restaurant by watching newspaper advertisements or through any print media. Television and other digital media helped business owners to reach the majority of the population in a short time. People have every option nowadays to access the internet with any of the gadgets like (Jee-Won & Young, 2019) Smartphones, laptops, and other devices at their fingertips (Law, Chan, & Wang, 2018). Various hotels and restaurants are already connected through internet connectivity to reach their customer. Due to the availability of sufficient broadband, wifi, and strong mobile internet connectivity at a lower price, today everybody can locate any restaurant at any destination (Julio & Claudia, 2019). Apart from locating any restaurant people can decide on different menus online and order the same. Some modern inventions like (Berezina, Ciftci, & Cobanoglu, 2019)

Figure 2. Challenges involved in Operating Restaurant Business
(Source: Self)

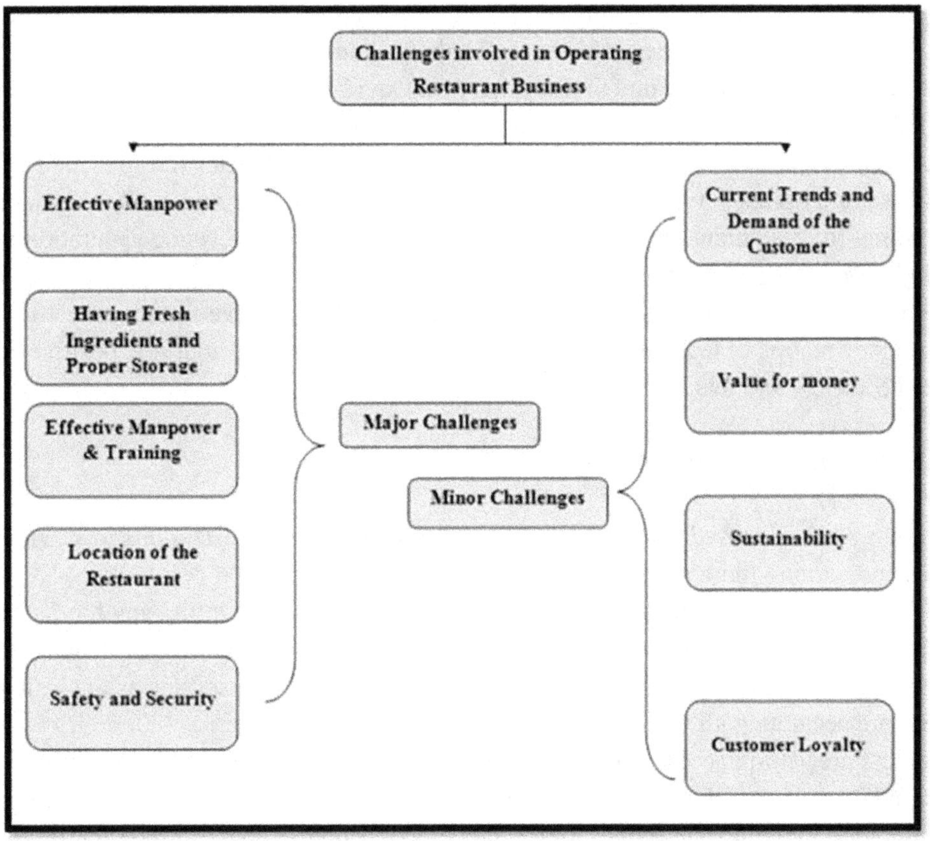

chatbots, robot order takers, and Cleaning robots are something that has improved the productivity of the restaurant business (Lee, Kwon, & Back, 2021). In today's world, people trust technology, social media, and digital medium than any other verbal or oral recommendation of any place (Kim, Jung, & Ryu, 2020). Hotels and restaurants are providing online promotions and discounts to get their customer. Every hotel or restaurant performs its marketing online to show its facilities and promotions. To support the growing demand restaurants are operating three types of services which are dining in, takeaway, and delivery. Most restaurants are now connected through the internet with different websites where they mentioned every detail of their establishments like how many cover restaurants they are, the number of outlets, the specialty of the restaurant or what cuisine they serve, the menu along with the price. Even most restaurants are encouraging their customer to make payments online to avoid any delay in service. Apart from online transactions and food service technology has achieved the benchmark with the innovation of virtual reality & holographic images. Virtual reality and holographic images help a customer to visualize a product or related material in a realistic form. For example, any restaurant making different pizzas for their customer can use holographic images of their pizza-making process to their customer to increase their desire to purchase pizza with specific toppings or ingredients. In terms of display, many Quick Service Restaurants (QSRs) are using Kitchen Order Ticketing Machines (KOTMs), Visual Displays, and Token detectors to attract their customer which is cost-effective in promotion and branding of the outlets. Inside the kitchen or at production units most restaurants and outlets are using high-capacity storage to keep their ingredients fresh and safe at specific temperatures for a long duration. Many developing countries are using semi or automated drone facilities to deliver food products at limited distances. Exceptional use of logistics and supply chain are happening daily to supply products through air transport and shipping unit in less time. Post Covid-19 (Ryu, Jarumaneerat, Promsivapallop, & Kim, 2023) many hotels have developed touches less food ordering system to avoid transmission of virus infection (Dirsehan & Cankat, 2021). Customer has the possibilities to order food from various automated or semi-automated vending machines (Marina, Liudmila, Alexander, Olga, & Natalyia, 2022) which is helping the restaurant owners to provide effective service in revenue generation without any further delay (María-Eugenia, Irene, & Gloria, 2014).

The future of the restaurant business and the concept of dine-in have improved with the development and implementation of technology (Khamidovich, 2021). But it has multiple positive and negative impacts on the society and mankind.

3.1 Positive Impact of Technology

1. Technology has improved the productivity standard which helps restaurants in serving more customers and earning more revenue in a short duration.
2. Helping customers to locate an outlet out of multiple options in ordering any food.
3. Helping customers in making online transactions for ordering any food without standing in the queue.
4. Improved the productivity standard along with personalized guest service as the waiting staff became more accurate with the specified remarks of the food orders.
5. Customers are getting access to choices while staying at home through online delivery.
6. Specifications for ordering have been achieved through online and virtual ordering systems.
7. Through food blogging and ratings, one can decide whether to visit a restaurant or not.
8. Customers can able to detect discounts, and offers through online ordering which provides them the option for value for money.

3.2 The Negative Impact of Technology on the Restaurant Business

1. There are possible chances of transaction failure during online ordering due to network and power failures.
2. There are so many especially able customers in society who have full or partial blindness and are not used to the usage of technologies and are not capable enough to use all these gadgets like Smartphones, laptops, or online transaction systems. As these facilities are not available in brogue features.
3. Human interactions are one of the necessary demands in the hospitality and restaurant industry which cannot be achieved through any machine, artificial intelligence, or other modern technologies.
4. At present, there are multiple food bloggers and food photographers who do not have any knowledge about food but do marketing for some restaurants only to get financial benefits.
5. Implementation of technology in the food business may impact on crisis of trained manpower.
6. Technology cannot replace human effort and emotion toward guest service.
7. Efficient training and surveillance are required to implement modern technologies in the food business. As compared to the old employees new employees require more time in learning the process of working with modern technologies.

4. CONCLUSION AND RECOMMENDATION

The concept of eating out has cropped in over the recent maybe 30 to 40 years. This started at a time when people in the travel industry used to travel for business. But today dining out is a way of life in most families which are neutral families and based in urban cities. People usually working husband and wife who are working do not have much time to cook. So there is a lot of dependency on package food and eating outside, visiting restaurants, dining outs, and takeaways. Numerous apps have come in and started a business online in suggesting and selling foods. With the ever-increasing demand of this segment as a result today technology is flourishing and delivering products to the customer both by dine-in and take away or through delivery (Oronsky & Chathoth, 2007). At present, there has to be control over what kind of food options we should select and which should be outlets we should decide. This is because there is a lot of artificial intelligence and behind-the-scenes story that common customers are not aware of. Therefore, whenever we search for any free or paid sites are the ones that pop up first likewise in these apps (Wu, Ngai, Wu, & Wu, 2020). These app manufacturers and companies promote marketing strategies (Li, Meng, & Pan, 2020) to reach out to the customers only to have their preferences and with better ratings so that apps will get sold for these better rating systems (Soon, 2020). There are many new professionals, software creators, food bloggers, and food photographers (Plotkina, Munzel, & Pallud, 2020) who do not know the food but for the sake of some income make unnecessary comments and suggest food and restaurant to the public (McGrew, Ortega, Breakstone, & Wineburg, 2017). In the hotel and restaurant business, there are a lot of positive as well as negative impacts of technology. Certain restaurants may be trying to sell their leftover or expired food products as an add-on to some main courses. But due to the influential comments and photos from food bloggers and food photographers, one is purchasing it online. There are maximum chances of discrepancies (Chen L. D., 2008) through online transactions (Patidar & Sharma, 2011). There are possibilities for disclosing valuable information about their bank accounts and other necessary details. That is why we should not be depending on any technology blindly or accept its capabilities unknowingly unless we have effective regulations and guidelines to make the best use of technology.

So, based on the above conclusion we must recommend that:

1. There have to be some strict regulations where the websites must not have any control over writing or modifying comments or feedback for any specific food products for any restaurants and food outlets.
2. All feedback and suggestions must be genuine from the customer and there should not be any influence from any individual owners to modify these comments for gaining financial benefits.
3. As this social media and different food websites are free platforms to make any comments or suggestions; food bloggers must be restricted to make comments unless they are fulfilling the criteria or certain eligibility on a similar field.
4. One should not be accepting any baseless recommendation in choosing any specific diet, or food products or deciding on a menu or restaurant just by watching on Google, Social Media, Websites, or verbal recommendation from any family member or friends.

5. SUGGESTED FURTHER READINGS

1. Restaurants and Dining Rooms, Edited By Franziska Bollerey & Christoph Grafe
2. Fine Dining: The Secrets behind the Restaurant Industry Paperback Edited by Jack Rasmussen

REFERENCES

Auty, S. (1992, July 1). Consumer choice and segmentation in the restaurant industry. *Service Industries Journal, 12*(3), 324–339. doi:10.1080/02642069200000042

Bendegul, O., Ahmet, B. O., & Anil, B. (2021). Generation Y's dining out behavior. *International Hospitality Review, 35*(1), 41–56. doi:10.1108/IHR-07-2020-0023

Berezina, K., Ciftci, O., & Cobanoglu, C. (2019). Robots, artificial intelligence, and service automation in restaurants. In Robots, artificial intelligence, and service automation in travel, tourism and hospitality, 185-219. doi:10.1108/978-1-78756-687-320191010

Bhatnagar, S. K. (2002). *Front Office Management* (2nd ed.). Frank Brothers and Company.

Bowie, D. (2018). Innovation and 19th century hotel industry evolution. *Tourism Management, 64*, 314–323. doi:10.1016/j.tourman.2017.09.005

Breffi, N. M. (2012). Applying Modern Robotics Technologies to Demand Prediction and Production Management in the Quice-Service Restaurant Sector. *Cornell Hospitality Quarterly, 53*(2), 122–133. doi:10.1177/1938965511434112

Burton-Hughes, L. (2020, February 7). www.highspeedtraining.co.uk. Retrieved from High Speed Training: https://www.highspeedtraining.co.uk/hub/improve-sustainability-in-restaurants/

Cha, S. S., & Seo, B. K. (2020). The effect of food delivery application on customer loyalty in restaurant. *Journal of Distribution Science, 18*(4), 5–12.

Chen, K. H., Hsieh, K. J., Chang, F. H., & Chen, N. C. (2015). The customer citizenship behaviors of food blog users. *Sustainability, 7*(9), 12502–12520. doi:10.3390u70912502

Chen, L. D. (2008). A model of consumer acceptance of mobile payment. *International Journal of Mobile Communications, 6*(1), 32–52. doi:10.1504/IJMC.2008.015997

Chen, S. L. (1804). R, Z. M., & Si, Y. (2021). Mobile communications for tourism and hospitality: A review for hospitality evolution, prsent status, and future trends. *Electronics (Basel), 10*(5).

Dirsehan, T., & Cankat, E. (2021). Role of mobile food-ordering applications in developing restaurants' brand satisfaction and loyalty in the pandemic period. *Journal of Retailing and Consumer Services, 102608*, 102608. Advance online publication. doi:10.1016/j.jretconser.2021.102608

Eun-Jung, K., & Loren, G. V. (2003). Consumers' Restaurant Choice Behavior and the Impact of Socio-Economic and Demographic Factors. *Journal of Foodservice Business Research, 6*(1), 3–24. doi:10.1300/J369v06n01_02

Goody, J. (2018). Industrial food: Towards the development of a world cuisine. In Food and culture. In C. Counihan, P. V. Estenik, & A. P. Julier, Food and Culture (pp. 263-282). Routledge.

Gupta, M. (2019). A Study on Impact of Online Food delivery app on Restaurant Business special reference to zomato and swiggy. *International Journal of Research and Analytical Reviews, 6*(1), 889–893. http://ijrar.com/

Ionel, M. (2016). Hospitality industry. *Ovidius University Annals: Economic Sciences Series*, *1*(1), 187–191.

Jee-Won, K., & Young, N. (2019). The role of personalization on continuance intention in food service mobile apps: A privacy calculus perspective. *International Journal of Contemporary Hospitality Management*, *31*(2), 734–752. doi:10.1108/IJCHM-12-2017-0783

Julio, N.-M., & Claudia, S.-S. (2019). Progress in wireless technologies in hospitality and tourism. *Journal of Hospitality and Tourism Technology*, *10*(4), 587–599. doi:10.1108/JHTT-07-2018-0061

Khamidovich, Z. K. (2021). Pandemic and trends in the restaurant business in 2021. *Central Asian Journal of Innovations on Tourism Management and Finance*, *2*(6), 54–61. doi:10.47494/cajitmf.v2i6.129

Kim, H., Jung, S., & Ryu, G. (2020). A study on the restaurant recommendation service app based on AI chatbot using personalization information. *International Journal of Advanced Culture Technology (IJACT)*, *8*(4), 263-270. doi:https://www.earticle.net/Article/A388080

Law, R., Chan, I. C., & Wang, L. (2018). A comprehensive review of mobile technology use in hospitality and tourism. *Journal of Hospitality Marketing & Management*, *27*(6), 626–648. doi:10.1080/19368623.2018.1423251

Lawrence, R. L., & Ken, M. W. (1988). The Effect of Multiple Measures of Age In Segmenting Hotel Markets. *Journal of Hospitality & Tourism Research (Washington, D.C.)*, *12*(2), 91–98. doi:10.1177/109634808801200210

Lee, H., Van Dolen, W., & Kolk, A. (2013). On the role of social media in the 'responsible'food business: Blogger buzz on health and obesity issues. *Journal of Business Ethics*, *118*(4), 695–707. doi:10.100710551-013-1955-0

Lee, M., Kwon, W., & Back, K. J. (2021). Artificial intelligence for hospitality big data analytics: Developing a prediction model of restaurant review helpfulness for customer decision-making. *International Journal of Contemporary Hospitality Management*, *33*(6), 2117–2136. doi:10.1108/IJCHM-06-2020-0587

Li, H., Meng, F., & Pan, B. (2020). How does review disconfirmation influence customer online review behavior? A mixed-method investigation. *International Journal of Contemporary Hospitality Management*, *32*(11), 3685–3703. https://www.researchgate.net/profile/Bing-Pan-5/publication/349913300_An_exploratory_study_on_Chinese_tourists%27_visitation_to_a_US_National_Park/links/62758c473a23744a72693115/An-exploratory-study-on-Chinese-tourists-visitation-to-a-US-National-Park.pdf. doi:10.1108/IJCHM-03-2020-0234

Lundberg, D. E. (1989). *The hotel and restaurant business* (Vol. 5). Van Nostrand Reinhold.

María-Eugenia, R.-M., Irene, G.-S., & Gloria, B.-C. (2014). Information and Communication Technology as a Differentiation Tool in Restaurants. *Journal of Foodservice Business Research*, *17*(5), 410–428. doi:10.1080/15378020.2014.967639

Marina, B., Liudmila, L., Alexander, S., Olga, B., & Natalyia, A. (2022). Innovative Engineering Developments of Equipment and Vending Machines. *Networked Control Systems for Connected and Automated Vehicles*, 1675–1682. doi:10.1007/978-3-031-11058-0_170

McGrew, S., Ortega, T., Breakstone, J., & Wineburg, S. (2017). The Challenge That's Bigger than Fake News: Civic Reasoning in a Social Media Environment. *American Educator, 41*(3), 4.

Oronsky, C. R., & Chathoth, P. K. (2007). An exploratory study examining information technology adoption and implementation in full-service restaurant firms. *International Journal of Hospitality Management, 26*(4), 941–956. doi:10.1016/j.ijhm.2006.04.001

Ottenbacher, M. C., & Harrington, R. J. (2009). The product innovation process of quick-service restaurant chains. *International Journal of Contemporary Hospitality Management, 21*(5), 523–541. doi:10.1108/09596110910967782

Patidar, R., & Sharma, L. (2011). Credit card fraud detection using neural network. [IJSCE]. *International Journal of Soft Computing and Engineering, 1*, 32–38.

Plotkina, D., Munzel, A., & Pallud, J. (2020). Illusions of truth - Experimental insights into human and algorithmic detections of fake online reviews. *Journal of Business Research, 109*, 511–523. https://www.elsevier.com/open-access/userlicense/1.0/. doi:10.1016/j.jbusres.2018.12.009

Rawson, K., & Shore, E. (2019). *Dining out: A global history of restaurants*. Reaktion Books.

Richard, H. P., Adams, L., & Walter, R. (2012). *Hotel Design, Planning and Development* (2nd ed.). Routledge., doi:10.4324/9780080967004

Roday, S., Biwal, A., & Joshi, V. (2009). *Tourism Opeartions and Management*. Oxford Publication.

Ryu, K., Jarumaneerat, T., Promsivapallop, P., & Kim, M. (2023). What influences restaurant dining out and diners' self-protective intention during the COVID-19 pandemic: Applying the Protection Motivation Theory. *International Journal of Hospitality Management, 109*, 103400. doi:10.1016/j.ijhm.2022.103400 PMID:36439897

Soon, M. J. (2020). Consumers' Awareness and Trust Toward Food Safety News on Social Media in Malaysia. *Journal of Food Protection, 83*(3), 452–459. doi:10.4315/0362-028X.JFP-19-415 PMID:32065648

Symons, M. (2013). The rise of the restaurant and the fate of hospitality. *International Journal of Contemporary Hospitality Management, 25*(2), 247–263. https://www.emerald.com/insight/content/doi/10.1108/09596111311301621/full/html. doi:10.1108/09596111311301621

Walker, J. R. (2021). *The restaurant: from concept to operation*. John Wiley & Sons.

Wu, Y., Ngai, E. W., Wu, P., & Wu, C. (2020). Fake online reviews: Literature review, synthesis, and directions for future research. *Decision Support Systems, 132*, 113280. doi:10.1016/j.dss.2020.113280

Chapter 5
Consumers' Attitudes Towards Using Food Delivery Apps

Deepa Sharma
https://orcid.org/0000-0003-4374-917X
Maharishi Markandeshwar (Deemed), India

ABSTRACT

The objective of this study is to find out consumer behavior toward food delivery apps. Analysis shows the most popular app consumers use to order food online, and there are several reasons why the authors think it's the best. Through this study, the authors focused on the factors that most contributed to this change and asked whether people still choose to eat out and if so why. This study aims to determine whether purchasing food online is as simple as downloading a restaurant delivery app and choosing the items you wish to get.

INTRODUCTION

A meal is a crucial source of survival, and the easiest and most practical way to get food is online through a food delivery service. Nowadays, individuals prefer ordering food online than going out to eat, making the conventional dining out a thing of the past. The development of digital technology has significantly altered the world today. Ten years ago there were no online food ordering services, so people had no choice but to stroll into restaurants and place their orders. With improvements in technology and delivery systems, online food ordering services that deliver customers' preferred food at their doorstep with the tap of a finger were introduced. Meal delivery services are dominating the industry because of their quick deliveries, ease of use, affordable food, simple payment methods, and range of alternatives. For certain parameters chosen by clients when placing an online food order, the trials were restricted to a certain geographic area. Different people enjoy different types of online food ordering. Due to the fact that meal delivery applications are only available in established urban areas, the study was limited to the attitudes of consumers regarding online food ordering. The perception of marketers or merchandisers was not studied in these investigations. As a result, the study solely includes residents of those locations and excludes a survey of individuals who reside in remote areas. In this digital age, we live. Over

DOI: 10.4018/978-1-6684-9094-5.ch005

the past 15 years, there has been a significant growth in internet availability. Geographically, it is contracted and geographic. The word is a small, close-to-you place. Social media networks, mobile apps, and a variety of other digital communications are now frequently used on the Internet. In this way, it has ingrained itself into the daily lives of billions of people. Apart from being a successful communication channel, the Internet is also an effective distribution channel. Electronic commerce also known as e-commerce, is a term used to describe digital platforms that use communication networks to enable both businesses and their customers to receive and relay information, establish and maintain relationships, sell products or enable payment-related transactions. E-commerce now includes the sale of services as well as the purchase of goods. Products focused on providing services, such as booking hotel rooms, flights, leisure tours, restaurant delivery, and residential services such as maintenance. This includes web-enabled platforms, such as company websites and online portals, which can be used to conduct business transactions. The current electronic meal ordering industry enables customers to place orders from a wide variety of restaurants with just a swipe on their mobile devices. Technology has significantly changed the industry of food delivery services. This can be a direct business-to-consumer strategy, or there are now aggregators that focus on the distribution of an entire product rather than making or manufacturing anything. A direct business-to-consumer distribution strategy is one option, but we also have aggregators that focus only on the distribution of a wide range of restaurant products. These food service firms do away with the difficulty of reaching out to the market and distribution thanks to the constantly shifting wants of the customers. Additionally, they specialized in customer understanding, forming partnerships with restaurants, hiring delivery personnel, and performing all other operations required to deliver cooked meals to clients' doorsteps. Customers around the nation are resorting to online buying more frequently, with convenience being the most enticing aspect of this door-delivery method. The COVID-19 epidemic has had a big impact on various industries, especially the one that deals with online food ordering, since customers are now expected to be cautious and watchful if they wish to keep doing so. Rapid technological innovation, a dynamic environment of fierce competition, and shifting consumer preferences and behaviours all contribute to the problems this industry faces. The pandemic adds to these difficulties the concern about security. In many other countries, there is a significant shift happening in how people are now starting to order food, which has already started. Managers are now adopting food app delivery systems to generate more customers each week for their restaurants without paying the extra cost of building even more dining space. When it comes to food, consumers have varying priorities and preferences. They might also have varying perspectives about the perceived value they place on utilizing particular meal delivery apps, as well as varying expectations about the features and quality of those apps. To our knowledge, no relevant research has examined these essential characteristics in regard to meal delivery apps. Furthermore, no research have examined the relationship between users' views of the quality of the products and services offered by meal delivery apps and their behavioural intends to use, re-use, or suggest those products and services to others. The current study analyzed the crucial factors related to these food apps such as perceptions, attitudes and desire to use. The fastest-growing household type for single-person households is one where food is delivered online. The fast increase in popularity of meal delivery apps may be primarily due to changing household demographics. Any country's population no longer consists primarily of households with more than one person. There are four causes for this: Each country has experienced an increase in population, a dip in domestic marriages, a notable drop in the overall fertility rate, and citizens moving to urban regions in quest of better economic opportunities. Because of these generational factors, it has become clear that some issues affect only certain social groups. These generational considerations make it clear that particular social groups from single-person households will face particular difficulties. Many nuclear families made up of older generation clients are particularly vulnerable and exposed to social and medical problems related to welfare, housing, and subsidized health care. The younger generation of consumers has inspired academia and hospitality industry to specifically recognize those needs and then develop strategies, and practices for a traditionally driven industry. The online businesses in India were minimal due to low internet penetration. But because of technological advances such as the expansion of the 4G services, the appearance of telecom operators who provide cheap data at quicker surfing speeds, even the middle and lower classes now have access to this expanding market. Local Indian citizens are also growing increasingly satisfied with app-based services and are willing to test out new ones. Indian consumers' lifestyles have seen a significant upheaval as a result of the swift adoption of smart phones and mobile applications. Consumer can purchase everything online, such as, lodging, groceries, food delivery, airline tickets, and real estate. Due to technological up gradations in the smart phone, a plethora of online food delivery companies have started their operations, and consumers have built a higher attraction for these start-up applications. The major differences in consumers' online food delivery application selection attributes and usage patterns across all nations can be attributed to changes in the lifestyles and family compositions of online food delivery application users as well as other socio-economic factors. Additionally, customers' shifting concerns and distinct preferences for online food delivery app selection and usage have persuaded them to use food delivery apps in search of value for invested money and convenience. The entire customer experience has been transformed with the introduction of online meal delivery applications. Customers can now browse all the closest eateries and food joints, study their menus and choose the things they want to eat with just a tap or click. The popular and widely utilized food delivery aggregator applications that are available to consumers have not been examined in the numerous prior research papers relating to online meal delivery applications in India. This study will examine the various and well-known online food delivery services available worldwide. Customers' ordering habits are influenced by a variety of elements, including price, on-time delivery, packaging, actions of other service providers, platform layout, *etc.* The way people order food is slowly changing in every country. The goal of this study is to identify the variables that impact consumer perception and find the most preferred food delivery app. The objective of this study is to thoroughly explore the dimensions influencing customers' purchase intentions to keep using a certain good or service. This study set out to thoroughly examine the factors affecting consumers' intentions when buying food online. This study's focus is on how individuals are turning to online food delivery services. In this study, we focused on the elements that influence this transition and investigated whether consumers still prefer eating out and, if so, why. Simply launching a food delivery app and selecting the food you wish to order is all it takes to place an online food order.

REVIEW OF LITERATURE

In India, applications have been created for online meal ordering and delivery. The Indian subcontinent has experienced the emergence of online meal ordering services for food delivery to homes due to the remarkable expansion of e-commerce (Pookulangara, 2022). The increasing use of smart technology such as smart phones, which offer real-time connectivity through mobile applications, and the rising ubiquity of online food delivery services among consumers seeking quick and practical food options make this worthy for study (Gupta, 2021). Customers select and use online food ordering and delivery services, and the perceived benefits and dangers involved (Khaliq, 2022). Consumers' level of hesitation and anxiety can be used to define perceived risk (Kaur, P., 2021) before making any decision about using and choosing an online food-ordering delivery application. These may include perceived functional risks, such as misinformation, incorrect product distribution, and worries about information accuracy; social risks like impersonal services, fake reviews, and ethical concerns; personal risks like a lack of personal touch, the high cost of Smartphone, and prior online experience (Akhtar,2021); and financial risks, such as the fear of online fraud, etc. financial risk, such as concerns about convenience fees and online scams (Tribhuvan,2020); psychological risks, such as concerns about trust and faulty food delivery (Singh, 2020); and perceived benefits, such as perceived savings and convenience (Bothra,2021). Consumers' buying behavior regarding online food delivering applications has been found to be strongly influenced by their perceptions of both risk and benefit (Cho, 2019). The mobile apps which are used to place meal orders through online food delivery platforms, includes restaurant-to-consumer delivery. It is crucial to know about the consumer's perception for using food delivery apps. The body of knowledge regarding consumer behaviour as it relates to online food delivery in general and specific food delivery applications in particular needs to be further expanded, even though researchers are increasingly interested in exploring various aspects of these applications. The more academic study would help inform practice in a field that is seeing amazing expansion in the Middle East, Europe, Asia and America and is associated with substantial funds and valuations (Dinu, 2022). Notably, little research has been done on customer attitudes and intentions about online meal delivery; the majority of the online-to-offline studies focused on individual behavior toward online products in general. The existing literature addresses issues such as the impact of processes on the preference to use online or offline food delivery apps (Garg, 2019), the effect of food quality and services on consumer trustworthiness towards online food delivery (Epuran, 2021), the impact of traffic while choosing options between online or offline food delivery services, and more. Kujur (2019) found that in Korea, the satisfaction of food delivery app users was influenced by food quality, receptiveness, dependability, pace of services, and, whereas in China, food quality, promptness, cost-effectiveness, and receptiveness were more significant. This underlines the geographic disparities in consumer behavior. Customer satisfaction and repurchase intentions were also highly associated in both countries. It is obvious that consumer behavior research on food delivery apps and online delivery apps, in general, is still in its infancy and is under-represented in the literature (Tribhuvan, 2020). To get a further understanding of the obstacles to and motivators of success for food delivery apps, which are fast becoming commonplace, more research is required. Such insights can be gained by analyzing customer behavior from the standpoint of consumption values. From the viewpoint of the hospitality industry as well, examining consumer behavior towards technological advancements is crucial. According to earlier research, the hotel industry must now take into account consumers' information system habits and hedonic motives as a result of technological advancements in digitization. Because so many people use smart phone to access apps for a multiple reasons, understanding clients' technology interactions is essential from the perspective of marketing for the hospitality industry in particular.

POPULAR ONLINE FOOD DELIVERY APPS WORLDWIDE

Applications for food delivery are increasingly widespread in the tech industry, and both their use and market share are growing. Our blog examines the industry's current state, including market size, growth, shopper trends, and projections for the future. It provides a complete analysis of the key rivals and their winning strategies for business owners looking to enter or expand the market. Here are a few of the most well-known international internet food delivery services.

Zomato

Zomato is an online restaurant discovery and delivery service first established in 2008 under the moniker "Foodiebay". The Indian company's domestic market success prompted Uber Eats to purchase the entire Indian operation for $400 million. Zomato is now available in over 10,000 cities and 24 countries across the globe. The platform's success reached a new peak in 2021 when a 39x oversubscribed IPO attracted shareholder bids worth US$46 billion.

Uber Eats

Uber Eats will be operational in 10,000 cities across the world by this year. They are market leaders so various food outlets want to partner with them.

FoodPanda, A Leader in International Food Delivery

FoodPanda is an online food ordering app that is working in 41 countries. The Germany based startup provides rapid delivery and has partnerships with about 50,000 surrounding restaurants.

Swiggy

The top online food delivery app in India, Swiggy, has received over 1,500,000 downloads and is currently available in practically all cities.

Grubhub: The Preferred Option for US Meal Delivery

One of the greatest apps for ordering meals are GrubHub, which has a large selection of more than 30,000 eateries, distributed throughout more than 800 U.S. locations. The headquarters of GrubHub are in Chicago, New York, and London. In 2004, the GrubHub app was released. Due to its early entry into the industry, the company was able to establish a wide range of alliances, and similar to Uber, market domination contributed to its continued success.

Deliveroo: One of Europe's Top Food Delivery Services

Food delivery startup Deliveroo is founded in London and operates in over 200 cities. It is the most widely used meal delivery app in Europe and allows users to order food from both well-known eateries and eateries without a typical eating area. Users are billed in accordance with their orders, and eateries receive a commission.

Domino's Pizza

A top-notch app is the well-known Domino's pizza delivery service. Unlike the other apps on this list, it delivers self prepared fast food.

Just Eat Takeaway: A Global Platform for Meal Delivery

A food delivery service in Europe called Just Eat Takeaway was started in 2001. More than 82,000 restaurants are partners, and it makes money via commissions.

DoorDash

One of the oldest and most reputable applications for meal delivery is DoorDash. In 32 markets, it services more than 300 cities. For the first time in its 60-year history, Little Caesars Pizza, the third-largest pizza company in the world, has partnered with DoorDash Inc. to offer delivery services.

Postmates: A Major US Player in Food Delivery

More than 90 cities across the United States offer the Postmates delivery app. With the exception of handling groceries and various delivery kinds, Postmates function similarly too many other food delivery applications.

REASONS OF POPULAR ONLINE FOOD DELIVERY APPS WORLDWIDE

The food business is once again growing and thriving. Both food delivery services and restaurants have benefited. Online food ordering is the act of placing an order for takeout or delivery from a nearby restaurant or food cooperative through a website or mobile application. Many of these sites allow users to maintain accounts with them to facilitate repeat orders, such as purchasing consumer goods online, etc. Various studies have found that 60 to 85% of respondents prefer mobile apps over mobile websites. The development of a food ordering app is more expensive than building a website. Some valuable reasons are described below:

1. **Menu Visible:** Though the restaurant's menu is already available on its website, the owner also makes sure that it is prominently displayed and easy to find within the app. Many customers primarily use the restaurant app for ordering, so they need to be able to peruse the customer menu first.
2. **Offer mobile ordering for takeout or delivery:** Some restaurants primarily use their apps to display their menus and other important company details such as location, hours of operation, phone numbers, *etc.* Customers can check the menu on the app and then use it to find their phone number to place an order. This is undoubtedly helpful, but a relatively straightforward website can achieve similar results. Instead, you want to think about including a mobile ordering feature within the app so that your users can place orders or even book reservations right away.
3. **Use location-aware Notifications to strengthen your relationship with Customers:** Mobile apps for restaurants can take advantage of location technologies to deliver better contextualized mobile

campaigns. Restaurants can leverage location-based services to their advantage by enabling order-ahead services, asking for online evaluations, and employing geo-located notifications to inform customers of special promotions and new menu items. There are countless options. Mobile apps that are aware of their location can utilize geofencing to provide highly targeted communications to only the clients who fit your criteria in a certain location at a specific time.

4. **Excellent performance:** Apps with good design typically load and operate more quickly than websites. Additionally, because they are simpler to explore, they offer a better user experience.
5. **Delivery fleet data can be provided in real time via an app:** Installing a GPS tracking system across the board can improve customer experience. We've added an Order Tracking System to the point-of-sale system for restaurants. This feature will keep track of any restaurant orders so that customers may see how much time has passed since placing their orders and learn about any delays in order preparation through color indications and other details.
6. **Push Notifications:** Push Notifications are a double-edged sword that has the power to make or kill your app. If you employ your app in a sophisticated and discrete manner, it will be successful. User retention is aided via push notifications. They can boost the returns you see from your app in addition to helping to increase interaction on the message's subject. Geographically-based notifications are another way to let users know about sales or promotions when they're nearby. Your app's ability to receive push notifications can make or destroy it.
7. **Consumer Benefits of Mobile Payment Systems and Convenience:** Mobile payment system streamline and simplify the payment process. Users may now make payments whenever they want, wherever they are, thanks to their mobile devices being connected to the Internet. Due to this, developers will need to create a mobile application that enables users to perform transactions without using physical debit/credit cards or cash. M-commerce includes things like in-app purchases, mobile banking, virtual marketplace apps like the Amazon mobile app, or electronic wallets like Android Pay, Apple Pay, and Samsung Pay.
8. **Locating locations near restaurants using GPS:** Sincere to say, several of these capabilities are also accessible online. However, they typically operate more quickly and effectively in native apps.
9. **More recognizable:** People should utilize a browser to access a website. For apps, all that is required is a single download and installation before your consumers sees a button with an icon on their home screen. Isn't that a brilliantly easy approach to reaching your customers? It is!
10. **Better user interface:** Website clutter is a common complaint among Internet users. Since food ordering apps feature numerous entrees with images, it is very crucial. Mobile apps typically have a clearer and more user-friendly user interface while being more condensed of course.
11. **Discount & Deals for Customers:** Discount deals apps have made it simpler for customers to go out and enjoy delectable meals at the most reasonable prices using discount deals that are supplied by various restaurants, in addition to the aforementioned apps. Deal News is one such app that collaborates with various eateries to attract new customers by providing discounts and deals on the establishment's goods.

STATUS OF POPULAR ONLINE FOOD DELIVERY APPS IN TODAY ERA

One thing is certain; technology is finding innovative ways to cater to specialized services and consumer demand. The following trends have emerged when it comes to new food delivery applications.

Apps and Delivery Fleets Operated by Restaurants

More restaurant businesses are likely to create apps and use their delivery fleets, similar to Domino's pizza. This concept has the benefit of giving the restaurant total control over all aspects of the client experience, from branding to app design. Because it is customized for their unique client journey, Domino's app experience is excellent. It is just impossible for Uber Eats, which collaborates with thousands of eateries, to create a universal app that keeps track of every aspect of the ordering process. Domino's is also far more suited than third-party apps to capitalize on cross-selling opportunities.

Increasing Growth of Grocery Delivery

Customers may order goods from their preferred supermarkets through services like Instacart, and this trend is gaining popularity. Since individuals preferred to avoid the stores during Covid-19, this lucrative market experienced a significant uptick. Customers are now addicted to the service, thus growth will probably continue. It's important to note that Instacart pricing strategy has come under scrutiny. When you order groceries online, it might be difficult to determine how much more you're paying because sometimes there is no markup signifying that the retailer has paid for the relationship, and other times the prices are higher. Furthermore, Instacart receives a monthly subscription charge. Maybe there's a market for an app that delivers groceries. A supermarket delivery app with more upfront pricing might find a market. Busy professionals are willing to pay a lot of money to have items delivered, that much is certain.

The Expansion of Business Models

Software as a Service corporation adores recurring income because it enables them to forecast earnings over time. It makes investors content and encourages more financial injections. Companies are then able to control their market thanks to that capital. Numerous of the top meal delivery apps on the list above feature monthly memberships with savings, which boosts user loyalty.

Deliveries of Prepared Meals and Meal

Pre-made, healthful meals that are simple to reheat in the microwave are available from companies like freshly. These are more affordable for the consumer than placing a takeout order from a restaurant, and they are ideal for a subscription business. Meal deliveries are an intriguing new development in the world of food app delivery. Companies like Blue Apron offer all the supplies and simple recipes needed for home cooks to prepare delicious meals. Not only does this save time, but it also teaches those who are hesitant about cooking how to navigate the kitchen.

The Effect of Robots and Artificial Intelligence

Future meal delivery apps may be dramatically impacted by artificial intelligence and robotics. Implementing chatbots, which can handle client inquiries and orders and enable a more effective and simplified ordering procedure, is one-way artificial intelligence may be employed. On the other hand, using robots for jobs like food preparation and delivery could possibly speed up and improve these procedures. The use of artificial intelligence and robotics in food delivery apps can also reduce costs for the business and enhance customer satisfaction.

The Function of Influencer Marketing and Social Media

Food delivery apps are increasingly relying on social media and influencer marketing to spread the word about their services and attract new users. Social media sites like Instagram, Facebook, and Twitter can highlight the services that meal delivery apps offer and disseminate details about sales and discounts. Influencer marketing is the practice of paying social media users to promote a brand's goods or services. Through sponsored articles, stories, or videos that highlight the food delivery apps and their products, they can successfully reach potential customers. Food delivery apps can raise brand awareness and stimulate sales by utilizing the social media platform's reach and impact, as well as influencer marketing.

FACTORS INFLUENCING CONSUMERS' ATTITUDE TOWARDS USING FOOD DELIVERY APPS

Four factors generally have an impact on consumer behaviour. These factors determine whether or not your target customer buys your products. They are cultural, social, cognitive, and personal. The conduct of consumers is influenced by a variety of things. Marketers need to be aware of the factors that affect consumer behaviour. The top 4 factors that influence consumer behaviour are listed above:

1. **Social factors:** Humans are social creatures; therefore they are always surrounded by people who can influence their purchasing decisions. The use of food delivery apps can also be influenced by various social factors. Here are some key social factors that can impact consumers' use of food delivery apps:
    a. **Reference groups:** Consumer behaviour is significantly influenced by social influence. When members of a social group, such as friends, coworkers, or family, use food delivery apps and share positive experiences, it may persuade other members of the group to follow suit. Customers may be persuaded to utilize meal delivery applications by recommendations and word-of-mouth from reliable sources.
    b. **Social norms:** The use of food delivery apps can be influenced by social norms, which are the acceptable actions and expectations within a society or social group. People might be more likely to utilise apps to order meals if it becomes commonplace or socially acceptable, in order to live up to the expectations of their perceived social circle.
    c. **Social Status:** Depending on one's social status, food delivery services may be viewed in different ways. While some people may view using meal delivery apps as a simple and time-saving solution, others may identify it with exclusivity or a particular social position. Consumer behaviour can be affected by the perceived image or status associated with utilizing these apps.
    d. **Generational Gaps:** Different generations can affect how people use applications that serve food. In comparison to earlier generations, younger generations, like millennial and Gen Z, who are more tech-savvy and habituated to digital platforms, may embrace meal delivery applications more readily. The use of meal delivery applications can be influenced by generational attitudes, interests, and digital fluency.

e. **Culture:** The use of meal delivery applications may be influenced by cultural variables. Dining out or getting food delivered may not be as common in some cultures while being an important aspect of society in others. Consumer attitudes and behaviours of using meal delivery apps might be influenced by cultural norms and acceptance surrounding convenience and food intake.

f. **Social Media:** The use of food delivery apps may be influenced by the expanding use of social media and digital connectivity. People may believe that utilising meal delivery apps is a socially acceptable behaviour as a result of seeing their peers and influencers post about their experiences with and recommendations for these services on websites like Facebook, Instagram, or Twitter. As a result, adoption of these services may rise.

2. **Cognitive factors:** Cognitive factors such as learning, attitude and motivations plays a crucial role in the adoption and usage of food delivery apps. These factors are described below:

a. **Learning:** Learning plays an essential role in adoption of food delivery apps. Users acquire the knowledge and skills necessary to effectively use the app through this learning phase. Users learn as they peruse the menus, read the descriptions, and find out about new foods the app offers. Through the app's features, they gain awareness of various culinary preferences, discover new cuisines, and increase their understanding of available local dining options. Users become familiar with the app's payment alternatives and the security precautions it has put in place to safeguard their financial information. Through this process of instruction, users gain trust and confidence in the app's payment system. When consumers share their opinions and evaluations about the meal delivery apps they've used, learning happens. They gain the ability to express their ideas, critique eateries or delivery services, and offer helpful criticism to app creators.

b. **Beliefs and attitudes:** Consumer views have a big impact on whether and how food delivery applications are used. Customers are more likely to view food delivery applications favourably if they place a high value on convenience and time savings. The adoption and continuous use of food delivery apps is influenced by consumers' positive sentiments towards trust and dependability because people are more confident that their meals will be accurately delivered and live up to their expectations. Food delivery app uptake and usage are influenced by consumer perceptions of the user experience and interface design. User-friendly, intuitive, and aesthetically pleasing apps are more likely to draw and keep users.

c. **Motivation:** The convenience and time-saving benefits that meal delivery apps provide are two of the main reasons people use them. Use of food delivery apps may be motivated by a desire to experiment with new flavors, try out new dining establishments, or obtain a particular sort of cuisine that may not be easily accessible in their area. People are more inclined to use food delivery apps as a handy alternative if they are driven by the desire to save time or to avoid the trouble of cooking or dining out. The need to save time, particularly while juggling several obligations or working long hours, can influence people's decision to use food delivery apps as a convenient way to get meals without disrupting their daily lives or routines.

3. **Personal factors:** Personal factors influences the use of food delivery apps by individuals, so organizations must focus on the same for making marketing strategies. Some of the individual factors are:

a. **Disposable Income:** The amount of discretionary income that a person has depends on their salary, which may have an impact on how likely they are to utilize meal delivery applications.

Higher earners might have more money available for non-essential purchases, such as ordering food using mobile apps. Contrarily, people with lesser incomes could priorities necessary spending and have less discretionary money, which makes using meal delivery apps a less common alternative.

b. **Age:** It plays a crucial role in usage of food delivery apps. Younger generations are typically more tech adept and more ease using smart phones and mobile apps, such as millennial and Gen Z. They are more likely to be among the first to adopt cutting-edge technology, such as meal delivery applications. They can quickly order food with a few touches on their smart phones thanks to food delivery applications, which provide convenience and time-saving advantages. Younger people typically feel more at ease revealing personal information and engaging in online commerce. They might have more faith in payment methods and app-based platforms.

c. **Lifestyle:** The use of food delivery apps may be influenced by urbanization and lifestyle changes, such as hectic schedules and time restraints. The effects of urbanization and changing lifestyles may increase the societal acceptance and use of meal delivery applications. Food delivery apps can be a practical answer for people who may not have much time to cook or eat out in busy urban surroundings.

d. **Occupation:** Occupational factors can have a variety of effects on how people use meal delivery applications. People who have jobs with lengthy hours, unpredictable schedules, or significant levels of work-related stress may rely on food delivery apps for their meals. The amount of time available for cooking or eating out may be constrained by jobs that entail a lot of travel or have strict deadlines. Food delivery services offer a practical solution by making it quick and simple to access a wide range of food selections. People who work from home or freelance may rely on food delivery apps for their meals.

4. **Economic factors:** The economic state of a nation or market has a substantial impact on consumer purchasing choices and behaviours. A healthy economy increases the quantity of money available on the market and raises consumer purchasing power. A faltering economy, however, reveals a market that is in danger due to rising unemployment and declining purchasing power. Consumer purchase decisions are significantly influenced by economic factors. There are numerous significant economic factors, including:

   a. **Affordability:** For many consumers, the price of using meal delivery apps can be a deciding factor, and income levels directly affect affordability. Delivery or service fees are often charged by some meal delivery apps; these prices might change based on the platform and the area. Customers might also need to think about the whole price of their food purchase, which includes menu prices and any applicable taxes or tips. Food delivery apps may be more affordable for those with higher incomes, making them more widely available to them.

   b. **Income Level:** The regularity with which people utilize meal delivery apps might also be influenced by income levels. Due to financial limitations, people with lower incomes may be more prone to utilize food delivery apps as an occasional treat or for special occasions. Contrarily, people who make more money might have more extra cash available, enabling them to use food delivery services more regularly as a regular eating alternative.

   c. **COVID-19 pandemic:** The food distribution sector was significantly impacted by the COVID-19 epidemic. As a result of lockdowns, limitations on dine-in services, and social distance-inducing policies, consumers have been turning more and more to meal delivery applications. The epidemic hastened the popularity of meal delivery services as a more practical and secure substitute for dining out.

d. **Competition:** A number of applications are fighting for market share in the fiercely competitive meal delivery sector. In order to draw and keep users, meal delivery app providers are using a variety of incentives, discounts, and promotional offers. These incentives are available, which has affected customer behaviour and promoted the usage of meal delivery applications.

e. **Consumer credit:** In times of need, when people may need fast access to food through delivery services, consumer credit might be very helpful. For instance, food delivery apps can help ensure access to meals during natural catastrophes, severe weather, or medical emergencies when traditional food supplies may be scarce or inaccessible. The essential financial flexibility to take care of such urgent needs can be provided by consumer credit.

CONSUMER ATTITUDES TOWARDS USING FOOD DELIVERY APPS

Online shopping is very popular in the minds of consumers because of its convenience and low cost. Online shopping saves people from visiting multiple stores and standing in long lines for holiday shopping. The Internet has successfully changed how consumers shop for products and has gained widespread acceptance. Many businesses have begun to use the Internet to compete with each other in an effort to reduce their marketing expenses and lower product prices. Many predict that online shopping will become more popular than traditional offline shopping. Consumers of the future will choose cash on delivery, and online marketing will have a wide range of applications. The article "The Impact of E-Commerce on Consumer Buying Behavior discusses the reasons why people shop for items online. According to their research, three factors influence whether or not consumers shop online. Product type, product quality and product pricing are characteristics of a shopping website. Users get the option to make a purchase or not. Additionally, there are other challenges that customers face while shopping online. As a result they move towards offline shopping. As a result they move towards offline shopping. However, online shopping is more convenient for customers who find it difficult to shop offline. However, the communities of elderly customers who do not live online are not involved in shopping as they do not view online shopping favorably and instead place more emphasis on physical shopping. Mobile e-payments have been accepted for many years. Many people also favour using their mobile devices to make payments. The digital wallet has grown in significance for customers. Numerous advantages of the digital wallet include its cost, simplicity, and security. The advancement of technology has made this payment method the most practical and straightforward method of transaction. Additionally, it is acceptable and accessible.

Customers are gravitating towards mobile payments because they provide advantages including flexible payment wallet brands and consumer ease. Trust, perceived ease of use, and express method of usage are all factors that encourage more people to use digital wallets as a form of payment. Shoppers benefit from the convenience of digital wallet payments by receiving additional payment options. Consumer Perception on Digital Payment evaluated the idea of acceptance and use of technology, which is the foundation for security, trust, social impact, and self-efficiency, as a model for how consumers accept mobile payments. Consumers can use digital wallets to pay conveniently without having to swipe their debit or credit cards. There is no distinction when assessing a customer's use of internet banking services.

FINDINGS

The overall findings of this study suggest that food apps are used by all consumers in the modern world, due to their quick response times. It improves understanding of people's preferences, effectiveness of time management, accessibility, food preferences, discounts available and door-to-door service without sacrificing quality. Attitude, subjective norms, control beliefs, and trust the four fundamental elements of the planned framework theory positively influenced users' intentions to continue using food delivery apps. These four fundamental constructs' predictors, namely perceived inventiveness, motivations, perceived utility, ease of use, and food safety risk perception, were also found to be confirmed.

CONCLUSION

As a result, this study on customer perceptions about food apps can conclude that most people use them because they are convenient and the biggest way to save time. Additionally, placing orders using the Food app requires accuracy. Online food app among respondents, and cash on delivery is the safest and most secure mode of payment. According to the report, food apps are used by people of all ages and socio-economic backgrounds, and are preferred for high levels of customer satisfaction, hygienic conditions, and packaging. The questionnaire actually had some intriguing responses, such as whether consumers still favor cash on delivery over the popular online payment method. And found that some people still favor ordering food the traditional way by phone while, in general, consumers are being persuaded.

FUTURE IMPLICATIONS OF THE STUDY

The study reveals consumer trends toward the use of food delivery apps. Also, in the future, this study can be further developed as a large geographical indication. Commerce online websites and other media channels may also be considered to integrate the study into exploration.

REFERENCES

Al Amin, M., Arefin, M. S., Sultana, N., Islam, M. R., Jahan, I., & Akhtar, A. (2021). Evaluating the customers dining attitudes, e-satisfaction and continuance intention toward mobile food ordering apps (MFOAs): Evidence from Bangladesh. *European Journal of Management and Business Economics*, *30*(2), 211–229. doi:10.1108/EJMBE-04-2020-0066

Bothra, K. (2021, December). Consumer's Perception Regarding Online Food Delivery Applications in India. *International Journal of Research Publication and Reviews*, *2*(12), 948–953.

Cho, M., Bonn, M. A., & Li, J. J. (2019). Differences in perceptions about food delivery apps between single-person and multi-person households. *International Journal of Hospitality Management*, *77*, 108–116. doi:10.1016/j.ijhm.2018.06.019

Dinu, A. (2022).The impact of food delivery applications on Romanian consumers' behaviour during the COVID-19 pandemic. *Socio-Economic Planning Sciences,* 82. *Part A*, (August), 101220.

Epuran, G. (2021). Enhancing Consumer Experience through Development of Implicit Attitudes Using Food Delivery Applications. *Journal of Theoretical and Applied Electronic Commerce Research, 16*(7), 2858–2882. doi:10.3390/jtaer16070157

Garg, V. (2019). The Study of Interest of Consumers In Mobile Food Ordering Apps. *International Journal of Recent Technology and Engineering* (IJRTE), *8* (1).

Gupta, V., & Duggal, S. (2021). How the consumer's attitude and behavioural intentions are influenced: A case of online food delivery applications in India. *International Journal of Culture, Tourism and Hospitality Research, 15*(1), 77–93. doi:10.1108/IJCTHR-01-2020-0013

Kaur, P., Dhir, A., Talwar, S., & Ghuman, K. (2021). The value proposition of food delivery apps from the perspective of theory of consumption value. *International Journal of Contemporary Hospitality Management, 33*(4), 1129–1159. doi:10.1108/IJCHM-05-2020-0477

Khaliq, A. (2022). Determinants of customers' intention to use online food delivery application through smart phone in Malaysia. *British Food Journal, 124*(3), 732–753. doi:10.1108/BFJ-01-2021-0075

Kujur, Rubi.(2019). Satisfaction of consumers by using online food services. *International Journal of Humanities and Social Sciences* (IJHSS) ISSN (P): 2319-393X; ISSN(E): 2319-3948 8(4) Jun - Jul 2019; 35-44.

Singh, A. (2020). Consumer Perception and Attitude towards Online Food Ordering. International Journal of Research in Engineering, Science and Management, 3(5), May-2020 www.ijresm.com | ISSN (Online): 2581-5792.

Tribhuvan, A. (2020). Analysis of customer attitude towards Electronic Food Ordering. *International Journal of Advance Research and Innovative Ideas in Education, 6*(4), 36.

Tribhuvan, Aditya (2020). A Study on Consumers Perception on Food Apps.2020 *International Journal of Advance Research and Innovative Ideas in Education*, ISSN (O)-2395-4396. 6(4).

Wen, H., Pookulangara, S., & Josiam, B. M. (2022). A comprehensive examination of consumers' intentions to use food delivery apps. *British Food Journal, 124*(5), 1737–1754. doi:10.1108/BFJ-06-2021-0655

Chapter 6
Robots in the Hotel Kitchen:
The Future Chef

Partho Pratim Seal
Manipal Academy of Higher Education, India

ABSTRACT

The adoption of robotics and automation tools in the hospitality sector is growing fast with robots programmed to perform a human-operated task. The shift towards the acceptance of robots in hotel kitchens with robot chefs as a part of the chefs team is a reality in some establishments. Robots are preferred by catering establishments as they reduce errors in food preparation and improve productivity; however, the initial investment for robots is high. Chefs do believe that humans could not be replaced by robots as the sensitivity of taste, and flavor may not be replicated by the robots even though standardization of recipes allows them to prepare distinctive food repeatedly. This study is about the customer's response to robots as chefs in various food establishments, and observation of robots in multi-cuisine establishments through secondary resources. The findings of the study will help hospitality chains to be better prepared for robot chefs and automation in the hotel kitchen.

INTRODUCTION

Staff shortage in hospitality has been stated in many parts of the world (Mathath and Fernando, 2015) followed by the impact of the pandemic which increased the interest of people in the area of robotics in the kitchen. Although robots were invented in 1920 (NPR, 2011) it was stated to be a thing of the future and not totally practical. With the passing of time, the intelligence of robots increased and they could be programmed so that they could perform human-operated tasks. Later in the 1980s robots initially were utilized in manufacturing before it was introduced in the service sector. The usage and adoption of robots in hospitality and tourism are found to be growing continuously in the last few years (Ivanov & Webster, 2019). Hospitality is mostly human-to-human interaction but it is slowly moving towards more human-to-robot interactions thereby challenging hospitality with a new service environment (Bowen & Morosan, 2018). Service robots intend to coexist with humans as they engage in relations for providing a better quality of life.

DOI: 10.4018/978-1-6684-9094-5.ch006

In hospitality services, personalization has an important role with the availability of personal service robot which provides a high level of social interaction and is the preferred option for many organisations (Murphy et al., 2017). In food preparations and serving the service robots can be effectively utilized for operations (Lukanova & Ilieva, 2019) with a new category of robots said to be catering robots. As per the McKinsey Global Institute Report in the future half of today's working activity would be automated, and by 2055 about 73 percent of jobs in hospitality and food establishments will be replaced by robots (McKinsey Global Institute Report, 2017).

The use of robots in restaurants is rapidly being accepted as service staff or waiters but also as chefs and kitchen staff. Kitchen robots are devices that help to create several pre-programmed meals as per the information provided by the users and in some robots the users have to also put in the necessary ingredients inside the robot. Initially, in the hospitality sector, it was believed when robots were introduced that they will possibly replace the jobs in the restaurant sector (Robinson, 2017).

Service robots have evolved a lot from the past and companies are likely to invest in robots as it improves hospitality productivity and profits. In the field of robotics having a robotic chef is a challenge as a robot chef may have to prepare many dishes as per the need of commercial establishment. The kitchen environment being semi-structured also has computational challenges as accuracy in the usage of ingredients, usage of different equipment for the process, and various cooking methods lead to the complexity of the entire process.

The current society is open to technology and new advances. New robots are been designed and manufactured by both research institutions and companies to fulfill the technological gap and to meet the need of the market. The culinary industry is accepting innovative technologies and many of the repetitive tasks in the kitchen such as stirring pasta, flipping the burgers, and assembling pizzas are been taken over by robots. The current robots are assisted by sensors onboard, optical cameras, and advanced AI technology to do multitasking thereby performing multi actions and movements which hotel chefs do in real-time. Now different kitchen robots are available in the market depending on the requirement and level of assistance necessary in cooking. AI is now powered by a robotic kitchen to prepare and abolish human interaction with the cooking process.

The purpose of the study is to examine the customer's views on the usage of robots in food service establishments particularly on the customer's perception of the food prepared by the robot chefs in food establishments and just not serving food. The study is conducted through secondary data obtained from online reviews of customers from restaurant portals and other websites. Considering the demand perspective the study findings especially disclose the meaning the guest have about robots as chefs. The study findings will help businesses both managers of restaurants and the manufacturers of robots to understand the challenges and provide the customers with best dining experience.

LITERATURE REVIEW

Robotics technology along with AI is the future with some researchers predicting that AI in the future might even exceed human intelligence (Callaghan et. al, 2017). The preference for hospitality towards automation tools is just not to reduce human error percentage but also to improve the productivity of the organization along with the quality of work.

In the United States, a California-based company developed a robot for preparing burgers that can prepare 120 orders in an hour (Troitino, 2018). The concept of robot chef has been found around the world with *Spyce*, a restaurant created by Chef Daniel Boulud in Boston which has a kitchen with a 14- foot full robotic kitchen (Doyle, 2018). In India also, a robot *'Julia'* was designed as a mechanized cooking pot that could prepare and dispense dishes in twenty minutes. Currently, there are many robots, CaliBurger created a robot *Flippy* which can cook hamburgers in Pasadena, California (Otero, 2018). A Wishdoing restaurant in Shanghai has a robot that can cook food in less than three minutes.

There is also a robot Bimby or Thermomix which has revolutionized cooking as it performs several functions which are done manually and automates the cooking process. The German-made food processor does prep work including (cutting, mixing, and kneading) and also cook (baking or steaming), and also seasons in a mixing bowl. The device could be programmed as per the recipe specifications (Fonseca et al., 2019). One of the world's fully automated multi-cuisine Robotic Chef is Nala Chef which uses machine learning to cook unlimited recipes with exact precision and can be personalized and customized as per the customer's liking. It has in build sensors that can check about 1200 parameters every microsecond which enhances productivity and adequate security. It has the capability to work 24/7 and can be a boon for pre-packed food during restaurant peak hours (Nalarobotics, n.d.).The Nala robotics have also developed a robot chef known as "Wingman" which is a fully automated system that can handle various cooking processes like breading of food products and frying, tossing of salads and preparing sauces, and packing food without much of an intervention. Another positive point of the robot is its multi-language compatibility which helps it to accept commands from the kitchen and respond to customer orders efficiently (Nalarobotics, n.d.).

Miso Robotics developed three robots named Flippy, Sippy, and Chippy to assist fast food restaurants in repetitive tasks (Reiley & Powell, 2022). These robots could cook fries, pour drinks and also prepare tortilla chips. Moley Robotics is a fully robotic kitchen with an AI autonomous system that makes the cooking process virtually automatic developed by a British robotic company stated as Moley Robotics MK1 and a professional kitchen. The robot chef cooks with the flair of a master chef and also has a recipe library that could also be accessed by professionals remotely The specialty of the robot is it is ceiling mounted and so can glide on the track and is accessible to the entire kitchen. The features of the robot are also distinct as it can adjust the temperature of equipment, can mix and pour ingredients into pans, stir ingredients for cooking and even use the sink like humans. It is pre-programmed with more than 5000 recipes which it can cook and even clean up the area after it is done. The robot also makes food with the help of sensors that are attached to the kitchen utensils which can monitor various parameters including touch, smell, seeing, and hearing with a learning loop similar to human beings. These robots are also equipped to have tactile, contact, and proximity sensors for recording tasks, capturing movements, and cooking recipes. The robots are programmed in a manner that they provide information when ingredients need replacing, suggest dishes, control calories, and adapt menus for different diets and lifestyles. The robot chefs could learn by storing information in the database and retrieving it when needed (Garcia-Haro et al., 2021).

A large amount of venture capital has funded companies that produce robots. The Creator earlier known as Momentum Machines did obtain $18 million in finance in 2017 (Robinson, 2017). The robot Spuce has also raised about $25 million from investors (Lucas, 2021).The catering robot requires the necessary skills and abilities to do multi-tasking which includes working with food and drinks, plate presentation and then transporting the prepared dish from the kitchen to the customer. Other than preparing food the robot chefs have other challenges which include working with humans in restricted space. The robots must also bear the heat and steam along with grease, smoke, and water. The robots clean and keep the area hygienic and also follow the food safety norms.

METHODOLOGY

The study objective was to find online reviews of customers regarding robot chef restaurants globally. A netnography approach was taken to obtain the maximum response of customers who had been to a restaurant catered by a robot chef. Netnography is a qualitative research methodology adapted to study cultures and communities with the help of computer-mediated communications (Kozinets,2002). The preference and success of the method are because of flexibility, less expensive, and pragmatic nature. In the present world where social life is influenced by the Internet, people feel free to share their experiences on various social media. The study was conducted considering two research questions- The diner's experience of food being prepared by robot chefs and the diner's preference towards visiting restaurants where robots cook.

The data was extracted from the reviews of popular robot chef restaurants around the world. Some of the reviews were also collected from Yelp as it is also found to be one of the platforms to collect online reviews on restaurants. Content analysis is chosen as it is an appropriate way to analyse the text as it offers both an objective and systematic way of studying the text (Kippendorff, 1980). A total of about 220 online reviews were collected from various restaurants' websites. A total of 183 reviews were considered for content analysis as some of the reviews were general and were not related to the research questions.

Result

The result of reviews led to the emergence of three categories with regard to acceptance, nonacceptance of robot chefs, and unsure of acceptance of robots as chefs even in the near future. The three categories are obtained from the reviews as part of the coding process. The maximum number of reviews stated that they are satisfied with the food and experience of a robot chef in a restaurant. However, some of the comments were negative and the customer was not satisfied with the food prepared by the robot chefs. While some of the reviews warned against replacing human chefs with robots as human-to-human interaction is a must for a dining experience.

The customers visiting various robotic restaurants have both positive and negative opinions about the restaurants.

The customers of the robotic restaurant stated that:

C1 "The robots can prepare the food as the customer wants the dish to be prepared. You can customize everything to be exactly how you like and in line with dietary restrictions."

Another customer shared the experience of viewing food prepared by robots which made it a memorable experience.

C2 "Though I usually order food and wait for the food to meet my expectations. I ordered the Bahn Mi bowl and it was delicious-- it was especially cool seeing the robots configure our food".

A customer's comments described the entire process of how the order is processed by the machine from placing the order to the finished dish and was quite happy with the food served.

C3 "The entire experience of ordering to the delivery of food was different. Interesting concept as you order at a machine, type in your name, pay, and then wait until your order is called and then you can also observe your order being processed as your name is displaced while the robot prepares the dishes".

By observing the process of each step of food being cooked in front of them the customer was spellbound by the entire dining experience.

C4 "Amazing watching the robots work their magic. Although much of it is behind the scenes, patrons can see robots moving horizontally, dropping ingredients into the different drums, and then the drums spinning, cooking everything up. The speed and accuracy by which the food is cooked watching is interesting".

As stated in the previous comment the customer is very pleased with the food being made with the ingredients and the entire cooking process made in front of them along with the efficiency by which the same is prepared.

C5 "Very cool concept of "robots making your food". It was very quick and precise (thanks to the robots). Even though the chef is a robot. All my specification for the dish was considered and the dish was customized for me."

The experience of a guest, the time taken for the dish to be cooked to the variation of different dishes the robot can cook were quite significant and the customer was quite happy with the customization of the dish according to the customer's need

C6 "A MUST-DO! This Place is Cool. The World's First Robotic. Restaurant Kitchen that cooks complex meals. It takes only 3 minutes for your meal to be prepared and you can see your name being displayed. Only garnish and accompaniments are added by the server".

Like previous reviews, the customer wants to repeat the visit to the robot restaurant considering its cooking time and food being processed in front of their eyes.

C7 "I will visit the place again I like the place the food is prepared and served fast. The great food, fresh tasting, and of course it's quite the show watching robotics and machines making your food. The time promised to prepare the order is adhered to by robots".

Like previous reviews, the customer is quite happy with the overall food experience including the quality of food, taste, and the time taken to prepare the dish.

C8 "There is no place for dialogue. Machine cannot do everything. Don't dare to use machines to replace real human work. This is not how the future of the culinary world suppose to be. Let the cook do the real precise and sophisticated culinary work. I will not go because I like human interaction to know about the chef and feel depressed if food is cooked by machines".

Contrary to most of the other reviews the customer is annoyed with the cooking process of food being prepared by a non-human and will not accept a deviation in the cooking

C9 "Sorry, but this place is absolute rubbish after you get past the "robot" novelty. The robots don't have common sense and hence provide a limited experience. Flavorless, boring fare. Summed up in one word. Bland. Scary if robots replace human beings.

Another customer was dissatisfied with the food being served and was very vocal about the poor taste of the food being served.

C10 "Barely edible. Visit this place if you want to take some dumb Instagram posts about how robots are the future of cooking. Do not recommend it to people who enjoy food. Missed the love the chef adds to the food. The food lacks soul and seems to be a more mechanized process."

Another not-very-positive comment by the customer stated that the food cannot be consumed at all and is more for the photographic purpose of how food is prepared with the robot.

C11 "As they are technologically developed they are able to manage information and data and provide improved services and be more sustainable. It can reduce waste, create just-in-time preparation and service, and reduce contamination, and risk".

A customer also was pleased by robots who can manage more information than chefs, be more efficient, reduce wastage, and can be more hygienic.

C12 "The robotic claw dropped and stamped flat by a robotic press manhandled crust a tiny bit, a little floppy and dripping with pepperoni oil. The crust and taste were not preferred. It didn't match the human touch. The pizza-making process was automated but the product was just satisfactory. Lacked interaction with the chef so could not state my specifications while ordering".

A customer also shared not so good experience with the pizza-making robot which made an undesirable pizza with a non-appetizing crust and taste.

The findings from the comments in Fig -1 do suggest that robot chefs are adequately fast and efficient as compared to human chefs. For diners, a robot restaurant is a memorable experience to watch the ingredients being sourced, the entire cooking process, and the dispensing of the food. As the food is not touched by humans so the chance of contamination is minimized. Some of the customers stated that the entire experience was mainly mechanical as pre-programmed robots could make only restricted dishes or fast food items and lacked love and affection. Most of the robots could not customize dishes as per the guest's needs with only slight variation considered to be a limitation. The dishes made were standardized to milligrams of ingredients and spices leading to same-taste dishes without any variation. The findings also suggest as in Fig -2 that the positive characteristics of robots are their accuracy of food preparation and timebound delivery of dishes while as compared to humans they are devoid of emotions and common sense to act accordingly.

Figure 1. A robot chef's characteristics

Figure 2. Customers' perception of a robot chef

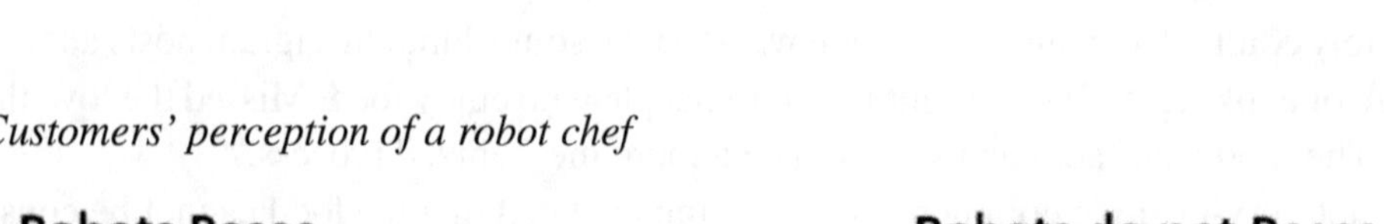

Though past studies have discussed chef robots in which survey data was collected the current study considers the guest comments in various robot restaurants for analysis. Robots are found to be machines that have additional skills than humans though they lack feelings. Robots are best suited for fast food joints dispensing burgers or pizzas or a single bowl meal but are not very effective in a specialty restaurant where food is prepared by skilled chefs and presented artistically.

DISCUSSION AND CONCLUSION

Most of the comments of the customer stated that robots are faster, and cleaner in operations in comparison to humans. The customer also must keep in mind and prioritize either the efficiency or interaction between humans to humans when opting for the dining experience in a restaurant. Some comments by the diners also state that a robotic restaurant may be a one-time visit experience for customers but the skills of a human chef could not be replicated by a robot. The food prepared by the robot and the customer experience is subjective in nature (Ivanov & Webster, 2019) though the study offers an understanding of the factors which influence the customer's perception of the robot chef in the overall dining experience. There is also a perception amongst consumers that is highlighted by various studies that food prepared by a robot chef will not taste like food prepared by humans (Xiao& Zhao, 2022).

In the future, it is expected that robots will assist the chefs and improve their work by enhancing precision and efficiency resulting in greater acceptance of robots as chefs (Fusté-Forné, 2021). So in conclusion we can state that automation will continue to happen in food preparation especially amongst the larger food establishment and fast foods where most of the jobs of the workers are repetitive in nature and the trend has mostly augmented by the post Covid pandemic (Morrison, 2022). The study highlights key challenges and opportunities linked with robot chefs. Though the presence of robot chefs provides novelty and experiential value to the customers but the long term viability considering the high cost of the robots is cause of concerns for the establishment. For the adaptation of robots extensively as chefs in restaurant operations, further research is required to understand various stakeholders' perspectives including owners, restaurant staff, and chefs in the kitchen globally. Future research can bridge the gap between academia and industry to utilize service automation thereby increasing the usage of robots effectively. However, the use of robots must be done along with the personalization which could be built on technology so that they can provide improvised food to the customers.

Limitations of Robot Chefs

The problem with some of the AI robot chefs is it is difficult for the robots to prepare the ingredients for cooking like peeling onions and garlic and cutting vegetables as per the specifications needed for preparing a dish. Currently, AI robots are very expensive and are out of reach for most people but the reports do suggest that the investment later will be comparatively cheaper than employing a human to perform the task. Cooking is natural for human beings and they may not like a non-human to prepare food for them.

REFERENCES

Bowen, J., & Morosan, C. (2018). Beware hospitality industry: The robots are coming. *Worldwide Hospitality and Tourism Themes*, *10*(6), 726–733. doi:10.1108/WHATT-07-2018-0045

Callaghan, V., Miller, J., Yampolskiy, R., & Armstrong, S. (Eds.). (2017). *The technological singularity: managing the journey*. Springer. doi:10.1007/978-3-662-54033-6

DigitalFoodLab. (2010). *NYMBLE LABS—The Smart Kitchen Appliance*. Digital Food Lab. https://www.digitalfoodlab.com/en/ foodtech-database/nymble-labs

Fonseca, E., Oliveira, I., Lobo, J., Mota, T., Martins, J., & Au-Yong-Oliveira, M. (2019). Kitchen Robots: The Importance and Impact of Technology on People's Quality of Life. In Á. Rocha, H. Adeli, L. Reis, & S. Costanzo (Eds.), *New Knowledge in Information Systems and Technologies. WorldCIST'19 2019. Advances in Intelligent Systems and Computing* (Vol. 931). Springer. doi:10.1007/978-3-030-16184-2_19

Francesc Fusté-Forné. (2021) Robot chefs in gastronomy tourism: What's on the menu? *Tourism Management Perspectives*, *37*. doi:10.1016/j.tmp.2020.100774

Garcia-Haro, J. M., Oña, E. D., Hernandez-Vicen, J., Martinez, S., & Balaguer, C. (2021). Service Robots in Catering Applications: A Review and Future Challenges. *Electronics (Basel)*, *10*(1), 47. doi:10.3390/electronics10010047

Ivanov, S., Gretzel, U., Berezina, K., Sigala, M., & Webster, C. (2019). Progress on robotics in hospitality and tourism: A review on the literature. *Journal of Hospitality and Tourism Technology*, *10*(4), 489–521. doi:10.1108/JHTT-08-2018-0087

Ivanov, S., & Webster, C. (2019). Perceived Appropriateness and Intention to Use Service Robots in Tourism. In J. Pesonen & J. Neidhardt (Eds.), *Information and Communication Technologies in Tourism*. Springer. (Original work published 2019), doi:10.1007/978-3-030-05940-8_19

Kozinets, R. V. (2002). The field behind the screen: using netnography for marketing research in outline communities. *Journal of Marketing Research.*

Krippendorff, K. (1980). Validity in content analysis. In E. Mochmann (Ed.), *Computerstrategien fur die kommunikations analyse* [Computer strategies for communication analysis]. (pp. 69–112). Campus.

Lukanova, G., & Ilieva, G. (2019). Robots, Artificial Intelligence, and Service Automation in Hotels. Ivanov, S. and Webster, C. (Ed.) Robots, Artificial Intelligence, and Service Automation in Travel, Tourism and Hospitality. Emerald Publishing Limited, Bingley. doi:10.1108/978-1-78756-687-320191009

Mathath, A., & Fernando, Y. (2015). Robotic transformation and its business applications in food industry. In Z. Luo (Ed.), *Robotics, Automation, and Control in Onustrial and Service Settings* (pp. 281–305). IGI Global. doi:10.4018/978-1-4666-8693-9.ch011

Moley Robotics. (2015). *MK1—The World's First Robotic Kitchen*. Moley Robotics. https://www.moley.com/ (accessed on 10 December 2020).

Morrison, R. (2022). Fancy a byte to eat? Chinese restaurant chain is forced to use ROBOT waiters priced at £ 14,500 each after struggling with staff shortages during the Covid pandemic. *Daily Mail Online*. https://www.dailymail.co.uk/sciencetec/ article-10402701/Chinese-restaurant-chain-forced-use-ROBOT-waiters-Covid-pandemic.html

Murphy, J., Hofacker, C., & Gretzel, U. (2017). Dawning of the Age of Robots in Hospitality and Tourism: Challenges for Teaching and Research. *European Journal of Tourism Research, 15*, 104–111. doi:10.54055/ejtr.v15i.265

Nalarobotics. (n.d.) *Nalarobotics-Nala Chef.* Nalarobotics. https://nalarobotics.com/nala-chef.html

Nalarobotics, (n.d.) *Nalarobotics-The Wingman*. Nalarobotics. https://nalarobotics.com/the-wingman.html

Naumov, N. (2019). The impact of robots, artificial intelligence, and service automation on service quality and service experience in hospitality. In S. Ivanov & C. Webster (Eds.), *Robots, artificial intelligence, and service automation in travel, tourism, and hospitality* (pp. 123–133). Emerald Publishing Limited. doi:10.1108/978-1-78756-687-320191007

NPR. (2011). *Science diction: the origin of the word "Robot."* NPR. https://www.npr.org/2011/04/22/135634400/science-diction-the-origin-of-the-word-robot

Otero, C. (2018). *Flippy, un automata ´ que hace hamburguesas a media jornada*. BeTech.

Reiley, L., & Powell, L. 2022. The robots are here. And they are making you fries. *The Washington Post*. https://www.washingtonpost.com/business/20 22/09/20/robots-automating-restaurant-industry/

Robinson, M. (2017). This robot-powered restaurant is one step closer to putting fast-food workers out of a job. *Business Insider*. https://www.businessinsider.com/momentum-machines-funding-robot-burger-restaurant-2017-6?r=US&IR=T

Spyce. (2018). *Culinary Excellence Elevated by Technology*. Spyce. https://www.spyce.com/

Troitino, C. (2018, June 21). Meet the world's first fully automated burger robot: Creator debuts the Big Mac killer. *Forbes*. https://www.forbes.com/sites/christinatroitino/2018/06/21/meet-the-worlds-first-fully-automated-burger-robot-creator-debuts-the-big-mac-killer/#7959e2926a89

Tuomi, A., Tussyadiah, I. P., & Steinmetz, J. (2020). Applications and implications of service robots in hospitality. *Cornell Hospitality Quarterly*, 1–16.

Xiao, C., & Zhao, L. (2022). Robotic chef versus human chef: The effects of anthropomorphism, novel cues, and cooking difficulty level on food quality prediction. *International Journal of Social Robotics, 14*(28), 1697–1710. https://www.mckinsey.com/featured-insights/future-of-work/jobs-lost-jobs-gained-what-the-future-of-work-will-mean-for-jobs-skills-and-wages. doi:10.100712369-022-00896-9 PMID:35910296

Chapter 7
The Use of Artificial Intelligence in Recipe Development:
How Technology Is Changing the Way We Create and Innovate in the Kitchen

Ceyhun Uçuk
Gaziantep University, Turkey & İstanbul Gelişim University, Turkey

Murat Doğdubay
Balıkesir University, Turkey

Sami Sonat Özdemir
Balıkesir University, Turkey

ABSTRACT

Artificial intelligence (AI) is rapidly reshaping recipe development by creating new recipes, enhancing existing ones, and providing personalized recommendations based on individual tastes. The key advantage of AI is its ability to swiftly generate and evaluate a large number of recipe variants, enabling more efficient experimentation and discovering novel flavor combinations. Additionally, AI's data analysis prowess allows it to identify patterns in ingredient combinations and cooking times for recipe enhancements. AI's influence extends to personalized recipe recommendations, where algorithms analyze culinary preferences, dietary restrictions, and cooking skills to offer tailored recipes, promoting culinary exploration. However, challenges lie in ensuring the practicality and creativity of AI-generated recipes. In conclusion, AI's integration in recipe development has the potential to revolutionize cooking and eating habits, and we can expect further exciting advancements in this field as technology advances.

INTRODUCTION

The application of artificial intelligence to the creation of recipes is an essential research topic that has garnered increasing interest in recent years. The rapid expansion of the food industry and shifting consumer preferences necessitate the development of novel and inventive recipes. Using complex algorithms and

DOI: 10.4018/978-1-6684-9094-5.ch007

data analysis techniques, artificial intelligence has tremendous potential in the recipe creation process. In this context, artificial intelligence-based methods can be applied in a variety of contexts, including the creation of recipes, the improvement of existing recipes, the optimization of flavor and nutritional content, and the discovery of distinct dishes for various cultures.

The use of artificial intelligence (AI) in recipe development is a rapidly expanding discipline that is revolutionizing our creative and innovative processes in the kitchen. AI technology is used to develop new recipes, improve existing recipes, and even suggest personalized recipes based on a person's taste preferences. One of the main benefits of using AI in recipe development is the ability to quickly create and evaluate multiple recipe variants. This allows for more efficient experimentation. It can also result in the discovery of previously unimaginable flavor combinations. In addition, AI can analyze large amounts of data such as ingredient combinations and October cooking times to identify patterns and suggest recipe improvements. This chapter highlights that artificial intelligence-based recipe creation is an area of research that is receiving increasing attention. The growth of the food industry and the changing consumer preferences reveal the necessity of new and creative recipes. Artificial intelligence brings great potential to the recipe creation process with complex algorithms and data analysis. These methods include creating recipes, improving existing recipes, optimizing nutritional value and discovering dishes for different cultures. In the chapter, it is stated that big data analysis using artificial intelligence is the basic step, and food preparation steps can be determined by matching the ingredients with natural language processing techniques. It is also stated that recipes can be created based on predetermined combinations of ingredients and flavors with deep learning models. It is emphasized that this technology will play an important role in the development of healthy recipes and the presentation of meals that meet consumer demands. In the future, it is predicted that the cooperation between artificial intelligence and the food industry will increase.

Artificial intelligence applications have a long history and have been considered by philosophers since ancient times. The foundations of modern artificial intelligence were laid in the 20th century. Artificial intelligence has made significant advances in various fields such as language and image recognition, natural language processing, robotics and driverless vehicles. This technology is seen in many successful applications, from early systems like Theseus and ELIZA to IBM's Deep Blue and Watson. As advanced AI is expected to become widespread, it is important to consider issues such as job losses and privacy concerns. In addition, there are various applications of artificial intelligence in the food and beverage industries. AI needs ethical and legal principles to understand and drive social consumers. While this technology can yield greater efficiency and benefits, the outcome of societal gain is an important consideration.

Personalized recipe suggestions are another area where artificial intelligence is influencing cuisine. AI algorithms can analyze a person's culinary preferences, dietary restrictions, and cooking skill level to suggest recipes that meet a person's specific needs. This not only makes it easier for people to discover recipes they will enjoy, but also encourages them to try new foods and cooking methods. However, there are challenges associated with using AI in recipe development. One of the biggest hurdles is ensuring that the resulting recipes are not only delicious but also practical and applicable to the average home cook. There is also the possibility that reliance on AI-generated recipes could lead to a lack of creativity and innovation in the kitchen.

Artificial Intelligence (AI) is a rapidly evolving field and has the potential to transform many aspects of us. However, with these developments, AI also brings ethical concerns. Bias can lead to unfair results as AI algorithms are trained on data. Privacy is a concern because of AI's ability to collect and analyze personal data. A lack of transparency can reveal the incomprehensibility of AI algorithms' decision processes and accountability issues. Also, the impact of AI on employment can increase the risk of job losses. In the kitchen, the use of AI can streamline the recipe development process and offer personalized recipes. However, difficulties may be encountered such as data quality and originality. Advanced AI-based recipes can drive culinary innovation and foster chefs' creativity, which can yield positive results. Global organizations are using AI tools to compile and analyze nutritional data, with data integration and standardization a key goal. However, taking action to manage the ethical and societal impacts of AI is vital. In the section, recipes were created with artificial intelligence and applications were included. As a result of the applications, it has been seen that artificial intelligence supported recipe development can have very effective results. It is predicted that time, energy and labor savings can be achieved in artificial intelligence supported kitchens.

Collecting and analyzing vast quantities of recipe data is the most fundamental step in the process of developing recipes using artificial intelligence. Using text mining, natural language processing, and machine learning techniques, recipes are analyzed and their contents are segmented in this procedure. By analyzing recipes with natural language processing techniques, it is known that there are systems that can determine the proper pairing of ingredients and the order of food preparation steps.

In the process of recipe development, artificial intelligence also plays an essential role in the creation of novel and original recipes. El-Allaly et al. (2021) conducted a study utilizing a deep learning model to autonomously generate recipes based on predetermined ingredient and flavor combinations.

Moreover, artificial intelligence can be employed to optimize nutritional values during the recipe creation process. By analyzing existing recipes, one can design a system that generates healthier alternatives. This technique has great potential for recommending healthy diets and eradicating nutrient deficiencies.

In the recipe development process, artificial intelligence is an essential research topic that is receiving increasing attention. The rapid expansion of the food industry and the shifting preferences of consumers necessitate the creation of innovative recipes. Using complex algorithms and data analysis techniques, artificial intelligence has tremendous potential in the recipe creation process.

Studies in this field indicate that AI-based methods can be applied in a variety of contexts, including the creation of recipes, the improvement of existing recipes, the optimization of flavor and nutritional content, and the discovery of new dishes for various cultures. The collection and analysis of massive quantities of recipe data is the most fundamental step in developing recipes using artificial intelligence. Using text mining, natural language processing, and machine learning techniques, the contents of recipes are analyzed and segmented. With the help of artificial intelligence, the order of food preparation stages can be determined by matching the recipes' ingredients. In addition, artificial intelligence plays a significant role in the development of new and unique formulas. Using deep learning models, recipes can be automatically created based on predetermined combinations of ingredients and flavors.

In addition, artificial intelligence can be used to optimize the nutritional value of recipes. Existing recipes can be analyzed to generate healthier alternatives, and therefore have great potential for promoting healthy diets and eliminating nutrient deficiencies. Therefore, artificial intelligence-based methods can contribute significantly to the recipe development process in the food industry. This technology enables the discovery of novel and delectable recipes in the food industry, thereby optimizing nutritional values and catering to consumer preferences. Future projections indicate that this synergy between artificial intelligence and the food industry will intensify.

USES OF ARTIFICIAL INTELLIGENCE

Artificial intelligence applications are applications that have been used for quite a long time. The history of artificial intelligence (AI) dates back to ancient times, when philosophers contemplated the existence or potential existence of artificial beings, mechanical humans, and other automata. Philosophers who attempted to describe the human thought process as the mechanical manipulation of symbols planted the seeds for modern AI. But it wasn't until the 20th century that artificial intelligence as a field of study began to take shape (Coşkun & Gülleroğlu, 2021; Öztürk & Şahin, 2018; Pirim, 2006; Kaul, Enslin & Gross, 2020; Muthukrishnan, Maleki, Ovens, Reinhold, Forghani & Forghani, 2020).

Before 1980, the first digital computers were developed (Russell & Norvig, 2010). Some computer scientists have endeavored to make machines as sentient as humans since the dawn of computer science (Benko & Lányi, 2009). The capabilities of artificial intelligence have advanced significantly over the past seventy years. The language and image recognition capabilities of artificial intelligence systems have advanced rapidly (Oztürk & ahin, 2018; Russell & Norvig, 2010; Benko & Lányi, 2009).

Theseus is the first system mentioned in the timeline. It was created by Claude Shannon in 1950 and was a remote-controlled rodent capable of navigating and remembering a maze. In the 1960s, MIT researchers created ELIZA, a program that could simulate human speech through pattern matching and substitution. Terry Winograd developed SHRDLU in 1974, a natural language processing program that can comprehend and respond appropriately to uncomplicated English sentences (Nihalani, Narang, Chandrasekaran, Eshraghi, and Bhatia, 2010; Zylberberg, 1989; Hof, 2021; Weizenbaum, 1966).

Terry Sejnowski and Charles Rosenberg introduced neural networks designed to replicate the structure and function of the human brain in 1985. Deep Blue of IBM defeated world chess champion Garry Kasparov in a six-game match in 1997. In 2011, IBM's Watson defeated two former Jeopardy! champions. AI is currently employed in a variety of applications, including speech recognition, image recognition, natural language processing, robotics, and autonomous vehicles. The future of artificial intelligence is promising, as numerous new applications are created daily. However, there are also concerns regarding the impact of AI on society, including job losses and privacy issues (Sejnowski & Rosenberg, 1985; Campbell, Hoane Jr. & Hsu, 2002; Lohr & Markoff, 2011; Dogan, Goru Dogan & Bozkurt, 2023; Ghaffari Laleh, Ligero, Perez-Lopez ve Kather, 2023; Cheng, Li, He, Guo, Lu, Gu ve Wu, 2023; Salvagno, Taccone ve Gerli, 2023).

There are also a variety of applications for artificial intelligence in the culinary and beverage industry. Some food manufacturers create software supported by artificial intelligence to make complex operations more understandable or to reduce the duration of excessively complex operations (Pallathadka, Ramirez-Asis, Loli-Poma, Kaliyaperumal, Ventayen & Naved, 2023; Kutyauripo, Rushambwa & Chiwazi, 2023; Nunes, Ribeiro, de Carvalho, Ferreira, de Oliveira & Pinheiro, 2023). There is documentation that simulates and predicts maintenance characteristics with the aid of AI, as well as storage recommendations for the automatic creation of aggregated models. In lieu of costly procedures, artificial intelligence-assisted simulators can be used, according to research. It's anticipated that artificial intelligence can now be supported by regulations such as culinary product composition, cooking time, and preparation costs. Current applications of artificial intelligence include language recognition, image recognition, natural language processing, robotics, and self-driving vehicles. The impact of artificial intelligence on society is a source of concern, despite the bright outlook for its future. With the proliferation of artificial intelligence, it is crucial to consider issues like employment losses and privacy concerns.

Additionally, there are numerous applications of artificial intelligence in the food and beverage industries. Food manufacturers are developing software supported by artificial intelligence to simplify or expedite complex processes. In addition, it is anticipated that artificial intelligence will play a facilitating role in regulations such as the composition of culinary products, cooking time, and preparation costs.

Artificial intelligence has demonstrated significant historical growth and continues to be utilized in numerous fields today. This technology may become more prevalent in the future and contribute to more efficient and effective operations across a variety of industries. However, it is also essential to develop ethical and legal frameworks in order to comprehend and manage the societal effects of AI.

AI AND ETHIC

Artificial Intelligence (AI) is a rapidly expanding field that has the potential to transform many facets of our existence. However, as AI becomes more advanced, ethical implications are becoming a concern.

Bias is one of the primary ethical concerns surrounding AI. AI algorithms are only as effective as the data they are trained on, so if the data is biased, so will the algorithm. This can result in unjust or discriminatory outcomes, such as less accurate facial recognition systems for people with darker skin tones (Thorat, Nayak, and Dandale, 2010; Gordon and Nyholm, 2021). To address this issue, it is crucial to ensure that AI algorithms are trained on diverse, bias-free datasets. Privacy is another ethical concern surrounding AI. As AI becomes more sophisticated, it will be able to collect and analyze enormous quantities of personal data. This information can be exploited for malicious purposes, such as identity theft or targeted advertising, by creating detailed profiles of individuals (Khan et al., 2022). To address this problem, AI systems must be designed with privacy in mind, and individuals must have control over their personal information.

The third ethical concern encircling AI is transparency. Many AI algorithms are "black boxes" because their decision-making processes are difficult or impossible to comprehend. This lack of transparency can make it difficult to hold artificial intelligence systems accountable for their decisions (Gordon & Nyholm, 2021). To address this issue, it is necessary to develop AI algorithms that are transparent and explicable. Concern exists regarding the impact of AI on employment. As AI becomes more advanced, there is a risk of human laborers being replaced in numerous industries (Frey & Osborne, 2017). To address this issue, it is crucial to ensure that workers are trained in the skills required to collaborate with AI systems and that policies exist to assist workers who may be displaced due to automation.

AI has the potential to revolutionize numerous aspects of human life. However, there are numerous ethical concerns regarding the development and use of this technology. It is essential to address bias, privacy, transparency, and employment in order to assure the development and use of artificial intelligence in an ethical manner.

USING ARTIFICIAL INTELLIGENCE IN THE KITCHEN: RECIPE DEVELOPMENT WITH ARTIFICIAL INTELLIGENCE

Today, technological advancements and advances in artificial intelligence (AI) have enabled remarkable innovations in the kitchen industry. Among the most intriguing applications in this discipline is the creation of recipes using artificial intelligence. Thanks to big data analysis and machine learning techniques, AI offers a novel approach that facilitates the creation of complex recipes. The contribution of artificial intelligence to the recipe development process is founded on the multidimensional analysis of data. AI suggests new recipes by accessing a large database of culinary recipes and analyzing variables such as ingredient combinations, cooking periods, temperature settings, and even flavor profiles. These suggestions encourage culinary innovation based on traditional recipes and allow chefs to serve distinct dishes. This technology also provides a significant advantage in its capacity to provide personalized recipes by analyzing the preferences and routines of its users. Using the user's health preferences, food intolerances, and dietary requirements, AI algorithms can provide users with individualized and personalized recipes. Nevertheless, the process of developing formulas with artificial intelligence is not without its challenges. The success rate of AI algorithms is significantly influenced by data quality. The lack of diversity and precision in data sets may hinder the applicability of proposed formulas in the real world. In addition, it is essential to showcase the creativity and individuality of chefs in original recipes while harmonizing them with those created by artificial intelligence. However, research indicates that AI-based recipe development yields positive results. IBM's Chef Watson initiative (IBM, 2021) demonstrates that AI can play a significant role in the creation of complex and original recipes.

Numerous global organizations compile high-quality nutrition databases for their respective regions. For instance, the USDA has a large archive of chronologically organized national dietary recommendations, allowing researchers to analyze the evolution of dietary recommendations over time. FAO, on the other hand, organizes its data according to global food systems, with a strong mission to combat malnutrition and famine and to integrate global UN programs. Currently, AI tools are utilized to connect and reconcile databases, replace data gaps, and develop standardized ontological frameworks (Eftimov et al., 2016; Eftimov et al., 2017; Ispirova et al., 2017; Dooley et al., 2018; Ispirova et al., 2019; Ispirova et al., 2019; Popovski et al., 2019; van Erp, 2021).

Figure 1. The image created by artificial intelligence, which is asked about what the food will be like in the future
Source: AIArtistic (2023)

Various disciplines have utilized artificial intelligence (AI) to increase efficiency and precision. In recent years, AI has been utilized in the creation of new and innovative recipes. According to a study by Van Erp et al. (2021), artificial intelligence can be used to develop healthier and more sustainable recipes than traditional ones. Additionally, AI can be used to create recipes based on an individual's preferences. Using machine learning algorithms is one application of artificial intelligence in the formulation of recipes. Algorithms capable of machine learning can analyze vast quantities of data and identify patterns that humans cannot see. This enables artificial intelligence to generate innovative and original new recipes. For instance, he developed an AI system that can generate new recipes by analyzing existing recipes and recognizing patterns in ingredients and culinary techniques (van Erp et al., 2021).

Figure 2. The image created by artificial intelligence, which was asked about the future neurogastronomy
Source: AIArtistic (2023)

Natural language processing (NLP) is another application of artificial intelligence in recipe creation. NLP enables computers to comprehend and generate text based on human language comprehension. This can be used to make recipe descriptions more engaging and simpler to comprehend. For instance, he devised a natural language processing system that can generate recipe descriptions based on user input. Additionally, AI can be used to create recipes based on an individual's preferences. For instance, it has developed an artificial intelligence system capable of generating customized meal plans based on the user's dietary preferences and restrictions. The system employs algorithms for machine learning to analyze the user's food preferences and prescribe nutritious meals (Dempsey, 2022, aiplusinfo, 2023).

In fact, artificial intelligence is transforming the domestic environment at a rapid rate. From advanced cooking devices to software that generates recipes, AI offers chefs new opportunities to produce delectable meals. Google Cloud created two entirely new culinary recipes using machine learning: a cake (cookie-cake mix) and a breakie (bread-cookie mix). They began by gathering numerous cookie, cake, and bread recipes. They then converted all their ingredients to ounces and added a few fundamental ingredients (yeast, flour, sugar, eggs, butter, etc.). According to Paul Hollywood, delicacies such as banana, zucchini, and pumpkin bread are more cake than bread. Consequently, they rearranged the ingredients.

They then used a Google Cloud utility called AutoML Tables to construct a machine learning model that analyzes the ingredient quantities in a recipe and predicts whether it is a cookie, cake, or bread recipe. In addition to accurately labeling breads, cookies, and pastries, the model was also able to identify recipes it deemed "hybrids" - for example, recipes containing 50% cake and 50% bread, or 50% cake and 50% cookies. Two such combinations were referred to as "breakie" (a bread-cookie; "brookie" is already in use) and "cakie" (a cake-cookie) (Robinson & Markowitz, 2021).

EXAMPLES OF RECIPES PREPARED USING ARTIFICIAL INTELLIGENCE

In recent years, artificial intelligence-prepared recipes have become an increasingly popular topic. AI-powered culinary appliances can expedite and optimize the food preparation process. Moreover, artificial intelligence can be utilized to enhance the flavor and quality of cuisine. For instance, an AI-based recipe recommendation system could provide chefs with novel and intriguing concoctions. Additionally, artificial intelligence can be used to monitor the inventory of kitchen ingredients. Thus, caterers are able to better manage ingredients and reduce food waste.1 McDonald's Turkey presented consumers with unique beverage concoctions created with artificial intelligence. McDonald's Turkey is offering My Dream Recipes beverages prepared with artificial intelligence at select McCafé locations for a limited time (webrazzi.com, 2023).

Within the scope of this section, the following question was asked to ChatGPT and Bing Chat, artificial intelligence tools developed by OpenAI: What are the most used products in Turkish cuisine?

ChatGPT: The cuisine of Turkey is rich and varied, with influences from Central Asia, the Middle East, and the Mediterranean. Numerous ingredients are commonly used to produce tasty and appetizing dishes. Some of the most common ingredients in Turkish cuisine include:

- Olive oil is the fundamental constituent in Turkish cuisine, used in sauces, dressings, and vegetable sautés.
- Tomatoes: Fresh tomatoes and tomato paste are fundamental ingredients in numerous Turkish dishes, including casseroles, stews, and salads.
- Eggplant: Eggplant is prevalent in popular dishes such as Imam Bayildi (filled eggplant) and Eggplant Moussaka.
- Yogurt is a ubiquitous ingredient used in sauces, dressings, and beverages such as ayran in Turkish cuisine.
- Bulgur is a variety of bulgur utilized frequently in dishes such as pilaf and salad.
- Roasted red bell peppers are frequently used as a base for appetizers and sauces.
- Lamb and Veal: Among the most popular meats in Turkish cuisine, lamb and veal are utilized in kebabs, casseroles, and grills.
- Salads and meat dishes are flavored with pomegranate seeds or jaggery to impart a sour and sweet taste.
- Fresh herbs such as mint and parsley are frequently used to garnish and flavor a variety of dishes.
- Garlic and Onion: Garlic and onion provide profundity of flavor to sauces, soups, and stews and are the foundation of many Turkish dishes.

- Cumin and Red Pepper Flakes: These spices are frequently used to add flavor and warmth to food in the kitchen.
- Popular delicacies such as Baklava and Borek are prepared with dough.
- Chickpeas and lentils are typically utilized in soups, stews, and salads.

Figure 3. Image created by Picsart, which was asked if you could make a new Turkish recipe with the 5 most used products in Turkish cuisine

Bing Chat: Turkish cuisine is a culinary culture full of flavor. Due to Turkey's position between Eastern and Western cultures, Istanbul is a place where civilizations meet culturally, linguistically, and gastronomically. Delicious products from Western cuisine, Mediterranean cuisine, and Eastern cuisine have contributed to Turkish cuisine. Here are some of the most commonly found ingredients in Turkish food:

- Cinnamon: Cinnamon is mostly used in Turkish cuisine.
- Cream: Although the cream is an English invention, it has found a place on Turkish breakfast tables.
- Feta Cheese: Feta cheese is served at every meal and accompanies almost every meal.
- Tomato: Turks use tomatoes in soups, salads, and delicious sauces.
- Pistachios: Pistachios may be the most Turkish of all the ingredients on this list. Pistachios are native to the Middle East, and archaeological evidence indicates that they were eaten in Turkey around 7000 BC.
- Onions: Onion is simply the main ingredient of every Turkish flavour.
- Tomato Paste: Tomato paste is one and only item of Turkish kitchens.
- White Flour: You can literally use a white floor with every Turkish flavour.
- Rice, Red Lentils, Olive Oil, Eggplant (Aubergine).

So, can you create a new recipe with the 5 most used products?

ChatGPT: Certainly! Let's prepare a basic and delicious Turkish dish with the five most common ingredients in Turkish cooking: olive oil, tomato, eggplant, yogurt, and bulgur.

Stuffed Eggplant in the Turkish Style with Tomato-Yogurt Sauce

Contents:

- 2 enormous eggplants
- 1 cup of buckwheat
- 2 cups warm water
- 3.0 ounces of olive oil
- 1 finely minced onion
- 2 minced cloves of garlic
- Two juicy tomatoes, diced
- 1 pint of basic yogurt
- Fresh parsley, chopped (as a garnish)
- Salt and pepper to flavor

Instructions

Preparation of Aubergines

Preheat the oven to 200 degrees Celsius (390 degrees Fahrenheit).
The eggplants are halved longitudinally, and the insides are drawn diagonally.
Salt the eggplant halves and allow them to sit for 15 minutes to absorb excess moisture.
Rinse and dry the eggplants with a paper towel.

Assembled Bulgur

In a basin, combine bulgur with hot water, cover, and let stand for 15 minutes, or until the water is absorbed and the bulgur has expanded.

In a pan, heat 1 tablespoon of olive oil over medium heat. Sauté the garlic and diced onion until translucent.

Combine the cooked bulgur with the onions and garlic in a pan. Add pepper and salt to suit.

Stuffed Eggplant With Rice

A grill pan or skillet should be heated over medium-high heat. Coat half of the eggplants with olive oil and position them cut-side down in the pan. Griddle for approximately 5 minutes, or until grill marks appear.

Flip the eggplants over, fill each half with the prepared bulgur, and lightly compress.

Tomato Sauce With Yogurt

In a separate basin, thoroughly combine diced tomatoes with plain yogurt by stirring. The salt should be added to flavor.

Cooking

Place the eggplant halves loaded with cheese on a baking sheet. Drizzle olive oil over the eggplants that have been filled.

Bake the eggplants in a preheated oven for 25 to 30 minutes, or until tender and faintly browned.

Figure 4. Visual created by artificial intelligence with service description in Turkish cuisine
Source: PicsArt (2023)

Service

After the filled aubergines have finished cooking, remove them from the oven.

The packed aubergines should be served with a tomato-yogurt sauce.

Use minced fresh parsley as a garnish for a burst of color and freshness.

This Stuffed Eggplant in Turkish Style with Tomato-Yogurt Sauce is a delectable and healthy combination of Mediterranean flavors. Enjoy the tangy tomato-yogurt sauce combined with tender and meaty eggplant stuffed with delectable bulgur. This dish encapsulates the essence of Turkish cuisine in a comforting and satisfying manner.

It can be said that recipe development with artificial intelligence can be used more in the future. More can be said about the future of recipe development, given that the above implementation with ChatGPT is quite logical, both in terms of ingredients and procedure.

Figure 5. Another image created by Picsart, which was asked if you could make a new Turkish recipe with the 5 most used products in Turkish cuisine

Bing Chat: Sure, I can create a recipe for you. Here is a recipe for Pistachio and Feta Cheese Salad:

Ingredients:

- 1/2 cup of pistachios
- 1/2 cup of feta cheese
- 1/2 cup of tomatoes
- 1/2 cup of onions
- 1/4 cup of olive oil

Directions

Preheat the oven to 350°F.
Spread the pistachios on a baking sheet and roast them in the oven for about 10 minutes, or until they are fragrant and lightly browned.
In a large bowl, combine the pistachios, feta cheese, tomatoes, onions, and olive oil.
Toss everything together until it is well combined.
Serve immediately and enjoy!

CONCLUSION

As can be seen, artificial intelligence has turned the development of technology in a different direction until today. Because, thanks to artificial intelligence, we are no longer just talking about a robotic

Figure 6. Turkish cuisine image according to artificial intelligence
Source: AIArtistic (2023)

technology that takes the responsibility of people for physical work. At the same time, we see that the cognitive responsibilities that people need to direct by thinking are also undertaken by artificial intelligence. This development, which was adopted in technology-intensive sectors, has now begun to show itself in more social areas. At this point, it is possible to say that the belief that people will make their lives easier is effective in their acceptance of this technology.

Studies on food, one of the most intimate areas of life, also predict the superiority of artificial intelligence. For a long time, artificial intelligence programs that we upload the materials to be used give us the recipes that can be made using them. Beyond that, it is possible to get the characteristics of ethnic and regional cuisines, the most used materials, and suggestions for their use.

Chat Gpt and Bing Chat AI applications were used in this study. Two questions were asked to both programs. Although the programs mostly approached the basic ingredients in Turkish cuisine through common elements, both programs highlighted different materials. Which program is more successful when referring to the materials used in Turkish cuisine is a matter that needs to be discussed over the literature. From another point of view, these findings can be evaluated in the context of the richness of Turkish cuisine and the diversity of the ingredients.

On the other hand, when they are asked to prepare a recipe using these ingredients, ChatGpt explains in detail by considering the stages such as preparation, cooking and serving separately; It is seen that Bing Chat combines all processes in a few sentences more simply. As a result, although they did not reach an academic level of knowledge, it is clear that the relevant AI programs have knowledge and transfer capabilities that can be applied to get basic information on a subject.

The following suggestions can be made to chefs regarding the use of artificial intelligence in kitchens:

- The application of artificial intelligence (AI) in the kitchen has the potential to revolutionize the work of caterers. Here are some recommendations for caterers who wish to use AI in their kitchens:
- AI can be overwhelming; therefore, it is essential to start small and focus on one area. For instance, you can begin by employing AI to assist with menu planning and ingredient selection.
- Choose the appropriate tools: There are a variety of artificial intelligence (AI) tools available for chefs, so it is crucial to select the ones that best suit your requirements. Look for instruments that are simple to use and compatible with your current systems.
- AI is only as good as the data it is trained on, so it is essential to train your employees on how to use AI tools effectively. This will aid in ensuring that the data collected is accurate and beneficial.
- Utilize AI to foster creativity: AI can be used to create new recipes or propose flavor combinations you may not have previously considered. This can help you create distinctive dishes that separate out from the crowd.
- Accept automation: Artificial intelligence can be used to automate a variety of kitchen duties, such as inventory management and recipe scaling. This can free up time for you to focus on more essential tasks, such as menu planning and customer service.
- Explore the Potential of Artificial Intelligence Chefs must investigate this technology in order to comprehend how AI can be applied in the kitchen. AI can aid chefs in multiple areas, including recipe formulation, ingredient selection, and inventory management. Understanding this potential can improve the efficiency and originality of a chef's work.
- Chefs should experiment with AI-based culinary tools and learn how they function and how to use them. By automating processes such as recipe suggestions, cooking times, and temperature control, these tools can save chefs time and enhance workflow in the kitchen.
- Encourage Creativity: AI should serve as a tool to foster chefs' creativity, not to replace them. Using artificial intelligence in their creative processes, cooks can discover new flavors and recipes. Additionally, they should bear in mind that AI can analyze data to help them comprehend customer preferences and provide personalized experiences.
- Consider Security and Privacy: Artificial intelligence systems used in kitchens may contain customer and company data. Chefs must therefore place a premium on security and seclusion. Protecting data and preventing unauthorized access are essential for the successful application of artificial intelligence.
- It is essential that chefs receive training and develop expertise in order to utilize AI technology effectively. Training can help chefs comprehend the potential of this technology and successfully implement it in their kitchen operations.

When conducting research on the application of artificial intelligence in kitchens, several considerations must be taken into account. Understanding the current status of technology in this field is essential. Reviewing the existing literature on the topic and identifying areas where additional research is required is the responsibility of researchers. The ethical implications of using AI in kitchens should be considered by researchers. For instance, they should consider privacy, security, and bias issues. Researchers should evaluate the practical implications of implementing AI in kitchens. For instance, they should consider how AI can be integrated into existing kitchen appliances to enhance cooking. Researchers should contemplate the potential societal effects of artificial intelligence. They must consider, for instance, how artificial intelligence can be used to reduce food waste and enhance food safety. There are numerous intriguing opportunities for researchers who wish to investigate the application of artificial intelligence in the kitchen. Researchers interested in the application of artificial intelligence in the kitchen should consider the following:

- Comprehend the Present State of AI in Kitchens: Prior to conducting research, it is necessary to have a thorough comprehension of the current applications of AI in the kitchen. Discover the current AI-powered culinary tools and technologies, as well as their potential impact on kitchen operations and customer experiences. This fundamental information will assist researchers in identifying research gaps and opportunities for further study.
- Define Precise Research Objectives: Define your research objectives and questions with precision. The field of artificial intelligence in kitchens is expansive; refining your focus enables you to conduct more in-depth and significant research. For instance, you can investigate the efficacy of AI-powered recipe creation, AI-powered content substitution, or the influence of AI on the efficiency of kitchen workflow.
- Cooperation with Culinary Professionals: Collaboration with professional chefs and culinary specialists is necessary to comprehend the practical implications and difficulties of implementing AI in actual kitchen environments. Working closely with industry professionals will provide valuable insights into chefs' requirements and preferences, allowing for the creation of AI solutions that are aligned with real-world demands.
- Explore the user experience and adoption of AI technologies among kitchen staff and clients. Evaluate how AI is perceived and adopted in kitchen environments, along with any potential barriers or resistance. Understanding user attitudes and requirements will contribute to the development of more effective and user-friendly AI systems.
- Ethical Considerations: The use of artificial intelligence in kitchens raises ethical concerns regarding data privacy, transparency, and equity. To ensure that AI technologies are developed and utilized in an ethical manner, researchers must meticulously address these issues. Transparency in AI decision-making processes and protection of sensitive data are essential for enhancing confidence in AI systems.
- Real-World Testing and Evaluation: When feasible, conduct real-world testing and evaluation of AI applications in actual kitchen settings. This empirical approach will provide practical insights into the efficacy and challenges of AI integration, as well as improvement opportunities.

AI technology is rapidly evolving, and researchers must remain current on the most recent advancements in this discipline. This includes monitoring the development of AI algorithms, hardware, and applications, as well as keeping abreast of emergent trends in the kitchen industry. Publish and Disseminate Results: Once research is complete, publishing results in reputable academic journals and disseminating findings to industry stakeholders will contribute to the development and widespread adoption of AI in the kitchen. Knowledge sharing can inspire additional research and facilitate collaboration.

The examples discussed in this study also demonstrate this striking development and that the data obtained is on the way to perfection. The next step is to create ideas that the perfecting of robotic technologies and combining them with the development in artificial intelligence will create a new dimension in the nature-technology relationship.

REFERENCES

Aiplusinfo. (2023). *AI and Recipes*. https://www.aiplusinfo.com/blog/ai-and-recipes/

Benko, A., & Lányi, C. S. (2009). History of artificial intelligence. In *Encyclopedia of Information Science and Technology* (2nd ed., pp. 1759–1762). IGI Global. doi:10.4018/978-1-60566-026-4.ch276

Cheng, K., Li, Z., He, Y., Guo, Q., Lu, Y., Gu, S., & Wu, H. (2023). Potential use of artificial intelligence in infectious disease: Take ChatGPT as an example. *Annals of Biomedical Engineering*, *51*(6), 1–6. doi:10.100710439-023-03203-3 PMID:37074486

Coşkun, F., & Gülleroğlu, H. D. (2021). Yapay zekânın tarih içindeki gelişimi ve eğitimde kullanılması. *Ankara University Journal of Faculty of Educational Sciences*, *54*(3), 947–966.

Dempsey, M. (2022). *AI: Why chefs are turning to artificial intelligence*. BBC. https://www.bbc.com/news/business-59651077

Dogan, M. E., Goru Dogan, T., & Bozkurt, A. (2023). The use of artificial intelligence (AI) in online learning and distance education processes: A systematic review of empirical studies. *Applied Sciences (Basel, Switzerland)*, *13*(5), 3056. doi:10.3390/app13053056

Dooley, D. M., Griffiths, E. J., Gosal, G. S., Buttigieg, P. L., Hoehndorf, R., Lange, M. C., Schriml, L. M., Brinkman, F. S. L., & Hsiao, W. W. L. (2018). FoodOn: A harmonized food ontology to increase global food traceability, quality control and data integration. *NPJ Sci. Food*, *2*(1), 23. doi:10.103841538-018-0032-6 PMID:31304272

Eftimov, T., Korošec, P., & Koroušić Seljak, B. (2017). StandFood: Standardization of foods using a semi-automatic system for classifying and describing foods according to FoodEx2. *Nutrients*, *9*(6), 121–148. doi:10.3390/nu9060542 PMID:28587103

Eftimov, T., Koroušić Seljak, B., & Korošec, P. (2016). Grammar and dictionary based named-entity linking for knowledge extraction of evidence-based dietary recommendations. In *Proceedings of the 8th international Joint conference on knowledge discovery, knowledge engineering and knowledge management*. SCITEPRESS. 10.5220/0006032401500157

El-Allaly, E., Khoshgoftaar, T. M., & Wald, R. (2021). An Intelligent Recipe Generation Model Using Deep Learning. In *2021 IEEE International Conference on Systems, Man, and Cybernetics (SMC)* (pp. 458-463). IEEE.

Frey, C. B., & Osborne, M. A. (2017). The future of employment: How susceptible are jobs to computerisation? *Technological Forecasting and Social Change*, *114*, 254–280. doi:10.1016/j.techfore.2016.08.019

Ghaffari Laleh, N., Ligero, M., Perez-Lopez, R., & Kather, J. N. (2023). Facts and hopes on the use of artificial intelligence for predictive immunotherapy biomarkers in cancer. *Clinical Cancer Research, 29*(2), 316–323. doi:10.1158/1078-0432.CCR-22-0390 PMID:36083132

Gordon, J. S., & Nyholm, S. (2021). Ethics of artificial intelligence. Internet Encyclopedia of Philosophy.

Hof, B. (2021). The turtle and the mouse: How constructivist learning theory shaped artificial intelligence and educational technology in the 1960s. *History of Education, 50*(1), 93–111. doi:10.1080/0046760X.2020.1826053

Ispirova, G., Eftimov, T., Korošec, P., & Koroušić Seljak, B. (2019). MIGHT: Statistical methodology for missing-data imputation in food composition databases. *Applied Sciences (Basel, Switzerland), 9*(19), 4111. doi:10.3390/app9194111

Ispirova, G., Eftimov, T., Koroušić Seljak, B., & Korošec, P. (2017). Mapping food composition data from various data sources to a domain-specific ontology. In *Proceedings of the 9th international Joint conference on knowledge discovery, knowledge engineering and knowledge management*. SCITEPRESS. 10.5220/0006504302030210

Kaul, V., Enslin, S., & Gross, S. A. (2020). History of artificial intelligence in medicine. *Gastrointestinal Endoscopy, 92*(4), 807–812. doi:10.1016/j.gie.2020.06.040 PMID:32565184

Khan, A. A., Badshah, S., Liang, P., Waseem, M., Khan, B., Ahmad, A., & Akbar, M. A. (2022, June). Ethics of AI: A systematic literature review of principles and challenges. In *Proceedings of the 26th International Conference on Evaluation and Assessment in Software Engineering* (pp. 383-392). 10.1145/3530019.3531329

Kutyauripo, I., Rushambwa, M., & Chiwazi, L. (2023). Artificial intelligence applications in the agrifood sectors. *Journal of Agriculture and Food Research, 11*, 100502. doi:10.1016/j.jafr.2023.100502

Lohr, S., & Markoff, J. (2011). Computer Wins on 'Jeopardy!': Trivial, It's Not. *The New York Times*. https://www.nytimes.com/2011/02/17/science/17jeopardy-watson.html

Mengucci, C., Ferranti, P., Romano, A., Masi, P., Picone, G., & Capozzi, F. (2022). Food structure, function and artificial intelligence. *Trends in Food Science & Technology, 123*, 251–263. doi:10.1016/j.tifs.2022.03.015

Muthukrishnan, N., Maleki, F., Ovens, K., Reinhold, C., Forghani, B., & Forghani, R. (2020). Brief history of artificial intelligence. *Neuroimaging Clinics, 30*(4), 393–399. doi:10.1016/j.nic.2020.07.004 PMID:33038991

Nihalani, J. C., Narang, R. H., Chandrasekaran, G., Eshraghi, M. Z., & Bhatia, S. (2010). *Artificial Intelligence in Gaming*. Academic Press.

Nunes, C. A., Ribeiro, M. N., de Carvalho, T. C. L., Ferreira, D. D., de Oliveira, L. L., & Pinheiro, A. C. M. (2023). Artificial intelligence in sensory and consumer studies of food products. *Current Opinion in Food Science, 50*, 101002. doi:10.1016/j.cofs.2023.101002

Öztürk, K., & Şahin, M. E. (2018). Yapay sinir ağları ve yapay zekâ'ya genel bir bakış. *Takvim-i Vekayi, 6*(2), 25–36.

Pallathadka, H., Ramirez-Asis, E. H., Loli-Poma, T. P., Kaliyaperumal, K., Ventayen, R. J. M., & Naved, M. (2023). Applications of artificial intelligence in business management, e-commerce and finance. *Materials Today: Proceedings*, *80*, 2610–2613. doi:10.1016/j.matpr.2021.06.419

Pirim, A. G. H. (2006). Yapay zeka. *Yaşar Üniversitesi E-Dergisi*, *1*(1), 81–93.

Popovski, G., Seljak, B., & Eftimov, T. (2019). FoodOntoMap: linking food concepts across different food ontologies. In *Proceedings of the 11th international Joint conference on knowledge discovery, knowledge engineering and knowledge management*. SCITEPRESS. 10.5220/0008353201950202

Robinson, S., & Markowitz, D. (2021). *Just desserts: Baking with AI-made recipes*. https://blog.google/products/google-cloud/just-desserts-baking-ai-made-recipes

Russell, S., & Norvig, P. (2010). *Artificial Intelligence: A Modern Approach* (3rd ed.). Prentice Hall.

Salvagno, M., Taccone, F. S., & Gerli, A. G. (2023). Can artificial intelligence help for scientific writing? *Critical Care (London, England)*, *27*(1), 1–5. PMID:36597110

Sejnowski, T. J., & Rosenberg, C. R. (1985). NETtalk: A parallel network that learns to read aloud. *Artificial Intelligence*, *134*(1–2), 57–83. doi:10.1016/S0004-3702(01)00129-1

Thorat, S. B., Nayak, S. K., & Dandale, J. P. (2010). *Facial recognition technology: An analysis with scope in India*. arXiv preprint arXiv:1005.4263.

van Erp, M., Reynolds, C., Maynard, D., Starke, A., Ibáñez Martín, R., Andres, F., Leite, M. C. A., Alvarez de Toledo, D., Schmidt Rivera, X., Trattner, C., Brewer, S., Adriano Martins, C., Kluczkovski, A., Frankowska, A., Bridle, S., Levy, R. B., Rauber, F., Tereza da Silva, J., & Bosma, U. (2021). Using Natural Language Processing and Artificial Intelligence to Explore the Nutrition and Sustainability of Recipes and Food. *Front. Artif. Intell.*, *3*, 621577. doi:10.3389/frai.2020.621577 PMID:33733227

Watson, C. (2021). *IBM Chef Watson*. IBM Research. https://www.research.ibm.com/artificial-intelligence/project-chef-watson.shtml

Webrazzi. (2023). https://webrazzi.com/2023/06/23/yapay-zeka-tarifleriyle-hazirlanan-icecekler-mcdonalds-turkiyede/

Weizenbaum, J. (1966). ELIZA - A Computer Program For the Study of Natural Language Communication Between Man And Machine. *Communications of the ACM*, *9*(1), 36–45. doi:10.1145/365153.365168

Zylberberg, T. (1989). THESEUS initiative in management. *Industry and Higher Education*, *3*(4), 243–244. doi:10.1177/095042228900300420

Chapter 8
The Use of Artificial Intelligence in the Food Industry:
From Recipe Generation to Quality Control

Hari Kishan Kondaveeti
https://orcid.org/0000-0002-3379-720X
VIT-AP University, India

Chinna Gopi Simhadri
https://orcid.org/0000-0002-8311-7495
VIT-AP University, India

Ganti Lakshmi Yasaswini
VIT-AP University, India

G. Kusuma Shanthi
VIT-AP University, India

ABSTRACT

The food industry is one of the most rapidly evolving sectors, with new technological advancements constantly emerging. Artificial intelligence (AI) has emerged as a crucial component in this industry, revolutionizing various aspects of food production, from recipe generation to quality control. In recent years, AI has been integrated into the food industry in a big way, and its impact on the industry has been profound. An important benefit of AI in the food industry is its capacity to automate monotonous tasks, freeing up human workers for more complex responsibilities. AI also aids in recipe generation, enabling food companies to create innovative products that cater to consumer preferences. Moreover, AI plays a crucial role in maintaining quality control, guaranteeing that food products adhere to safety and quality requirements before they are delivered to consumers. To provide insights into the impact of AI in the food industry, this chapter will present case studies highlighting successful AI implementations.

DOI: 10.4018/978-1-6684-9094-5.ch008

INTRODUCTION TO AI IN FOOD INDUSTRY

The food industry is a rapidly evolving sector, constantly embracing new technological advancements (Hassoun et al.,2022). One such technology that has gained significant importance is AI, which plays a vital role in various aspects of food production, ranging from recipe creation to quality assurance. Over the past few years, AI has been extensively integrated into the food industry, resulting in a significant impact on its operations.

Due to the expected increase in global population, food demand is projected to rise significantly by 2050, with estimates ranging from 59 to 98% (Elferink et al.,2016). As a result, the use of AI has become increasingly prevalent in various aspects of the food industry which encompasses various areas such as supply chain, sorting food, optimizing production processes, improving quality, and maintaining industrial hygiene (Mavani et al.,2022).

AI brings several benefits to the food industry, including the automation of routine and repetitive tasks, which allows human workers to focus on more complex and critical tasks. With AI, food companies can also generate new and innovative recipes that appeal to consumers. Additionally, AI can help with quality control, making sure the safety and quality of food products before they reach consumers. By leveraging AI technologies, the food industry can enhance productivity, reduce waste, and improve the overall customer experience.

Potential Benefits of Using AI in the Food Industry

AI is widely employed in the food industry for multiple purposes such as detection, safety, prediction, quality analysis, control, and classification. The use of AI in the food industry has been steadily increasing for several years, driven by its potential to address a range of challenges and benefits. These include food sorting, parameter classification and prediction, quality control, and food safety. Various AI techniques have been employed in the industry, encompassing expert systems, fuzzy logic, Artificial Neural Networks (ANNs), Adaptive Neuro-Fuzzy Inference systems (ANFIS), Machine Learning (ML), etc. These methods have proven to be effective in improving food production processes, reducing waste, and enhancing the overall standard of the products (Strong, A. I., 2016).

AI technologies assist the food industry in effectively promoting its products by implementing global strategies and efficient planning based on current food trends. With the help of AI, food processing becomes more adaptable since machines can now differentiate between fundamental issues such as sorting different fruits, to more complicated problems like identifying the difference between low saturated fats and high unsaturated fats. By utilizing AI in food processing, food companies can optimize their production processes, reduce waste, and improve the quality of their products. Moreover, AI technologies can assist in predicting and meeting consumer demands, likewise introducing innovative products that satisfy the changing consumer preferences.

Challenges

Automation, Internet of Things (IoT), and AI integration in the food and processing sector is expected to bring about significant changes, but there are also drawbacks. Despite the incredible technological advantages for food processing facilities and industries, there are many difficulties that must be addressed in order for food manufacturing facilities to achieve growth and increased revenue. Effective implementation of AI and automation in industries and food processing units requires significant changes, and companies must carefully consider the type of AI model that is best suited for their units. It also requires a significant effort to fully comprehend and install these technologies (Mor et al., 2022).

Main obstacles of AI to the food sector:

- **Cost Management:** Organizations and food processing facilities face challenges while upgrading their technology systems due to potential significant expenses, including initial purchases and potential repair costs.
- **Technological Challenges:** Increased use of AI and IoT in the food sector brings about transformative changes, leading to higher profits. However, the integration of new technology can be complex, especially for small businesses still reliant on manual labor.
- **Expertise and Skills:** Successful implementation of AI systems requires skilled personnel who can effectively manage and update them. Training employees to maintain and utilize AI tools is crucial for sustained success.
- **Integration with Existing Systems:** Harmonizing AI with current quality assurance systems poses compatibility and data integration challenges. A seamless transition without disrupting ongoing operations is key.
- **User Acceptance and Trust:** Building confidence among industry professionals and consumers in AI-based quality assurance is vital for successful adoption, as some stakeholders may prefer traditional methods.

The following sections discuss about the Usage of AI in Recipe Generation, Quality Control, Applications of AI in the food industry and Impact of AI in the food industry.

RECIPE GENERATION

Recipe generation using AI involves using algorithms to create new recipes or modify existing ones. There are several ways to go about this, but one popular alternative is to employ machine learning algorithms (Marın et al., 2021) which examine massive datasets of recipes and ingredients to spot trends and come up with new combinations.

The general steps involved in Recipe generation using AI are:

- **Data collection:** Collect a large dataset of recipes and ingredients from various sources.
- **Preprocessing:** Clean and preprocess the data, removing any duplicates or irrelevant information.
- **Data analysis:** Use machine learning algorithms to analyze the data, identify patterns, and create models that can generate new recipes.
- **Recipe generation:** Use the trained model to generate new recipes, either from scratch or by modifying present recipes.
- **Evaluation:** Evaluate the generated recipes using metrics such as flavor, texture, and nutritional value.
- **Iteration:** Refine the model based on the evaluation results and repeat the process to generate better recipes.

There are also several platforms and tools available that can assist with recipe generation using deep learning to generate recipes from a list of ingredients, and also Chef Watson is an AI-powered cooking technology created by IBM (Chef Watson, IBM, 2016). It uses machine learning algorithms to analyze flavor compounds in various ingredients and suggest novel ingredient combinations, recipe ideas and can suggest recipe variations based on user preferences.

Deep learning models like the character-level Long Short-Term Memory (LSTM) used which is used in Ratatoullie (Goeal et al., 2022) and forage (Kiddon et al., 2016) word-level LSTM model, and transformer-based pre-trained GPT-2(Generative Pre-trained Transformer 2) model are used to generate new recipes from a given ingredient list (Team, K. (no date) Keras Documentation). The models were built using OpenAI, have 1.5 billion parameters, and were trained on a variety of themes from 8 million web pages (Batra et al., 2020). To create a corpus of recipes with titles, ingredients, and cooking directions, an AI training model uses the RecipeDB dataset. The goal is to produce unique and varied recipes using the input list of ingredients provided to the model.

Usage of AI in Recipe Generation

Producing cooking instructions that match a particular food image is the task of recipe generation. This requires creating text that appropriately describes the cooking procedure for the provided food using a language decoder to produce coherent and pertinent text. The decoder may better understand the relationships between various components and culinary stages with the aid of unsupervised recipe tree structures, leading to more accurate and pertinent instructions (Bien et al.,2020).

In order to integrate the knowledge of the tree structure into the training and inference phases, a unique framework called the Structure-aware Generation Network (SGN) is utilized, as illustrated in the accompanying image. A target structure inference module is added to the recipe creation process using SGN. The recipe tree structures are also created using Recurrent Neural Network (RNN) and food image data. The graph attention networks (GAT) are used to embed the trees based on the created trees in an effort to provide the model with additional direction when creating recipes. By using tree structure embeddings, it is possible to substantially boost generation performance while maintaining the same length of the generated recipes as the actual recipes.

Evaluation of the cross-modal food retrieval task using unsupervised learned recipe tree structures, where the tree representations are converted into recipe characteristics as illustrated in Figure 1 (Wang, H.,2022).

Benefits of Using AI in Recipe Generation

AI can quickly analyze vast amounts of data to generate new and unique recipes in a matter of seconds. It can tailor recipes to meet specific dietary needs, preferences, and restrictions of individual users, such as gluten-free, vegan, or low-carb diets (Salvador et al.,2017). It can generate new and innovative recipes by identifying and combining unusual or unexpected flavor combinations and ingredients that humans may not have considered before. It can calculate precise measurements and cooking times, reducing the chances of errors and ensuring consistent results every time. The generated recipes can be easily accessed and shared online, making it possible for people all over the world to discover and try new cuisines and cooking techniques.

The author (Salvador et al.,2019) explores the concept of generating recipes from food images using a technique called inverse cooking. The authors propose a deep learning approach to infer the ingredients

Figure 1. Image to recipe generation

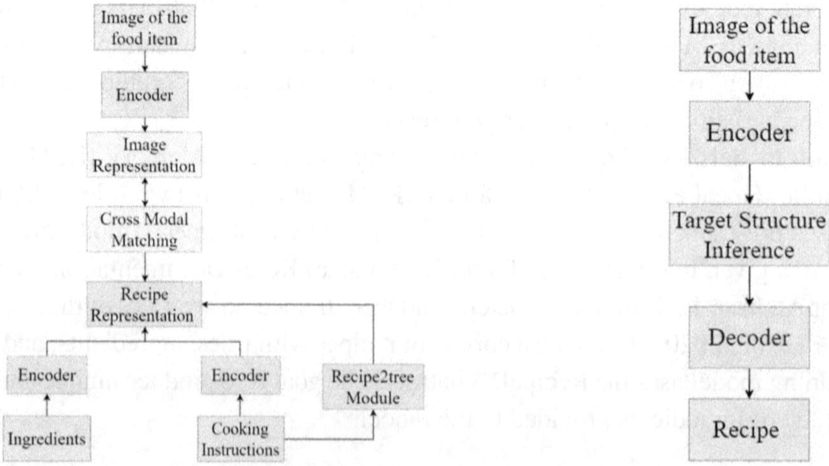

and cooking instructions of a dish by analyzing its visual features. The goal is to enable machines to understand and recreate recipes based on visual cues alone. The study introduces a large-scale dataset of food images and corresponding recipe information to train and evaluate the model. It discusses the methodology, experimental results, and potential applications of recipe generation from food images using AI techniques. This system is able to generate more interesting recipes than retrieval-based methods. It makes cooking accessible to a wider audience, including individuals who may not have prior cooking knowledge or experience. It allows users to explore new dishes and cooking ideas easily.

Challenges of Using AI in Recipe Generation

Lack of creativity is one of the biggest challenges in recipe generation. While AI can generate new and innovative recipes, there is a risk that the generated recipes may lack the creativity and flair that comes from human creativity. Quality of input data matters here as the accuracy and quality of input data can significantly affect the quality of the AI-generated recipes. If the data used to train the AI is incomplete or biased, the recipes generated may not be suitable for a broad audience (Goel et al., 2022).

Limited understanding of cultural and regional differences with AI may not have a full understanding of the cultural and regional differences in cuisines, which could result in recipes that are not culturally appropriate or do not align with the culinary traditions of a particular region. Difficulty in replicating complex techniques with AI may result in difficulty replicating complex cooking techniques that require a high level of skill and expertise, making it challenging to generate recipes that are truly authentic. Lack of sensory experience is also the challenge as AI does not have the ability to taste or smell, making it challenging to generate recipes that fully take into account the sensory experience of cooking and eating.

Another highly demanding aspect of recipe generation involves managing the intricate constraints among the input ingredients. Previous studies have attempted to simplify the problem by considering the inputs as independent entities and generating recipes that incorporate as much information as feasible. Consequently, there is a need for a systematic approach to navigate content selection while adhering to internal restrictions. To address this, a routing Routing-enforced Generative Model (RGM) can generate suitable recipes based on the provided ingredients and user preferences.(Yu et al.,2020).

QUALITY CONTROL

Quality control in the food industry refers to the process of ensuring that food products meet specified quality standards and regulatory requirements (Henson et al.,1999). It involves monitoring and evaluating various aspects of the production, processing, and distribution of food to maintain consistent quality and safety for consumers.

The goals of quality control in the food industry include:

- **Ensuring food safety:** This involves identifying and controlling potential hazards that can affect the safety of food, such as microbial contamination, chemical residues, and foreign objects.
- **Maintaining product quality:** Quality control measures are implemented to ensure that food products meet predetermined quality attributes, such as taste, texture, appearance, nutritional value, and shelf life.
- **Compliance with regulations:** Food manufacturers must adhere to local, national, and international regulations and standards to ensure that their products meet legal requirements and consumer expectations.

Quality control activities in the food industry include:

- **Raw material inspection:** Assessing the quality and safety of incoming ingredients and materials before they are used in food production.
- **Process monitoring:** Continuously monitoring and controlling critical points in the production process to ensure consistency and adherence to quality standards.
- **Testing and analysis:** Conducting laboratory tests and analysis to evaluate the composition, microbiological safety, and sensory characteristics of food products.
- **Packaging and labeling checks:** Verifying that the packaging and labeling of food products are accurate and compliant with regulatory requirements.
- **Traceability and recall systems:** Implementing systems to track and trace products throughout the supply chain and enabling efficient recalls if necessary.

AI has become an integral element of food quality assurance since it is efficient in every manner. AI is a human-machine partnership that has taken manufacturing and quality assurance to the next level. AI ensures precise, low-cost, sanitary, efficient, large-scale, high-quality production. Aspects of technical and economic significance for assessing food quality and safety improvements are also examined, as well as the effectiveness of various AI techniques for enhancing food quality, the best AI techniques for improving food quality at various stages, and the effect of farm scale on the costs associated with the implementation of AI techniques.

Some of the applications of AI in food security are food accessibility, availability, use, and sustainability. While AI has been slower to gain adoption in the food sector compared to other industries, it is being implemented worldwide across all four pillars of food security. As such, it is essential to examine the potential of AI and its current impact on the food system. The use of AI technology has a crucial role to play in the future of agriculture. (Sahni et al.,2021).

As quality traits are regarded more highly by enterprises, consumers, and governments, quality assurance is becoming more and more crucial. The rising value placed on food processing characteristics has prompted food corporations to consciously improve quality control and governments to tighten restrictions. Regulations are also being examined more closely on a national and international level. Governments are increasingly forced to utilize risk assessment and cost-benefit analysis to determine whether current or new food laws improve public welfare as demands for regulatory responsibility have risen.

Usage of AI in Quality Control

The highest priority in food production and processing is ensuring the safety and quality of food items. While standard guidelines are in place to achieve this goal, processed foods are still vulnerable to contamination, resulting in compromised product quality and foodborne illnesses. To address this issue, the food industry is switching from manual to automated processing using AI. However, this requires significant investments in technology, expertise, and resources.

To explore the potential of AI in food processing and presenting various technologies and methods for implementing automated technology-aided processing. A conceptual design for an automated processing line with many operating layers and procedures is elaborated, with the emphasis being on liquid goods like milk and beverages. The plan intends to improve liquid foods' microbiological safety and quality assurance (Kyaw, et al.,2022).

By implementing automated processing using AI, the food industry can ensure the safety and quality of food products, reducing the risk of foodborne illnesses and recalls. AI-powered systems can detect and prevent contamination, monitor processing conditions, and improve product quality in real-time. However, the implementation of automated processing requires significant investments and expertise, making it a challenging but necessary step towards safer and higher quality food production.

Historically, determining the quality of food has relied on a number of disruptive approaches that necessitate sample destruction and substantial postharvest losses. A particular emphasis is placed on the power of AI applications in assessing food quality to provide high and ideal yields. The best results for modeling and efficient real-time monitoring strategies were shown to be provided by ANN (Ali et al.,2021). Future applications of AI for quality inspection show promise and may enable quick and real-time examination of various foods and agricultural products.

By utilizing AI methods and Fractal Theory, Automated Quality Control in the Food Industry (AQCFI) seeks to reduce the number of microbiological and chemical procedures required to detect dangerous substances and harmful bacteria in food. The research suggests employing a technique for microbe identification based on the fractal dimension, which can speed up identification. This approach simulates the process of expert judgment to determine the production quality. A module for fractal identification and an intelligent module for running the expert simulation are both included in the computer program created to implement this strategy. Based on data regarding the identifications of microorganisms and chemicals, the latter module uses statistical calculations and heuristics to determine the level of production quality. The suggested method can significantly cut down on the time and work required to find poisonous chemicals and harmful microbes in food, which is essential for guaranteeing consumer food safety and quality (Castillo et al.,1970).

The authors (Sharma et al.,2021) provided a comprehensive review of the application of AI in food quality assurance. The review paper states that AI can enhance quality control measures by automating processes, detecting contaminants, monitoring production parameters, and predicting shelf life. It explores the potential of AI in different stages of the food supply chain, including raw material inspection, processing, and packaging. In this case the author explains the process of making ketchup using AI methods in quality control.

Picking Using Computer Vision

Initially the tomatoes undergo harvesting and washing processes. Once cleaned, they are transported via a conveyor belt where they undergo scanning. Computer vision technology is employed in this scanning process to conduct a quality check before proceeding to the next stage. Through computer vision, all defective tomatoes, including raw, rotten, or undersized ones, are identified and separated. The algorithm utilized in this process examines various criteria such as diameter, color, and discoloration caused by bacterial impact. If any of these criteria indicate a poor quality tomato, it is discarded. Previously, performing this task required a significant amount of manual labor and time, resulting in increased production costs. Similarly, computer vision is employed to ensure the quality of ketchup by scanning all the raw materials used in its production.

Processing Using Deep Learning

In the processing of tomato ketchup, several common image pre-processing techniques are employed, particularly for color processing of the final sauce. Image segmentation plays a significant role in image processing as it determines whether the analysis focuses on the desired sample. Additionally, regression prediction models utilizing both 2D and 3D image features have been utilized to estimate the mass of tomato fruits based on their volume. These AI methods establish a connection between the 2D size and mass of the tomato fruit, leveraging multiple techniques to derive the physical relationship between these characteristics. The models are trained and evaluated using the tomato dataset, utilizing the output from the geometry module (estimated dimensions) to estimate the final mass.

Packaging Using Computer Vision

Once the tomatoes are processed into ketchup, the next crucial step is packaging and palletizing. In the packaging industry, the application of computer vision is of utmost importance. To ensure the quality of the final product, it is necessary to verify that all bottles are properly filled and clean. This task is efficiently carried out using computer vision or machine vision technology. In this process, a camera, in conjunction with a sensor, captures digital images of the bottles. By analyzing these images, any bottles that are not properly filled can be identified and removed from the conveyor belt before the cap is applied. This ensures that only bottles meeting the required standards proceed in the packaging line.

Benefits of Using AI in Quality Control

The AI-powered technologies can identify flaws and anomalies in the food products in real-time, which can help prevent contaminated products from reaching consumers. This ensures that the final product is of high quality and safe for consumption, reducing the risk of foodborne illnesses and product recalls. AI can also help to optimize the production process, reducing waste and increasing efficiency. AI-powered systems can monitor various parameters such as temperature, humidity, and ingredient levels, adjusting them in real-time to ensure that the process is running at optimal levels.

Using AI in quality control of food production can lead to higher quality, safer products, increased efficiency, and cost savings, making it a valuable investment for the food industry. AI can help predict and prevent potential issues before they occur. By analyzing data from sensors and other sources, AI can generate models that can predict potential issues, allowing manufacturers to take corrective actions before any problems arise and using AI in quality control can help save time and resources.

Challenges of Using AI in Quality Control

There are several challenges associated with the implementation of AI in food quality assurance. They include:

- **Insufficient Data:** One of the main challenges is the lack of data or incomplete data sets, which can limit the effectiveness of AI algorithms. Another challenge is the need for high-quality data, as AI systems heavily rely on accurate data for effective decision-making. The implementation of AI systems in food production may require significant investment in terms of resources, expertise, and infrastructure.
- **Data quality and availability:** AI algorithms require large amounts of high-quality data to train and make accurate predictions. Obtaining relevant and representative data sets can be challenging, especially in the food industry where data collection and standardization may vary across different stages of the supply chain.
- **Regulatory compliance:** The food industry is subject to various regulations and standards to ensure food safety and quality. Implementing AI systems that comply with these regulations can be complex. Ensuring that AI solutions meet regulatory requirements and can provide transparent and auditable results is crucial.

APPLICATIONS OF AI IN THE FOOD INDUSTRY

AI has seen increasing application in the food industry over the years, driven by various factors like food sorting, parameter classification, quality control, prediction, and safety. Before the implementation of AI, extensive studies on food have been conducted to enhance public knowledge about food and improve outcomes related to food properties and production. Along with AI, Computer vision techniques (Kakani et al., 2020) and Machine learning algorithms are currently being utilized by numerous businesses and startups across various industries, including agriculture, food processing, and livestock management. These technologies are employed to enhance processes, optimize operations, and improve outcomes in these specific domains.

AI Based Methods for Processing of Food

The availability of food-related data has led researchers to study how AI might be used in the food industry. Machines could recognize food from images as early as 2015. By the year 2016, the AI developed by Massachusetts Institute of Technology (MIT) was able to predict the components and nutritional values of food that was presented to it (Kakani, et al., 2020). This technology was made accessible to the public as a smart phone application. These AI advancements are assisting the food industry in effectively marketing their products by employing food trend strategies and strategic planning methods. The application of advanced algorithms has enhanced the food processing industry by enabling the identification of low-saturated fats in contrast to high-unsaturated fats, thereby increasing the flexibility of food processing operations. Figure 2 illustrates the application of deep learning in categorizing food utilizing the LeNet framework (LeCun et at., 1998).

Figure 2. Food classification: Deep learning architecture

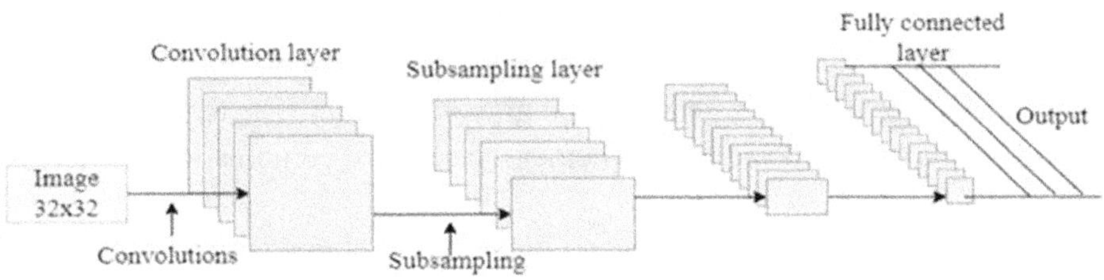

Sorting Techniques

The variable supply of feedstock is one of the biggest issues that food processing facilities are dealing with. Manual sorting of vegetables in these plants often leads to inefficiencies and increased costs. However, the implementation of automation in food sorting can be made possible by AI. Food processing companies can enhance the efficiency of food sorting processes by using sensors, cameras, lasers and machine learning algorithms (Nosratabadi, et al., 2020). For example, sensor-based optical sorting solutions powered by AI can eliminate time-consuming sorting processes for fresh produce, resulting in higher yields, improved quality, and reduced wastage (Agbai, 2020). AI plays a crucial role in calibrating machines to effectively handle different product sizes, thereby reducing waste and cutting down costs.

Management of Supply Chain

AI is extremely helpful in enhancing supply chain management in the food sector, contributing to safety of food, accurate forecasting, efficient procurement, and increased consumer confidence. Ensuring effective management and transparency of the supply chain is an important objective for food manufacturers. To address this, the food industry is leveraging AI technology to enhance supply chains in various ways. One important application is using AI for food safety monitoring and testing. This method ensures food safety by closely monitoring products at each level of the supply chain to make sure they meet industry and consumer standards. AI also enables more accurate forecasting, which is crucial for managing pricing and inventory (Jayashankar P, et al., 2020). By analyzing large volumes of data, AI systems can generate insights and predictions that assist in optimizing pricing strategies and maintaining appropriate inventory levels.

Image recognition technologies powered by AI play a significant role in the procurement of produce. By accurately analyzing images, AI systems can assist in identifying and selecting high-quality produce more efficiently, improving the overall procurement process. AI also makes it possible to track produce effectively from the farm to the customer. This end-to-end traceability builds consumer confidence as they can easily track and verify the journey of the food they consume, ensuring its authenticity and quality.

Consumer Needs

Food companies are increasingly utilizing AI to better understand and predict consumer flavor preferences. Food producers may predict and model the dietary needs of their target consumers by utilizing AI-based predictive analytics, enabling the creation of new food products that closely match consumer needs.

An example of AI implementation in the food business is the technology launched by The Kellogg Company in 2017 (Kuo TC, et al.,2018). Customers can choose from 50 ingredients with the help of this AI-enabled technology to produce a personalized granola product. The technology suggests which ingredients or components to add. It evaluates whether the selected combination is likely to result in a good taste experience.

The advantages of AI in this industry go beyond individual customization. The data gathered from flavor combinations, actual consumer selections, and repeated orders create a valuable feedback loop. This data helps refine and improve flavor preferences, providing useful insights to the parent company when making decisions about introducing new products within their larger brands.

By leveraging AI technology, food manufacturers can enhance their understanding of consumer tastes and preferences, enabling them to develop innovative and personalized food products. This not only improves the overall consumer experience but also provides valuable market insights for the company's future product development strategies.

Innovating New Products

AI technology and machine learning that make use of predictive algorithms, plays a crucial role in modeling consumer flavor preferences and predicting their response to new tastes. This enables companies to gain insights into consumer preferences and develop products that align with the tastes of their target audience. By segmenting the data into demographic groups, manufacturers can better understand the preferences of different consumer segments and tailor their product development accordingly, increasing the likelihood of success in the market.

One of the examples of AI application in the beverage industry is Coca-Cola's implementation of self-serve soft drink fountains at numerous sites (Babu Loganathan, G.,2021). Through the addition of various flavors to their base beverages, these machines enable users to customize their own drinks, resulting in hundreds of possible combinations. AI is being used to analyze the enormous quantity of data produced by these self-service machines, which include thousands of freestyle drink fountains that dispense hundreds of different drinks every day.

Coca-Cola's AI analysis of consumer preferences led to the creation of a new product called CHERRY SPRITE. The AI algorithms identified that customers were frequently adding cherry flavor to Sprite, indicating a significant demand for a cherry-flavored variant. As a result, CHERRY SPRITE was introduced as a standalone product, leveraging the insights gained from AI analysis of consumer preference data.

Companies like Coca-Cola can identify emerging trends, anticipate consumer demands, and develop new products that are more likely to succeed in the market with the help of AI. This demonstrates how

AI-driven analysis of consumer preferences can inform product development strategies and contribute to the creation of successful and tailored food and beverage offerings.

Self-Ordering AI Based Kiosks

Self-ordering AI-based kiosks, such as the ones used by McDonald's, have the potential to significantly improve the experience for customers by reducing waiting times and removing the need to wait in the queue to pay. These machines allow customers to place their orders directly, making the process more efficient and convenient. With the integration of card readers, customers can make payments directly at the kiosk without requiring human assistance. This streamlined payment process not only saves time but also provides a seamless and hassle-free transaction experience for customers.

By leveraging AI technology, self-ordering kiosks can accurately capture and process customer orders, ensuring accuracy and reducing the chances of errors or miscommunication that can occur during traditional order taking. Customers have the flexibility to customize their orders and explore the menu at their own pace, enhancing their overall satisfaction. Overall, self-ordering AI-based kiosks in fast-food establishments like McDonald's exemplify how technology can improve customer experiences by optimizing order placement and payment processes, resulting in reduced waiting times and greater convenience for customers (Writer, S.,2022).

Food Delivery

AI-based food delivery mechanisms have gained significant traction among various companies, revolutionizing the entire process from ordering to delivery. Uber Eats, Grubhub, and DoorDash are some of the successful examples of successful meal delivery services. In the food industry robots are also being utilized not only in food processing and manufacturing but also in the delivery of food directly to customers' doorsteps. The use of AI techniques has enabled these developments in the field. Several major companies are actively operating in the AI in Food & Beverages Market. These companies leverage AI technologies to enhance various aspects of the food and beverages industry, such as sorting solutions, automation, and intelligent systems (Pandey, S. 2021).

Some of these prominent players include:

- TOMRA Sorting Solutions AS
- Rockwell Automation Inc.
- ABB Ltd
- Honeywell International Inc.
- Key Technology Inc.

IMPACT OF AI ON THE FOOD INDUSTRY

AI-driven methodologies and equipment have revolutionized various aspects of processing, and production techniques. In addition to simply showcasing food images, computers now possess the ability to recognize and offer calorie information for different food items.

The fields of image processing and computer vision have undergone a significant revolution due to deep learning and machine learning, both of which are AI techniques. Previously, image processing enabled computers to understand the contents of images for decision-making purposes. However, the introduction of deep learning and machine learning has expanded the capabilities of image processing to include tasks such as detection, object tracking, and analysis of various data types, including photos, videos, and linguistic sequences.

Possible sectors in the food industry where AI has made significant process improvements.

Food Safety

For the whole supply chain, stricter hygiene standards have been set by the Food Safety Modernization Act (FSMA), particularly for non-refrigerated food products such as cereals and spices, which are more susceptible to contamination (Dolci, R., 2017). In the past, these products were relatively safe from contamination, but the situation has changed. AI-based systems can help address these challenges by eliminating the risk of human-transmitted illnesses and offering simpler maintenance. According to a report by Technavio, to fulfill government demands, the usage of robots in the food processing industry is anticipated to rise by 30%. There are also promising new inventions in food safety that use AI, aiming to reduce the occurrence of foodborne diseases.

Supply Chain

Supply chain management plays a crucial role in food companies, driven by the increasing demand for transparency. Integration of AI in the supply chain includes food safety tracking and analysis throughout the supply chain to ensure compliance with the industry and consumers' requirements (Kumar, I.,2021) Precise forecasting is necessary to manage pricing and supplies effectively. Effective food production, food maintenance energy management, and food retail packaging labeling control are examples of recent advancements in the subject. Additionally, AI facilitates the efficient monitoring of products from producers to consumers, which in turn increases consumer confidence.

AI technology plays a crucial role in optimizing supply chain operations, minimizing delays, and maximizing profit margins. It enables accurate forecasting for better pricing and inventory management. AI is also instrumental in addressing transparency issues by effectively tracking products from the farm to consumers. To reduce waste and enhance inventory management, suppliers and retailers are increasingly relying on technology. The increased visibility and management in the food supply chain is able to benefit from AI, leading to waste reduction and improved product freshness.

Food Waste Management

Food waste is a problem that has to be addressed, according to the US Department of Agriculture, which calculates that it accounts for between 30 and 40% of the world's food supply. This equates to almost 133 billion and $161 billion pounds worth of food in 2010, causing significant societal impacts (*Food Waste Faqs USDA*).

AI has the potential to address these food waste challenges and significantly reduce waste by 2030. By implementing regenerative leisure farming practices and leveraging AI technologies, substantial progress can be made in reducing food waste.

To achieve this, smarter farming methods can replace traditional farming approaches. Sensors are deployed to collect data, which is then processed using machine learning algorithms to make faster and more accurate decisions. A few approaches to reduce food waste with AI include assessing fruit maturity, identifying beneficial microorganisms for fruit and vegetable growth without artificial fertilizers, eliminating unnecessary ground inspections to save costs, utilizing computer vision technology for farm-based management of food in supply chain, and implementing AI-based food tracking technologies to sell food before it becomes waste. These solutions enable greater connectivity between farmers and consumers, reducing food waste.

CONSIDERATIONS FOR THE FUTURE DEVELOPMENT AND INTEGRATION OF AI IN THE FOOD INDUSTRY

Experts and scientists anticipate that AI will play a pivotal role in fueling various organizations as a result of recent advancements in the field. By 2020, the exponential increase of digital data is projected to reach 44 trillion gigabytes per year. The Fourth Industrial Revolution (4.0 IR) is characterized by innovative approaches that leverage vast amounts of data and trainable AI technologies to address existing challenges in diverse fields (Lezoche, M.,2020).

The concept of industrial revolutions, denoted by the terms IR, has been used to describe significant shifts in human history driven by technological advancements. The first industrial revolution (1.0 IR) emerged with the introduction of industrial machinery and equipment. The development of radio, electricity, and airplanes contributed to the start of the second industrial revolution (2.0 IR) (Nicholas, 2019). The development of electronics and the internet sparked the third industrial revolution (3.0 IR), which brought about globalization.

Now, with recent developments in data analytics, computer vision, and AI, we are on the brink of witnessing the 4.0 IR revolutionize every aspect of production and various industries. IR 4.0 involves the integration of intelligent systems and technologies into traditional industries. One of the key technologies in IR 4.0 is AI, which aims to create intelligent machines capable of performing tasks similar to humans.

The integration of Fourth Industrial Revolution (4IR) technologies, including AI and computer vision, is evident in different sectors such as automotive, medicine, smart robotics, vision-based virtual agents and AI systems. This widespread adoption has led to a significant increase in investments in AI-based services (Evans, G. W., 2017). Startups like Numenta (Home (2023) Numenta), Clarifai (Clarifai 2023) and others are using 4IR technologies, including computer vision and AI, in industries like healthcare, unmanned vehicles, cognition and core AI, and smart robotics. The investments in computer vision-based startups are depicted as 3.5 billion USD, while AI machine learning-based startups have received nearly double that amount, totaling 7 billion USD (Kakani et al,2020). Particularly, entrepreneurs investing approximately 2 billion USD in natural language processing, intelligent robots, and unmanned vehicles use AI trainable methodologies and exploit computer vision's sensory data for autonomous navigation.

Nations such as China and India are currently implementing government initiatives or programs aimed at enhancing food supply efficiency through the utilization of technologies like deep learning and information analysis. Notably, companies like Microsoft and Google are contributing their technologies to these nations, actively supporting the development of sustainable global economies. For instance, the Microsoft Cortana Smart Suite was developed in partnership with the ICRISAT organization of the Indian government in order to collect and analyze agricultural data using deep learning methods (Babu

Loganathan, G., 2021). In India, trial sites were set up in 13 districts to explore Labs specializing in soil analysis, smart systems for water management, techniques for Integrated Modular Optical Design (IMOD), promoting farmers through public-private partnerships and state investments. These initiatives showcase the power of technology and collaboration in addressing food production challenges in developing nations.

CONCLUSION

This chapter presents various facts about the integration of AI technologies in the sector of food industry along with its potential benefits and challenges. It explains about AI's significance in the food industry and its contributions including management of food waste, market prediction, Quality control, recipe generation, Food safety, cost management, and increased revenue. Ongoing research aims to address AI-related challenges and further enhance its adoption in the food industry. AI has proven to be beneficial in enhancing business strategies through sales prediction and increased yield. The widespread recognition of AI in the food industry is attributed to its simplicity, accuracy, and cost-saving capabilities.

REFERENCES

Agbai, C. M. (2020). Application of artificial intelligence (AI) in the food industry. *GSC Biol. Pharm. Sci, 13*(1), 171–178. doi:10.30574/gscbps.2020.13.1.0320

Ali, M. M., Hashim, N., Abd Aziz, S., &Lasekan, O. (2021). Quality inspection of food and agricultural products using artificial intelligence. *Advances in Agricultural and Food Research Journal, 2*(2).

Babu Loganathan, G. (2021). Recent Scope for AI in the Food Production Industry Leading to the Fourth Industrial Revolution. *Webology, 18*(2), 1066–1080. doi:10.14704/WEB/V18I2/WEB18375

Ban, Y., Kearney, K., Sundahl, B., Medina, L., Bonnecaze, R. T., & Chopra, M. J. (2021, February). Fast etch recipe creation with automated model-based process optimization. In *Advanced Etch Technology and Process Integration for Nanopatterning X* (Vol. 11615, pp. 66–76). SPIE. doi:10.1117/12.2583868

Batra, D., Diwan, N., Upadhyay, U., Kalra, J. S., Sharma, T., Sharma, A. K., & Bagler, G. (2020). Recipedb: A resource for exploring recipes. *Database (Oxford)*, 2020. PMID:33238002

Bień, M., Gilski, M., Maciejewska, M., Taisner, W., Wisniewski, D., & Lawrynowicz, A. (2020, December). RecipeNLG: A cooking recipes dataset for semi-structured text generation. In *Proceedings of the 13th International Conference on Natural Language Generation* (pp. 22-28). IEEE.

Castillo, O., & Meliif, P. (1970). Automated quality control in the food industry combining artificial intelligence techniques with fractal theory. *WIT Transactions on Information and Communication Technologies, 10*.

Clarifai. (n.d.) The *world's AI: Clarifai Computer Vision AI & Machine Learning Platform*. Clarifai. https://www.clarifai.com/

Dolci, R. (2017, July). IoT solutions for precision farming and food manufacturing: artificial intelligence applications in digital food. In *2017 IEEE 41st Annual Computer Software and Applications Conference (COMPSAC)* (*Vol. 2,* pp. 384-385). IEEE. 10.1109/COMPSAC.2017.157

Elferink, M., & Schierhorn, F. (2016). Global demand for food is rising. Can we meet it. *Harvard Business Review*, *7*(04), 2016.

Evans, G. W. (2017). *Artificial intelligence: where we came from, where we are now, and where we are going*.

Food Waste Faqs (no date) USDA. https://www.usda.gov/foodwaste/faqs

Goel, M., Chakraborty, P., Ponnaganti, V., Khan, M., Tatipamala, S., Saini, A., & Bagler, G. (2022, May). Ratatouille: A tool for Novel Recipe Generation. In *2022 IEEE 38th International Conference on Data Engineering Workshops (ICDEW)* (pp. 107-110). IEEE. 10.1109/ICDEW55742.2022.00022

Hassanien, A. E. (2022). *Artificial Intelligence: A Real Opportunity in the Food Industry* (Vol. 1000). Springer Nature.

Hassoun, A., Jagtap, S., Trollman, H., Garcia-Garcia, G., Abdullah, N. A., Goksen, G., ... Lorenzo, J. M. (2022). Food processing 4.0: Current and future developments spurred by the fourth industrial revolution. *Food Control*, 109507.

Henson, S., & Caswell, J. (1999). Food safety regulation: An overview of contemporary issues. *Food Policy*, *24*(6), 589–603. doi:10.1016/S0306-9192(99)00072-X

Jayashankar, P., Johnston, W., & Nilakanta, S. (2020, July). A Conceptual Introduction to an AI-Based Food Marketing System: Navigating Assortment Expansion from Farm to Fork. In *the 45th Annual Macromarketing Conference* (*Vol. 35*).

Kakani, V., Nguyen, V. H., Kumar, B. P., Kim, H., & Pasupuleti, V. R. (2020). A critical review on computer vision and artificial intelligence in food industry. *Journal of Agriculture and Food Research*, *2*, 100033. doi:10.1016/j.jafr.2020.100033

Kiddon, C., Zettlemoyer, L., & Choi, Y. (2016, November). Globally coherent text generation with neural checklist models. In Proceedings of the 2016 conference on empirical methods in natural language processing (pp. 329-339). ACL Anthology. 10.18653/v1/D16-1032

Kumar, I., Rawat, J., Mohd, N., & Husain, S. (2021). Opportunities of artificial intelligence and machine learning in the food industry. *Journal of Food Quality*, *2021*, 1–10. doi:10.1155/2021/4535567

Kuo, T. C., & Smith, S. (2018). A systematic review of technologies involving eco-innovation for enterprises moving towards sustainability. *Journal of Cleaner Production*, *192*, 207–220. doi:10.1016/j.jclepro.2018.04.212

Kyaw, K. S., Adegoke, S. C., Ajani, C. K., Nwabor, O. F., & Onyeaka, H. (2022). Toward in-process technology-aided automation for enhanced microbial food safety and quality assurance in milk and beverages processing. *Critical Reviews in Food Science and Nutrition*, 1–21. doi:10.1080/10408398.2022.2118660 PMID:36066463

LeCun, Y., Bottou, L., Bengio, Y., & Haffner, P. (1998). Gradient-based learning applied to document recognition. *Proceedings of the IEEE*, *86*(11), 2278–2324. doi:10.1109/5.726791

Lezoche, M., Hernandez, J. E., Díaz, M. D. M. E. A., Panetto, H., & Kacprzyk, J. (2020). Agri-food 4.0: A survey of the supply chains and technologies for the future agriculture. *Computers in Industry*, *117*, 103187. doi:10.1016/j.compind.2020.103187

Marın, J., Biswas, A., Ofli, F., Hynes, N., Salvador, A., Aytar, Y., Weber, I., & Torralba, A. (2021). Recipe1m+: A dataset for learning cross-modal embeddings for cooking recipes and food images. *IEEE Transactions on Pattern Analysis and Machine Intelligence*, *43*(1), 187–203. doi:10.1109/TPAMI.2019.2927476 PMID:31295105

Mateo-Babiano, I., & Palipane, K. (2020). Reimagining place through the sandbox studio pedagogy: An introduction. Placemaking sandbox. *Emergent approaches, techniques and practices to create more thriving places*, 1-14.

Mavani, N. R., Ali, J. M., Othman, S., Hussain, M. A., Hashim, H., & Rahman, N. A. (2022). Application of artificial intelligence in food industry—A guideline. *Food Engineering Reviews*, *14*(1), 134–175. doi:10.100712393-021-09290-z

Mor, R. S., Kumar, D., & Singh, A. (Eds.). (2022). *Agri-Food 4.0: Innovations, Challenges and Strategies*. Emerald Publishing Limited. doi:10.1108/S1877-6361202227

Nosratabadi, S., Khazami, N., Abdallah, M. B., Lackner, Z., Band, S. S., Mosavi, A., & Mako, C. (2020). Social capital contributions to food security: a comprehensive literature review. *Foods*, *9*(11). Home (2023) Numenta. https://www.numenta.com/

Pandey, S. (2021, June). Artificial Intelligence and Food Industry. *In Proceedings of the 22nd International RAIS Conference on Social Sciences and Humanities* (pp. 72-76). Scientia Moralitas Research Institute.

Sahni, V., Srivastava, S., & Khan, R. (2021). Modelling techniques to improve the quality of food using artificial intelligence. *Journal of Food Quality*, *2021*, 1–10. doi:10.1155/2021/2140010

Salvador, A., Drozdzal, M., Giró-i-Nieto, X., & Romero, A. (2019). Inverse cooking: Recipe generation from food images. In *Proceedings of the IEEE/CVF Conference on Computer Vision and Pattern Recognition* (pp. 10453-10462). IEEE. 10.1109/CVPR.2019.01070

Salvador, A., Hynes, N., Aytar, Y., Marin, J., Ofli, F., Weber, I., & Torralba, A. (2017). Learning cross-modal embeddings for cooking recipes and food images. In *Proceedings of the IEEE conference on computer vision and pattern recognition* (pp. 3020-3028). IEEE. 10.1109/CVPR.2017.327

Sera, T., Izukura, S., Hashimoto, I., Motegi, T., & Motohashi, Y. (2020, April). A case study of Food Production Using Artificial Intelligence. In *Proceedings of the 2020 Symposium on Emerging Research from Asia and on Asian Contexts and Cultures* (pp. 33-36). IEEE. 10.1145/3391203.3391211

Strong, A. I. (2016). Applications of artificial intelligence & associated technologies. *Science [ETEBMS-2016]*, *5*(6).

Team, K. (no date) *Keras Documentation: Character-level text generation with LSTM*. Keras. https://keras.io/examples/generative/lstm_character_level_text_generation/

Wang, H., Lin, G., Hoi, S. C., & Miao, C. (2022). Learning structural representations for recipe generation and food retrieval. *IEEE Transactions on Pattern Analysis and Machine Intelligence*, 1–15. doi:10.1109/TPAMI.2022.3181294 PMID:35687622

Writer, S. (2022, February 17). *McDonald's & Sonic Self Order Kiosk Ai Technology*. Kiosk Industry. https://kioskindustry.org/mcdonalds-sonic-optimizing-with-self-order-ai-technology/

Yu, Z., Zang, H., & Wan, X. (2020, November). Routing enforced generative model for recipe generation. In *Proceedings of the 2020 Conference on Empirical Methods in Natural Language Processing (EMNLP)* (pp. 3797-3806). IEEE. 10.18653/v1/2020.emnlp-main.311

ADDITIONAL READING

Bottani, E., Murino, T., Schiavo, M., & Akkerman, R. (2019). Resilient food supply chain design: Modelling framework and metaheuristic solution approach. *Computers & Industrial Engineering*, *135*, 177–198. doi:10.1016/j.cie.2019.05.011

Kudashkina, K., Corradini, M. G., Thirunathan, P., Yada, R. Y., & Fraser, E. D. (2022). Artificial Intelligence technology in food safety: A behavioral approach. *Trends in Food Science & Technology*, *123*, 376–381. doi:10.1016/j.tifs.2022.03.021

Linko, S. (1998). Expert systems—What can they do for the food industry? *Trends in Food Science & Technology*, *9*(1), 3–12. doi:10.1016/S0924-2244(97)00002-2

Murrell, C., & van Rooyen, F. (2021). Next-tech, agtech: How the Fourth Industrial Revolution is intersecting with agriculture. *Red Meat/Rooivleis (AFR & ENG)*, *12*(5), 22-26.

Rawat, R. M., Rana, A., Toppo, A. J., & Beck, A. (2021, May). AI based Impact of COVID 19 on food industry and technological approach to mitigate. In *2021 5th International Conference on Intelligent Computing and Control Systems (ICICCS)* (pp. 1743-1748). IEEE. 10.1109/ICICCS51141.2021.9432152

Sharma, T. (2019). *How Is AI Optimizing Food Supply Chain?* Global Tech Council. https://www.globaltechcouncil.org/artificial-intelligence/how-is-ai-optimizing-food-supply-chain/

KEY TERMS AND DEFINITIONS

Artificial Intelligence: AI is a computer science field focused on developing intelligent machines capable of performing tasks that typically require human intelligence that include learning, problem-solving, perception, and decision-making.

Artificial Neural Network: Artificial Neural Networks are computational models composed of interconnected processing elements called neurons that receive inputs, process them using predefined activation functions, and produce outputs.

Computer Vision: Computer vision is a branch of AI that enables machines to interpret and understand visual data. It involves the development of algorithms and techniques that allow machines to see, perceive, and recognize objects, patterns, and features in images and videos.

Deep Learning: DL is a subset of machine learning that utilizes artificial neural networks to model and train these networks on large datasets, allowing them to learn patterns and make predictions or classifications and solve complex problems.

Machine Learning (ML): ML is a subset of AI that trains algorithms to make predictions and decisions based on input data, rather than being explicitly programmed to perform a specific task.

Quality Control: Quality control refers to a set of procedures implemented to ensure that manufactured products or performed services meet predefined quality criteria or client/customer requirements.

Supply Chain Management: Supply Chain Management involves the management of the flow of goods and services from raw material acquisition to final product delivery.

Chapter 9
The Use of Augmented Reality in the Food Industry:
Enhancing the Dining Experience

Melike İmre
Tekirdağ Namık Kemal University, Turkey

Seydi Yıkmış
https://orcid.org/0000-0001-8694-0658
Tekirdağ Namık Kemal University, Turkey

ABSTRACT

In this chapter, augmented reality (AR) and its use in the food industry are evaluated under important headings. AR applications that aim to improve the customer or user experience and the contents of these applications are explained. The use and effects of these applications have been examined separately in important steps of the food industry, such as food safety, food cooking stage, food education, food promotion, and user/consumer experience. The aim is to deeply understand the use of AR and to show that the power of the technological opportunities provided by AR can be used at every step of the food industry. In addition, the perspectives of the applications developed and considered to be developed in the literature are examined to present the aspects that should be considered when planting seeds for new applications.

INTRODUCTION

Contemporary advancements in food science have successfully integrated diverse disciplines and cutting-edge technologies, resulting in convenient availability of a wide range of nourishing and delectable food options. The food system has embraced numerous technologies, reflecting a growing and pervasive trend towards embracing novel approaches (Antonucci et al., 2019; Suprem et al., 2013).

As companies constantly compete for the acceptance of consumer products in today's highly innovative and global food environment, the need for sensory evaluation in the food industry is becoming increasingly complex. The latest technological developments in virtual and augmented reality provide the food

DOI: 10.4018/978-1-6684-9094-5.ch009

industry with new opportunities to create more reliable insights into consumers' sensory perceptions of food and beverages and contribute to the design and development of new products with optimized consumer benefits (Crofton et al., 2019). In the upcoming years, the food technology industry holds significant potential for growth, and we are already witnessing the emergence of visible connections between augmented reality (AR) technology and the food industry (Casari et al., 2021).

With the rapid adoption of the food and beverage industry, augmented reality (AR) technology has become an industry investment to improve the customer experience and has had a huge impact (Fritz et al., 2022). Considering its relationship with the food industry, it is known that augmented reality can affect the eating and drinking experiences of individuals and includes a series of technological innovations that improve the quality of the experience in food and beverage businesses. The use of augmented reality applications in gastronomy experiences is considered a valuable process with the potential to provide unforgettable memories. (Aydın & Uçkan Çakır, 2022).

Academic research often highlights the advantages of augmented reality (AR) technology for the food industry. Beck et al. have noted that utilizing AR technologies can offer real-time inventory location information, instant food identification, heightened situational awareness, enhanced visibility of preparation areas, and identification of potential food hazards to food businesses (Beck et al., 2016).

Moreover, the utilization of mobile applications incorporating AR technology is believed to empower users to make healthier dietary decisions, explore novel taste experiences, enhance their cooking and food shopping endeavors, boost workplace productivity, and refine precision agriculture applications (Xie et al., 2022).

In this review, it is aimed to understand the reflections of augmented reality technologies in the food industry. In this direction, AR technology and its use in restaurants, as well as the effects of this technology under the headings of food safety and education, food supply chains, food marketing, customer satisfaction, which are important components of the food industry, are discussed.

AR Technologies and the Food Industry

AR Technology

Augmented reality is a technology that enriches the user experience by overlaying digital information onto real-world objects or locations (Berryman, 2012). The increasing prevalence of this technology is leading to its growing popularity and widespread adoption in society (Johnson et al., 2010).

AR technology offers an interactive and three-dimensional (3D) experience by blending real and virtual objects. Paul Milgram and Fumio Kishino introduced Milgram's Continuum of Virtuality, which describes a spectrum ranging from the physical environment to the virtual environment: "Augmented Reality (AR) falls closer to the real world on this continuum, while Augmented Virtuality (AV) leans towards the virtual environment" (Milgram & Kishino, 1994). These concepts are expressed in Figure 1.

Figure 1. Augmented reality sprectrum

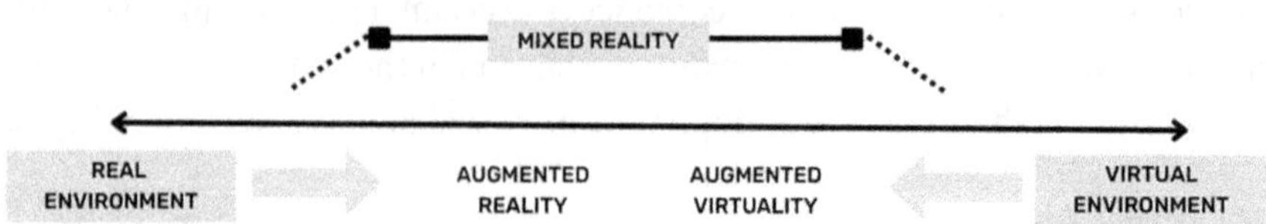

Augmented reality (AR) refers to the dynamic utilization of real-time information, including text, graphics, sound, and other virtual enhancements, seamlessly integrated into the physical environment. AR can be regarded as a digital expansion of an individual's five senses. Particularly useful on mobile devices, AR allows the user to give faster responses or make decisions by enhancing their senses through digital tools (Nguyen, 2013). Augmented reality applications are divided into image-based and location-based. These and their elaboration are shown in Figure 2.

Figure 2. AR Applications defined by (Tzima et al., 2018)

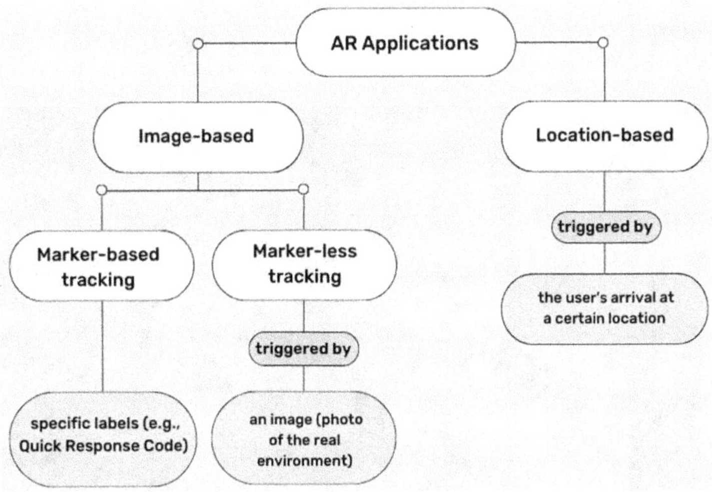

Augmented reality (AR) encompasses three primary elements: the fusion of real and virtual worlds, facilitating real-time user interaction, and presenting the experience within a three-dimensional environment (Jagtap et al., 2021). However, despite all this, it does not isolate the user from the physical environment, and this feature is perhaps one of the main factors in the increasing popularity of AR (Kounavis et al., 2012). The broad usage of hardware devices, primarily smartphones and tablets, capable of supporting application development and utilization, contributes significantly to the proliferation of AR technology (Bottani & Vignali, 2019).

Use of AR Technology in Restaurants

In recent years, with the advancement in science and technology, changes have been observed in restaurants as well as food and beverages. It has come to the fore to offer experiences aimed at satisfying sensory and emotional quests rather than meeting customer needs. In addition to providing products and services, providing extraordinary experiences to consumers is defined as "experience marketing" and the number of companies that are examples of this is increasing (Akbulut & Aksoy, 2016). In this section, some of the restaurants using AR technology are given.

Inamo. Located in London, the restaurant named Inamo is known for its Far Eastern service experience and as a place that offers an extraordinary experience. With the development of hospitable virtual environments, Inamo offers services such as e-table, chef cam, local area and taxi booking. In addition, customers can order food and drink thanks to the AR interactive menus placed on the tables in the restaurant. The desks feature electronic sales points powered by four electronic video cards, while waterproof touch panels, high-quality projectors and interactive AR-supported computers are also features of the desks (Figure 3). There are main servers, main server printers working in the bar and kitchen, and various tools to manage the operational aspects of the restaurant. Inamo restaurant management conducted a semi-structured interview to gather guest opinions and gain insights into the impact of AR on the hospitality industry. The findings indicated that the technology was well-received in London, and guests expressed enjoyment while dining with AR-enhanced experiences (Shabani et al., 2019).

Figure 3. Inamo restaurant
(Inamo, 2022)

KabaQ Augmented Reality Food. KabaQ, an AR technology company based in New York, offers AR-powered 3D menus for restaurants. In this way, customers can view restaurant menus in real size. KabaQ employs cutting-edge scanning technologies to generate exceptionally high-quality 3D models, which can be utilized in various augmented, virtual, and mixed reality encounters. KabaQ states that thanks to AR, all the information desired to be shared with the customer is provided in a digital layer, thus creating the opportunity to share more with the customer. KabaQ use cases are summarized in Figure 4 (KabaQ, 2021).

Ultraviolet by Paul Pairet. Ultraviolet is an innovative restaurant concept by renowned French chef Paul Pairet. Located in Shanghai, China, the restaurant aims to provide a dining experience that engages all the senses. Conceptually, the restaurant creates a multi-sensory dining experience where food, music, lighting and aroma are carefully selected to take the diners on a unique culinary journey. The restaurant has a single table for ten and the menu consists of a set meal that changes regularly (Figure 5) (Paul Pairet, 2022).

Figure 4. KabaQ use cases
(KabaQ, 2021)

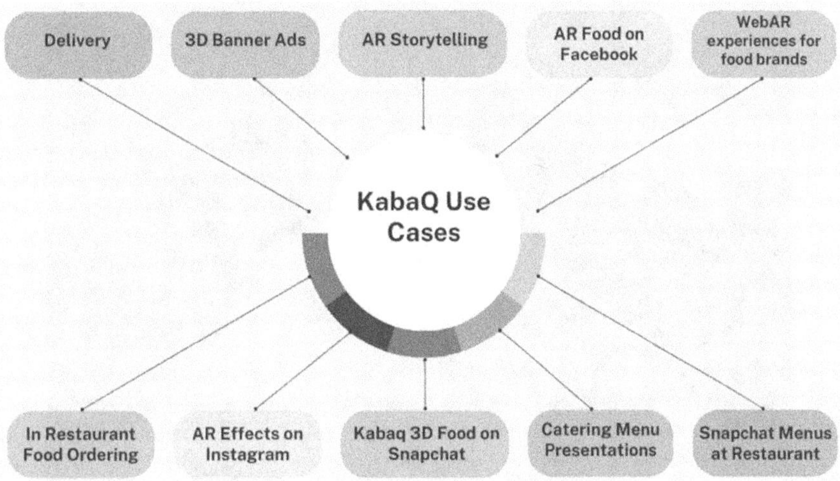

Figure 5. Ultraviolet by Paul Pairet

Sublimotion. Sublimotion, a restaurant located on the Spanish island of Ibiza, is also a concept that offers an exclusive dining experience. The restaurant was designed by Michelin-starred chef Paco Roncero. Sublimotion offers a gastronomy and show experience that combines food with art. The restaurant, which uses advanced technology, draws attention with its AR applications and creative concept. Decorative elements, special lighting and artistic touches were used to create a different atmosphere in the interior design of the restaurant. The restaurant serves only a certain number of people every night with a special beer and a special menu. During the experience, the space can change with different visual and auditory effects, Figure 6 shows a dining experience from Sublimotion.

Figure 6. Sublimotion, Ibiza
(Veebrant, 2022)

Use of AR Technology in Kitchens

It is thought that the use of augmented reality in kitchen design can provide an attractive and interactive experience (Phan et al., 2010). Thanks to this technology, when designing the kitchen, the measurement errors can be minimized compared to the traditional way, and the virtual kitchen result can be superimposed, which can fit the real kitchen environment. In addition, this technology can reveal more than human limited imagination in design (Muhammad Nizam et al., 2018).

AR continues to be the subject of many studies due to its various uses in kitchens and its features such as facilitating the food preparation process, providing educational materials and enriching the visual experience. For example, in a rapidly developing and technological world, he developed a sound-enhanced kitchen prototype system called Chop Chop, which recommends spending quality time on cooking (Halupka et al., 2012). In this system, which encompasses a computer or mobile device, headphones, a cutting board, and a knife, users can enhance their cooking experience while performing everyday tasks like chopping food. Through the application of filters and composite events, the sound of chopping is enriched, enabling users to encounter a novel and thrilling sensation. The study concluded that integrating this augmented reality system in a kitchen environment does not undermine traditional tools and techniques, but rather enhances the overall user experience in a positive manner (Halupka et al., 2012).

(Bonanni et al., 2005) used AR technology for Counter Intelligence, a kitchen they augmented with information projection onto their objects and surfaces to guide users, coordinate between multiple tasks, and increase trust in the system. In this kitchen, the augmented reality projection is arranged to display information and project interfaces directly on the task being performed. For example, thanks to Augmented Cabinetry, which is one of the components of the kitchen, the LEDs embedded in the translucent cabinet handles are illuminated in a certain order by the virtual system in order to make it easier to find the materials in the cabinet. If the desired items are located far from the user, the last location of the item is shown by projecting an arrow centered between the user and the item (Figure 7).

Figure 7. Augmented cabinetry in counterintelligence by (Bonanni et al., 2005)

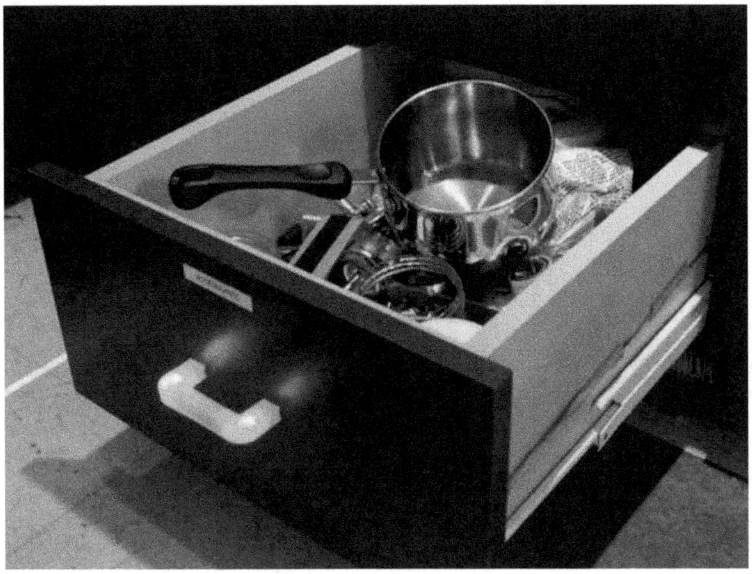

The Relationship Between Important Concepts in the Food Industry and AR Technology

In this section, separate titles have been created to examine the impact of AR on the food industry. Under these headings, the results of the researches made in recent years and the possible benefits of AR are discussed.

Food Promotion

Food marketing is a process used to introduce food products to consumers and to be effective in their purchasing behavior. It includes various strategies and techniques to raise awareness, generate interest and trigger purchasing decisions. Key elements such as branding, packaging, advertising and promotions affect the success of food marketing. The food and beverage industry employs food and beverage marketing as a crucial strategy to enhance the attractiveness and acceptance of their products in the market (Kickbusch et al., 2016).

The changing content of the ingredients in the food supply chain with the increasing digitalization of the world requires the creation of appropriate strategies for these changes. According to (Vanderroost et al., 2017), the emergence of augmented reality in the sales and consumption stages of the food package lifecycle is attributed to the development of new technologies and methods. These advancements offer opportunities for the stakeholders involved in food marketing strategies to adapt to changes, fulfill consumer demands, and address evolving market dynamics. These components, described in Figure 8, are summarized.

The significance of packaging in food marketing should not be underestimated. As the final element influencing purchase decisions in the consumer's decision-making process, packaging plays a crucial role. A well-designed package can enhance the brand's connection with the buyer by adding value, enhancing product attributes (such as freshness or convenience), creating a distinctive

Figure 8. Sales and consumption components of food supply chain AR applications

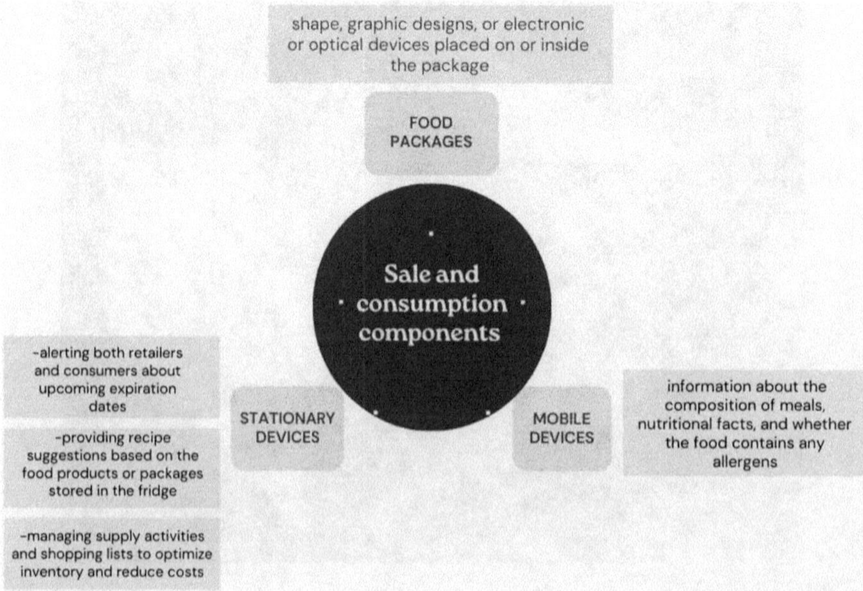

shelf presence, positioning the brand effectively, and gaining a competitive edge. It should serve as a "visual magnet," as described by senior designer Michael Peters, enticing customers to choose a specific brand and fostering long-term brand loyalty (Smith & Zook, 2016). In a recent study conducted by (Gu et al., 2023), on the impact of packaging on consumers' negative perception of purchasing and consuming packaged food, it was concluded that augmented reality technology offers a more convenient user experience compared to traditional packaging. This finding suggests that augmented reality has the potential to enhance consumer satisfaction and alleviate negative reviews associated with packaged food. According to the study findings, augmented reality technology has improved consumers' perception of interaction, perceived vitality and innovation experience, and has achieved the purpose of promoting packaged food retail sales by improving negative reviews by users. As a result of their study, Gu et al. showed that interactive packaging developed using AR technology can improve consumers' negative perceptions of packaged food and improve their evaluations on this subject (Gu et al., 2023).

Augmented reality packaging serves as a catalyst that brings attention to the product by providing additional information crucial for decision-making or offering supplementary benefits to the customers. In simple terms, with technological developments, customers have the opportunity to further examine the market for products that suit their needs and desires. In this scenario, augmented reality packaging gives companies the ability to present and highlight their products and fulfill more needs and desires (Edelman & Singer, 2015).

The digitalization of marketing has led to the emergence of digitally empowered consumers who can be effectively engaged by businesses in digital environments. Recognizing the potential, many businesses are embracing digital platforms and investing in digital marketing opportunities. Consequently, the widespread adoption of technology in marketing activities has been driven by the influence of digital media (Tuncer, 2020).

Creating a unique marketing strategy in today's world is a challenging process, but augmented reality can help with marketing campaigns by providing innovative and engaging ways to showcase brands. Integrating AR into daily marketing activities has the potential to increase possibilities (Todorović et al., 2019). According to (Rauschnabel et al., 2022), AR marketing can be defined as the deliberate integration of AR experiences, either independently or in conjunction with other media or brand-related signals. The purpose is to attain comprehensive marketing goals by generating value for the brand, its stakeholders, and society as a whole, while considering ethical implications.

Many studies have been conducted to determine the power of AR to support food promotion. (Weking et al., 2020) aimed to develop a mobile augmented reality application to promote Indonesian food. The application, created using the Unity3D program, has succeeded in presenting information about traditional Indonesian food in a more interesting way using mobile AR technology. As highlighted by (Wang et al., 2019), technology offers the potential to create enjoyable and interactive experiences that allow consumers to see and hear a product while consuming it.

Utilizing AR technology in marketing is known to be more cost-effective compared to traditional marketing approaches. Print advertising and marketing campaigns tend to be relatively more expensive, requiring ongoing resource investment. In contrast, AR, being a computer-based technology, proves to be less expensive as it can be utilized on digital platforms or online. Although the initial development costs of AR-based marketing campaigns may be higher, the ongoing operating costs remain largely consistent over time. (Shabani et al., 2019).

Food Safety and Education

Occupational safety must be ensured in the kitchens, which is one of the most important stages of the production phase in the food industry. For this purpose, it is an important step to raise the awareness of the kitchen staff about the possible risks and to provide the necessary precautions.

Injuries in the kitchen, particularly burns and scalds caused by hot surfaces, food, liquids, and caustic chemicals, encompass a variety of potential hazards. (Le et al., 2014). When it comes to workplace accidents and injuries, the social and economic costs are often complex and cannot be attributed to a single factor. Instead, worker safety is influenced by a multitude of interconnected elements, such as the physical and psychosocial work environment, job-related factors, and individual differences (Wilkins & Mackenzie, 2007).

Food safety training is commonly imparted to employees in the food industry to equip them with essential knowledge on preventing foodborne illnesses. However, despite evidence indicating the effectiveness of current training methods in enhancing employee knowledge, observed behaviors often fail to undergo significant changes. Consequently, the risk of foodborne illnesses persists without a corresponding decrease (Beck et al., 2016). Food businesses have a keen interest in exploring AR technology as technology opens up many opportunities (Rejeb et al., 2021).

By integrating virtual information, such as live video streams, images, or instructions, into the real-world environment, augmented reality (AR) systems aid users in executing tasks and operations during human-machine interactions. (Vignali et al., 2018). AR systems have features that can function in ensuring and controlling food safety. For example, by providing AR food traceability, it is of great importance in protecting consumers by precisely determining the origin of the food product and disclosing relevant information (Chai et al., 2022).

With the help of developing AR technology, people can experience a reality where virtual items are placed over the real environment. Researchers have been interested in this technology, which has been applied in a number of educational contexts (Franco-Mariscal, 2018). To address the issue of complex operations faced by users when accessing food safety information, Min-Jing and colleagues introduced a smartphone-based augmented reality information disclosure model specifically designed for food safety. In this proposed model, "real objects" and "virtual objects" distinguish between food recommendations and safety information and inform the user in a more effective way (Min-jing & Bo, 2013). Juan and his team have developed an augmented reality application that focuses on food packaging, particularly boxes or cans, to aid users in interpreting the nutritional information of actual packaged foods (Juan et al., 2019). Studies and applications like this enable user to gain information about the food they consume, providing awareness in many aspects such as food safety, health, and more.

In a study by (Clark et al., 2018), it was concluded that food safety education in the kitchen can be an effective way to implement information transmission and behavior when carried out with wearable technology (Figure 9).

Figure 9. Video recordings were captured to train food handlers in proper handwashing techniques, presenting two perspectives: first-person perspective (on the left) and third-person perspective (on the right) (Clark et al., 2018)

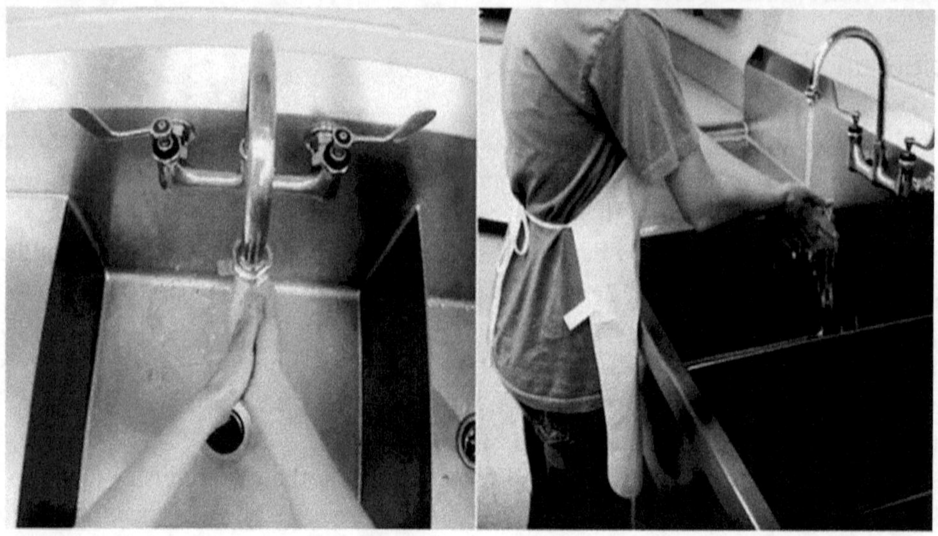

The implementation of augmented reality (AR) headsets in food and beverage manufacturing and processing facilities can offer assembly line workers comprehensive information required for ensuring proper preparation and packaging of food (Beck et al., 2016). (Luque et al., 2017) suggest that augmented reality (AR) brings several advantages to the food supply chain. These include enhanced visualization of food processes, mitigating risks of contamination, facilitating food education, bolstering food marketing efforts, and optimizing food logistics, such as warehousing activities.

Over the past decade, the operations and structures of food systems and food supply chains have experienced notable transformations due to the expanding globalization, which has increased the range and accessibility of food options.As food supply chains lengthen and food trade relations develop; food origin, growing methods, harvesting, processing and working conditions, transparency about sustainability and food confidence are declining (Keogh et al., 2020). As a result, appropriate technological developments to ensure safety at every step in the transportation of food to the consumer are closely followed and implemented with different applications as mentioned.

Traditional food safety education designs may not always effectively bring about behavioral changes. However, it is believed that the utilization of augmented reality applications, such as headphones or head-mounted action cameras, can revolutionize the delivery of food safety education within the food industry (Clark et al., 2018). The concept behind using AR technologies is to enhance the integration of real-world and digital learning resources, thus broadening the possibilities for educational experiences (Wu et al., 2013). In order to provide an interactive experience in food safety training, the functional features of AR technology used while performing this function should be understood.

Meta-perceptual knowledge involves the amalgamation of haptic and pictorial knowledge acquired within the AR environment. This knowledge is subsequently merged with pre-existing mental models and schemas, forming a consistent internal representation of spatial information. This representation is continuously updated through the incorporation of new knowledge and experiences into existing ones. With this type of information encoding process, understanding of the environment is enhanced (Haken, 1991). Through its integration of visual, spatial, and sensorimotor feedback in response to interface manipulations, augmented reality (AR) presents users with distinct advantages in terms of interacting with the physical world and engaging in various actions. While multi-sensory exploration of objects is beneficial for gathering and triangulating perceived information, visual perception becomes particularly effective during movement. As a result, AR provides opportunities for users to experience information in a comprehensive manner, combining visual perception with spatial understanding and sensorimotor engagement (Hung et al., 2017).

In comparison to studies conducted on more established technologies in education, such as multimedia and web-based platforms, research focusing on the applications of augmented reality (AR) in education is still in its nascent stages. Consequently, the existing evidence regarding the impact of AR on teaching and learning outcomes is limited. Over the years, empirical research on computer or internet use in education has made progress in unraveling the connections between learner characteristics and technology utilization. This research has already shown some effectiveness in utilizing technologies to support positive learning outcomes (Lee et al., 2011). To establish concrete evidence regarding the educational value of augmented reality (AR), it is imperative to conduct rigorous and comprehensive evaluation studies that encompass a sizable sample and employ valid instrumentation. Such studies would help solidify the understanding of the benefits AR can offer in an educational context (Wu et al., 2013).

Agriculture and Food Supply Chains

The increasing demand for modern investments in agricultural management paves the way for a new era in the agro-technical industry sector, where technology meets agriculture. Since the turn of the millennium, the agricultural landscape has undergone a significant transformation as a result of numerous inventions. These innovations, spanning automation, wireless communications, geo-satellite research, and various other fields, have turned the agricultural sector into a platform for addressing existing challenges through cutting-edge solutions (Kakani et al., 2020; Morgan & Murdoch, 2000).

Although augmented reality is not the subject of much research yet in agricultural sector, it offers various uses. AR can increase the efficiency of agricultural operations, provide growers with important information to make better decisions, and improve agricultural education and consultancy. It can also provide transparency in detecting plant diseases and pests and marketing agricultural products. Caria et. all demonstrate the benefits of employing augmented reality (AR) via smart glasses for obtaining real-time information on various aspects such as feeding, milking, breeding, and animal health. Through this advanced technology, farmers can conveniently scan QR codes attached to specific subjects in the field, granting them access to comprehensive files containing pertinent information about the respective animals. The significance of utilizing AR technologies in this context lies in the ability for farmers to access information seamlessly without disrupting their ongoing tasks (Caria et al., 2019).

Special transportation, preparation and logistics are important in making food safe for human consumption, which is a basic necessity for human life. In recent times, food supply chains have undergone a transformation, becoming increasingly global, fragmented, and intricate. As a result, stakeholders within the food industry are reassessing conventional practices and embracing new approaches to effectively manage these complex food chains while capitalizing on the latest trends in technological innovation (Rejeb et al., 2021).

The success of developing new technology hinges on more than just cutting-edge features and significant investments from renowned brands. A complete and well-supported ecosystem plays a crucial role in driving growth. This entails collaborative efforts from various stakeholders, including core hardware and accessory manufacturers, technology and solution providers, content creators, distribution channels, and an existing user base. Such collective endeavors contribute to the overall success and advancement of the technology (Chang & Chen, 2017). AR technology, with its structure that allows to concentrate almost unlimited amount of information in a limited area (Logaldo, 2014) is adapted to all stages of the food industry and provides benefits for the process to operate safely, efficiently and quickly.

The powerful impact of smart technologies on core business processes goes beyond just increasing productivity and efficiency. This development lays new foundations for both economic and social sustainability (Porter & Heppelmann, 2014). However, it should not be forgotten that digital applications such as AR can also create additional demand by increasing general consumption. Additional resources and energy are needed to run digital manufacturing processes and produce the necessary equipment. In addition, direct digital manufacturing processes may result in higher scrap rates than standard mass production, as they require less skilled labor and can potentially contain more suitable materials (Ford & Despeisse, 2016).

There is a growing interest and acceptance among food industry stakeholders, including food suppliers, retail establishments, and food service providers, regarding the advantages of augmented reality (AR). They are increasingly recognizing the potential benefits that AR can bring to their respective sectors. Although the integration of AR technologies into food chains is challenging and the development costs are high, it seems that this integration is necessary. Due to its ability to provide dynamic information and experiences that cannot be obtained from existing technologies, the use of AR is believed to have the potential to revolutionize research in the food industry (Chai et al., 2022).

User/Consumer Experience

Recently, it is seen that augmented reality applications are widely used in the marketing and advertising sector. Food and beverage businesses also use these applications, especially for the purpose of creating a positive image for their consumers and promoting their products. In this way, existing businesses can gain sustainable competitive advantages by increasing their original values. In this direction, food and beverage businesses have the opportunity to use augmented reality applications to speed up the communication between the customer and the business through location services, on their tables via projectors, on their menus in virtual environment and through electronic devices (Cankül et al., 2018). In addition, the use of such technologies is also important in terms of ensuring customer satisfaction, which is a factor of great importance for the food industry (Çöl et al., 2023).

With the advancements in Android and iOS platforms, augmented reality (AR) has found its way into the everyday lives of consumers. Consequently, the emergence of mobile augmented reality (MAR) technologies has become a prominent subject in the realm of digitalization (Penco et al., 2021). AR applications play a pivotal role in broadening consumers' understanding of products, enabling them to make more informed assessments, particularly in terms of the products' functional aspects such as perceived price, quality, utility, and reliability (Poushneh & Vasquez-Parraga, 2017).

Customer compliance with AR technology and the impact of AR on user experience depend on many factors. These can be summarized as:

- Customers must have the necessary knowledge of the product prior to a particular AR technology experience.
- Products are designed to have user-friendly interfaces and provide easy access.
- Customers' technology experience and competencies are appropriate.
- Experienced technology provides access to valuable and interesting information, rather than just entertainment or ostentation.

According to a study conducted by (Fritz et al., 2022), the integration of augmented reality in a real-time environment enhances consumers' capacity to mentally simulate consuming a depicted food. As a result, their desire to purchase the food and the likelihood of actually making the purchase are heightened. Furthermore, (Hilken et al., 2017) highlight that one of the advantages of AR is its ability to provide entertainment to a broad audience.

In their study, (Albayrak et al., 2019), focused on the development of a training program using augmented reality glasses for fast-food restaurant employees. The program aimed to enhance the onboarding experience of new employees by incorporating gamification, personalization, and shortened training sessions to make the process more enjoyable and efficient. As a result, they reported that they expected an increase in employee satisfaction and service quality. (Albayrak et al., 2019). In another study, the effect of Pokemon Go, a popular mobile game using AR technology, on local restaurants was examined.

The results indicated that restaurants connected with Pokemon Go experienced increased consumer engagement and a more favorable perception among consumers (Pamuru et al., 2021).

In the food industry, restaurants are adopting electronic digital menus that offer diverse interactive functionalities. For instance, the Domino Pizza ordering application (Figure 10) allows customers to personalize their pizza by selecting toppings and bases, providing them with a virtual visualization of their customized pizza (Yim & Yoo, 2022).

With the new Pizza Chef feature in this ordering application, users can now visually craft their favorite pizza by handpicking ingredients and placing them onto a virtual pizza. They have the freedom to choose their preferred side, sauce, cheese, and other desired ingredients, and witness the creation process firsthand (Santos, 2019).

Figure 10. Domino's Pizza AR app
(Santos, 2019)

Similar to this innovation, Pizza Hut has planned a fun AR-based game that innovatively introduces its delivery services. PCCW Solutions was employed in this program, which aims to enable customers to experience fast delivery services virtually and to feel intertwined with the environment in terms of augmented reality. In the project, interactive board games are offered to customers using AR technology (Figure 11). Customers can play games and have a fun experience through AR projections on restaurant tables. By harnessing this technology, Pizza Hut aims to enhance customer engagement, attract and retain a larger customer base, and solidify its position in the market (Pccw Solutions, 2022).

Figure 11. Pizza Hut AR based pizza box game
(8th Wall, 2021)

When all these technological applications are examined, it can be said that augmented reality (AR) technology will continue to play a major role in the future of customer experience. The fact that AR creates an environment that connects customers more by providing them with interactive and interactive experiences simplifies, saves and improves the shopping experience of customers.

CONCLUSION AND FUTURE WORK

The food industry is a complex and global network that involves many steps from processing the raw materials to delivering the final meal to the customer. It has important activities to be considered at every stage. These can be summarized as the production, processing, distribution and sale of food. The framework of a new technology such as AR can benefit from its use in the large food industry at all these stages. Utilizing technology to overcome the challenges in food industry applications will help to ensure sustainability in this industry. As the interest in AR increases, there is a growing number of AR-based applications being developed in the food industry. It is believed that careful analysis of the effects of these applications on users or customers will guide researchers in identifying the missing or ineffective aspects, leading to the development of new applications.

A meal experience blended with AR technology will provide customers with memorable experiences that will satisfy them. Restaurants aim to enrich their customers' experiences by using AR applications such as interactive menus, virtual dining experiences and visual presentations of food. In addition, thanks to AR, restaurants have many opportunities to promote their services and brands in an innovative way, and to offer tailor-made experiences to customers.

In addition, the presence of restaurants that have kept up with such technologies in a region, city, or country can lead to gastronomic mobility. The frequency of using tools that test customer or user experiences should increase during the development and renovation of such applications. By emphasizing the aspects that customers or users are satisfied with and improving the aspects they are uncomfortable with, significant steps can be taken to elevate the meal experience to the highest level.

Food manufacturers and retailers are increasingly using AR technology to market their products and tell brand stories. By using AR labels or QR codes, companies have become able to provide consumers with detailed information such as product content, source, and sustainability information. In the future, AR technology is expected to be used in areas such as more personalization, interactive experiences and product information provision in the food industry. Thus, consumers can be more connected to the product and brand loyalty can be improved. It has been observed that most of the studies carried out so far in the use of AR technology in the food industry have focused on food sales and customer satisfaction. Considering that the use of AR technology in the early stages of the food industry can provide great benefits for businesses in terms of providing occupational safety and saving time, it may be beneficial to increase the number of studies supporting the use of AR at these stages. Therefore, it is understood that more studies are needed to test the use of this technology.

REFERENCES

8th Wall. (2021). *Pizza Hut turns special edition pizza boxes into an augmented reality arcade*. 8TH Wall. https://www.8thwall.com

Akbulut, B. A., & Aksoy, M. (2016). Restoranlardaki Teknolojik Yeniliklerin Deneyim Pazarlaması Açısından Değerlendirilmesi (Assessment of the Technological Innovations in Restaurants in terms of Experience Marketing. *International Congress on Cultural Heritage and Tourism*, 1–10.

Albayrak, M. S., Oner, A., Atakli, I. M., & Ekenel, H. K. (2019). Personalized training in fast-food restaurants using augmented reality glasses. *Proceedings - 2019 International Symposium on Educational Technology, ISET 2019*, (pp. 129–133). IEEE. 10.1109/ISET.2019.00035

Aydın, Ş., & Uçkan Çakır, M. (2022). Gastronomi ve Dijitalleşme. *Nevşehir Hacı Bektaş Veli Üniversitesi SBE Dergisi*, *12*(4), 2143–2159. doi:10.30783/nevsosbilen.1123324

Beck, D. E., Crandall, P. G., O'Bryan, C. A., & Shabatura, J. C. (2016). *Taking food safety to the next level—An augmented reality solution*. Taylor and Francis.. doi:10.1080/15378020.2016.1185872

Berryman, D. R. (2012). Augmented Reality. *RE:view*, *31*(2), 212–218. Http://Dx.Doi.Org/10.1080/02763869.2012.670604. doi:10.1080/02763869.2012.670604 PMID:22559183

Bonanni, L., Lee, C.-H., & Selker, T. (2005). CounterIntelligence: Augmented Reality Kitchen. MIT Media Laboratory, 2239(45).

Cankül, D., Doğan, A., & Sönmez, B. (2018). Innovation And Augmented Reality Applications in Food And Beverage Companıes. *İşletme Araştırmaları Dergisi Journal of Business Research--Turk, 10*, 576–591.

Caria, M., Sara, G., Todde, G., Polese, M., & Pazzona, A. (2019). Exploring Smart Glasses for Augmented Reality: A Valuable and Integrative Tool in Precision Livestock Farming. *Animals 2019, 9*(11), 903. doi:10.3390/ani9110903

Casari, F. A., Navab, N., Hruby, L. A., Kriechling, P., Nakamura, R., Tori, R., de Lourdes dos Santos Nunes, F., Queiroz, M. C., Fürnstahl, P., & Farshad, M. (2021). Augmented Reality in Orthopedic Surgery Is Emerging from Proof of Concept Towards Clinical Studies: A Literature Review Explaining the Technology and Current State of the Art. *Current Reviews in Musculoskeletal Medicine*, *14*(2), 192–203. doi:10.100712178-021-09699-3 PMID:33544367

Chai, J. J. K., O'Sullivan, C., Gowen, A. A., Rooney, B., & Xu, J. L. (2022). Augmented/mixed reality technologies for food: A review. *Trends in Food Science & Technology*, *124*, 182–194. doi:10.1016/j.tifs.2022.04.021

Chang, S. N., & Chen, W. L. (2017). Does visualize industries matter? A technology foresight of global Virtual Reality and Augmented Reality Industry. *Proceedings of the 2017 IEEE International Conference on Applied System Innovation: Applied System Innovation for Modern Technology, ICASI 2017*, (pp. 382–385). IEEE. 10.1109/ICASI.2017.7988432

Clark, J., Crandall, P., & Shabatura, J. (2018). Wearable technology effects on training outcomes of restaurant food handlers. *Journal of Food Protection*, *81*(8), 1220–1226. doi:10.4315/0362-028X.JFP-18-033 PMID:29963922

Çöl, B. G., İmre, M., & Yıkmış, S. (2023). Virtual reality and augmented reality technologies in gastronomy: A review. *EFood*, *4*(3), e84. doi:10.1002/efd2.84

Crofton, E. C., Botinestean, C., Fenelon, M., & Gallagher, E. (2019). Potential applications for virtual and augmented reality technologies in sensory science. *Innovative Food Science & Emerging Technologies*, *56*, 102178. doi:10.1016/j.ifset.2019.102178

Edelman, D. C., & Singer, M. (2015). Competing on Customer Journeys. *Sales and Marketing*. https://hbr.org/2015/11/competing-on-customer-journeys

Ford, S., & Despeisse, M. (2016). Additive manufacturing and sustainability: An exploratory study of the advantages and challenges. *Journal of Cleaner Production*, *137*, 1573–1587. doi:10.1016/j.jclepro.2016.04.150

Franco-Mariscal, A. J. (2018). Discovering the Chemical Elements in Food. *Journal of Chemical Education*, *95*(3), 403–409. doi:10.1021/acs.jchemed.7b00218

Fritz, W., Hadi, R., & Stephen, A. (2022). From tablet to table: How augmented reality influences food desirability. *Journal of the Academy of Marketing Science*, 1–27. doi:10.1007/S11747-022-00919-X/FIGURES/2 PMID:36589144

Gu, C., Huang, T., Wei, W., Yang, C., Chen, J., Miao, W., Lin, S., Sun, H., & Sun, J. (2023). The Effect of Using Augmented Reality Technology in Takeaway Food Packaging to Improve Young Consumers’ Negative Evaluations. *Agriculture 2023, 13*(2), 335. doi:10.3390/agriculture13020335

Haken, H. (1991). *Synergetic Computers and Cognition: A Top-Down Approach to Neural Nets*. Springer Series In Synergetics. https://books.google.com.tr/books?hl=en&lr=&id=-e6JfVRtLMwC&oi=fnd&pg=PA1&ots=iLJZne3UIh&sig=qDwqQH_BasNLUU0f1nLbWhlMK1g&redir_esc=y#v=onepage&q&f=false

Halupka, V., Almahr, A., Pan, Y., & Cheok, A. D. (2012). Chop chop: A sound augmented kitchen prototype. Lecture Notes in Computer Science (Including Subseries Lecture Notes in Artificial Intelligence and Lecture Notes in Bioinformatics), 7624 LNCS, 494–497. doi:10.1007/978-3-642-34292-9_43/COVER

Hilken, T., de Ruyter, K., Chylinski, M., Mahr, D., & Keeling, D. I. (2017). Augmenting the eye of the beholder: Exploring the strategic potential of augmented reality to enhance online service experiences. *Journal of the Academy of Marketing Science*, *45*(6), 884–905. doi:10.100711747-017-0541-x

Hung, Y. H., Chen, C. H., & Huang, S. W. (2017). Applying augmented reality to enhance learning: A study of different teaching materials. *Journal of Computer Assisted Learning*, *33*(3), 252–266. doi:10.1111/jcal.12173

Inamo. (2022). *Inamo London*. Inamo. http://www.inamorestaurant.com

Jagtap, S., Saxena, P., & Salonitis, K. (2021). Food 4.0: Implementation of the Augmented Reality Systems in the Food Industry. *Procedia CIRP*, *104*, 1137–1142. doi:10.1016/j.procir.2021.11.191

Johnson, L., Levine, A., Smith, R., & Stone, S. (2010). The 2010 Horizon Report. In *New Media Consortium*. New Media Consortium. http://www.nmc.org

Juan, M. C., Charco, J. L., García-García, I., & Mollá, R. (2019). An Augmented Reality App to Learn to Interpret the Nutritional Information on Labels of Real Packaged Foods. *Frontiers of Computer Science*, *1*, 1. doi:10.3389/fcomp.2019.00001

KabaQ. (2021). *Augmented Reality Food.* kabaq.io

Kakani, V., Nguyen, V. H., Kumar, B. P., Kim, H., & Pasupuleti, V. R. (2020). A critical review on computer vision and artificial intelligence in food industry. *Journal of Agriculture and Food Research*, *2*, 100033. doi:10.1016/j.jafr.2020.100033

Keogh, J. G., Dube, L., Rejeb, A., Hand, K. J., Khan, N., & Dean, K. (2020). The future food chain: digitization as an enabler of Society 5.0. *Building the Future of Food Safety Technology*, 11–38. doi:10.1016/B978-0-12-818956-6.00002-6

Kickbusch, I., Allen, L., & Franz, C. (2016). The commercial determinants of health. *The Lancet. Global Health*, *4*(12), e895–e896. doi:10.1016/S2214-109X(16)30217-0 PMID:27855860

Kounavis, C. D., Kasimati, A. E., & Zamani, E. D. (2012). Enhancing the Tourism Experience through Mobile Augmented Reality: Challenges and Prospects. *International Journal of Engineering Business Management*, *4*, 1–6. doi:10.5772/51644

Le, S., Bazger, W., Hill, A. R., & Wilcock, A. (2014). Awareness and perceptions of food safety of artisan cheese makers in Southwestern Ontario: A qualitative study. *Food Control*, *41*(1), 158–167. doi:10.1016/j.foodcont.2014.01.007

Lee, S. W. Y., Tsai, C. C., Wu, Y. T., Tsai, M. J., Liu, T. C., Hwang, F. K., Lai, C. H., Liang, J. C., Wu, H. C., & Chang, C. Y. (2011). *Internet-based Science Learning: A review of journal publications.* doi:10.1080/09500693.2010.536998

Logaldo, M. (2014). *Promoting Augmented Reality in the Food Industry: A Linguistic Analysis of Web-Based Communication.* Food and Identity in Literature and Culture.

Luque, A., Peralta, M. E., de las Heras, A., & Córdoba, A. (2017). State of the Industry 4.0 in the Andalusian food sector. *Procedia Manufacturing*, *13*, 1199–1205. doi:10.1016/j.promfg.2017.09.195

Milgram, P., & Kishino, F. (1994). A Taxanomy of Mixed Reality Visual Displays. *IEICE Transactions on Information and Systems*, *12*.http://vered.rose.utoronto.ca/people/paul_dir/IEICE94/ieice.html

Min-jing, P., & Bo, L. (2013). An Augmented Reality Based Disclosure Model of Information on Food Safety and Its Application. *International Journal of U- and e- Service. Science and Technology*, *6*(4), 69–80.

Morgan, K., & Murdoch, J. (2000). Organic vs. conventional agriculture: Knowledge, power and innovation in the food chain. *Geoforum*, *31*(2), 159–173. doi:10.1016/S0016-7185(99)00029-9

Muhammad Nizam, S. S., Lam, M. C., Arshad, H., & Suwadi, N. A. (2018). A Scoping Review on Tangible and Spatial Awareness Interaction Technique in Mobile Augmented Reality-Authoring Tool in Kitchen. *Advances in Multimedia*, *2018*, 1–14. doi:10.1155/2018/5320984

Nguyen, T. (2013). *Innovation Insight: Augmented Reality Will Become an Important Workplace Tool.* Gartner Research.

Pamuru, V., Khern-Am-Nuai, W., & Kannan, K. (2021). The Impact of an Augmented-Reality Game on Local Businesses: A Study of Pokémon Go on Restaurants. *Information Systems Research*, *32*(3), 950–966. doi:10.1287/isre.2021.1004

Paul Pairet. (2022). *uvbypp.cc*. http://paulpairet.com/

Pccw Solutions. (2022). *Pizza Hut: Enhancing Customer Engagement with Augmented Reality Game - Lenovo PCCW Solutions*. PCCW. https://www.pccwsolutions.com

Penco, L., Serravalle, F., Profumo, G., & Viassone, M. (2021). Mobile augmented reality as an internationalization tool in the "Made In Italy" food and beverage industry. *The Journal of Management and Governance*, *25*(4), 1179–1209. doi:10.100710997-020-09526-w

Phan, V. T., Seung, P., & Choo, Y. (2010). A Study on Tangible AR for interior design. *International Journal of CAD/CAM, 10*(1), 21–28. https://scholar.google.com/scholar_lookup?title=A

Porter, M. E., & Heppelmann, J. E. (2014). How Smart, Connected Products Are Transforming Competition - Article - Faculty & Research - Harvard Business School. *Harward Business Review, 92*(11), 64–88. https://www.hbs.edu/faculty/Pages/item.aspx?num=48195

Poushneh, A., & Vasquez-Parraga, A. Z. (2017). Discernible impact of augmented reality on retail customer's experience, satisfaction and willingness to buy. *Journal of Retailing and Consumer Services*, *34*, 229–234. doi:10.1016/j.jretconser.2016.10.005

Rauschnabel, P. A., Babin, B. J., tom Dieck, M. C., Krey, N., & Jung, T. (2022). What is augmented reality marketing? Its definition, complexity, and future. *Journal of Business Research*, *142*, 1140–1150. doi:10.1016/j.jbusres.2021.12.084

Rejeb, A., Rejeb, K., & Keogh, J. G. (2021). *Enablers of Augmented Reality in the Food Supply Chain: A Systematic Literature Review.* doi:10.1080/15378020.2020.1859973

Santos, K. (2019). *Domino's ordering app to acquire further customer insight with new augmented realityfeature*. QSR Media Australia. https://qsrmedia.com.au/technology/news/dominos-ordering-app-acquire-further-customer-insight-new-augmented-reality-feature

Shabani, N., Munir, A., & Hassan, A. (2019). E-Marketing via Augmented Reality: A Case Study in the Tourism and Hospitality Industry. *IEEE Potentials*, *38*(1), 43–47. doi:10.1109/MPOT.2018.2850598

Smith, P. R., & Zook, Z. (2016). *Marketing Communications: Offline and Online Integration, Engagement and Analytics*. Kogan Page.

Todorović, V., Milić, N., & Lazarević, M. (2019). Augmented Reality in Food production traceability - Use case. *EUROCON 2019 - 18th International Conference on Smart Technologies*. IEEE10.1109/EUROCON.2019.8861734

Tuncer, I. (2020). Customer Experience in the Restaurant Industry: Use of Smart Technologies. Handbook of Research on Smart Technology Applications in the Tourism Industry, 254–272. doi:10.4018/978-1-7998-1989-9.ch012

Tzima, S., Styliaras, G., Zaragas, H., & Bassounas, A. (2018). The Effects of an Educational Augmented Reality Application on the Interpretation and Communication of Cultural Assets: The "Greek Type" Watermill's Case. *Proceedings of EdMedia: World Conference on Educational Media and Technology*, (pp. 909–925).

Vanderroost, M., Ragaert, P., Verwaeren, J., De Meulenaer, B., De Baets, B., & Devlieghere, F. (2017). The digitization of a food package's life cycle: Existing and emerging computer systems in the logistics and post-logistics phase. *Computers in Industry*, *87*, 15–30. doi:10.1016/j.compind.2017.01.004

Veebrant. (2022). *Dining at Sublimotion in Ibiza Is a Mind-Blowing Multisensory Experience*. veebrant.com.

Vignali, G., Bertolini, M., Bottani, E., Di Donato, L., Ferraro, A., & Longo, F. (2018). Design and testing of an augmented reality solution to enhance operator safety in the food industry. *International Journal of Food Engineering*, *14*(2). doi:10.1515/ijfe-2017-0122

Wang, Q. J., Mielby, L. A., Thybo, A. K., Bertelsen, A. S., Kidmose, U., Spence, C., & Byrne, D. V. (2019). Sweeter together? Assessing the combined influence of product-related and contextual factors on perceived sweetness of fruit beverages. *Journal of Sensory Studies*, *34*(3), e12492. doi:10.1111/joss.12492

Weking, A. N., Suyoto, & Santoso, A. J. (2020). *A Development of Augmented Reality Mobile Application to Promote the Traditional Indonesian Food.*

Wilkins, K., & Mackenzie, S. G. (2007). Work injuries. *Health Reports*, *18*(3), 25–42. https://pubmed.ncbi.nlm.nih.gov/17892250/ PMID:17892250

Wu, H. K., Lee, S. W. Y., Chang, H. Y., & Liang, J. C. (2013). Current status, opportunities and challenges of augmented reality in education. *Computers & Education*, *62*, 41–49. doi:10.1016/j.compedu.2012.10.024

Xie, J., Chai, J. J. K., O'Sullivan, C., & Xu, J. L. (2022). Trends of Augmented Reality for Agri-Food Applications. *Sensors (Basel)*, *22*(8333), 8333. doi:10.339022218333 PMID:36366030

Yim, M. Y. C., & Yoo, C. Y. (2022). Are Digital Menus Really Better than Traditional Menus? The Mediating Role of Consumption Visions and Menu Enjoyment. doi:10.1016/j.intmar.2020.01.001

KEY TERMS AND DEFINITIONS

Augmented Reality: A tool that offers richer and more interactive experiences using technology that combines the virtual and real worlds.

Digitalization: It is the process of replacing traditional methods in the food industry with technology-oriented and digital solutions.

Food Industry: The food and beverage industry that includes processes such as production, processing, distribution, and consumption.

Food Marketing: It is the process of promoting, branding, and delivering food and beverage products to the consumer.

Food safety: Measures and practices that ensure that food conforms to health standards and is delivered safely to consumers.

Mobile Applications: They are software used on mobile devices such as smartphones and tablets and can be used for various purposes in the food industry.

Technology: Various innovative tools and methods used in the food industry.

Chapter 10
Tracing the Future: Blockchain and IoT's Role in Revolutionizing Food Supply Chain Transparency

Swapnil Morande
https://orcid.org/0000-0003-4720-2980
University of Naples, Italy

Tahseen Arshi
https://orcid.org/0000-0002-5244-7862
American University of Ras Al Khaimah, UAE

Kanwal Gul
University of Naples, Italy

Mitra Amini
Bennett University, India

Veena Tewari
https://orcid.org/0000-0003-4025-9909
University of Technology and Applied Sciences, Omanv

ABSTRACT

The chapter presents a comprehensive evaluation of the transformative role of blockchain and internet of things (IoT) in augmenting food supply chain transparency, primarily focusing on inventory management and food safety tracking. The global food supply chain, fraught with myriad challenges such as economic adulteration, microbial pathogens, and non-permitted additives, necessitates the integration of cutting-edge technology to ensure safety and sustainability. Blockchain and IoT, with their inherent qualities of traceability and transparency, can revolutionize the food industry by enabling real-time tracking, fostering responsible sourcing, minimizing waste, and bolstering consumer satisfaction. The chapter underscores these advantages while highlighting some existing challenges within the food supply chain and how these technologies can address them.

DOI: 10.4018/978-1-6684-9094-5.ch010

INTRODUCTION

In an increasingly interconnected world, food safety is a global concern that transcends borders and cultures. Our global food supply chain is more complex than ever before, with food items sourced from different parts of the world to meet the growing demands of a booming population, projected to reach 9.8 billion by 2050 (United Nations, 2017). This complexity, while necessary, presents an array of food safety issues that require focused management strategies. From microbial pathogens to heavy metals, undeclared allergens, and foreign material contamination, the range of potential food safety hazards is broad (Djekic, Jankovic and Rajkovic, 2017). Consumers today, more than ever, are aware and concerned about what goes into their food. Issues such as non-permitted additives, avian influenza, and the use of biotechnology have raised regulatory and genuine consumer concerns (Pigłowski, 2020). These incidents underline the need for enhanced transparency and traceability in the food supply chain. Although bodies like FDA (Food and Drug Administration) are in place to ensure food safety, we are still far from achieving the desired level of food safety assurance. The food supply chain is a series of processes, operations, and entities that take the food from its raw material state to our plates (Haji *et al.*, 2020; Abideen *et al.*, 2021).

Pursuant to Nosratabadi *et al.* (2020) this chain includes producers, processors, retailers and distributors, consumers, and the hospitality industry. Each segment of this chain has its challenges and potential for improvement. For instance, producers face increasing uncertainties due to changing climatic weather patterns, scarcity of water, and soil degradation caused by industrialization and urbanization. In contrast, food processors need to reduce waste and extend the shelf life of their products while maintaining consumer expectations. Regardless of whether they are organized or unorganized, retailers and distributors have to manage inventory, waste, and technology. There is a dire need for practical solutions that can help in managing these issues while ensuring 'auditability' in the food supply chain. In such a scenario the advent of technology, specifically the Internet of Things (IoT) and blockchain, can be significant in revolutionizing the food supply chain (Kaur *et al.*, 2022). The Internet of Things refers to a system where common objects have identifying, sensing, networking and processing abilities allowing them to communicate with each other and other devices over the Internet to achieve a goal (Sethi and Sarangi, 2017).

Integrating this technology into the food supply chain gives companies visibility into what is happening upstream in the supply chain and communicates this information both internally and externally (Tan and Sidhu, 2022). Furthermore, IoT can enable traceability in the food supply chain, i.e., using technology to trace each raw material in a product, how the product is used and where it is discarded. Employing IoT to track food products in the supply chain can significantly enhance provenance, supporting sustainability by providing visibility of inputs and source information to trace its origin and enabling comprehensive tracking of environmental, economic and social consequences of agricultural production (Dutta *et al.*, 2020). While traceability allows companies to 'know' what is happening in the supply chain, transparency allows them to 'show' this information to stakeholders, including consumers. To achieve sustainability in the food supply chain, this chapter explores how IoT and blockchain technology are being used for inventory management and food safety tracking.

Background

There are several critical challenges facing the global food supply chain today, including food safety, inventory management, and sustainability. This review delves into the role of emerging technologies such as Blockchain and the Internet of Things (IoT) in transforming the food supply chain to improve safety, sustainability, and transparency (Kumar *et al.*, 2023).

Food Safety Issues

Food safety, undeniably a multi-layered concern, encompasses a broad range of potential hazards that could compromise the quality and safety of the food we consume daily. These hazards not only include microbial pathogens and undeclared allergens, but also heavy metals and contamination from foreign material, each of which could seriously impair the health of the consumer (Young and Waddell, 2016). With the presence of such dangers, the exigency for the establishment of rigorous food safety management strategies cannot be overstated. In recent years, trends in food safety have started to lean towards increased use of automation in the food processing stages, the promotion of healthier dietary options, an enhanced focus on sustainability, and a heightened emphasis on overall food safety (Knorr, Augustin and Tiwari, 2020).

These trends have largely been molded by a triad of factors - consumer demands, environmental impacts, and the economic standing of nations. Consumer demands, for instance, have evolved in response to the growing awareness of the health implications of food choices, leading to a call for healthier, safer, and more sustainably produced food options (Miller *et al.*, 2021). Concurrently, the advent of climate change and the environmental toll of food production processes has spurred the industry towards more sustainable practices. Lastly, the economic status of nations influences food safety regulations and the feasibility of implementing innovative food safety strategies.

Food Supply Chain Dynamics

The dynamic nature of the food supply chain poses numerous challenges and calls for efficient and effective supply chain strategies that ensure the food reaches the consumer in a safe and timely manner. Simultaneously, In 2023, the food market's revenue is reported to be US$9.24 trillion, with a Compound Annual Growth Rate (CAGR) of 7.25% expected from 2023 to 2028 (Statista, 2023). This growth, however, is not without its trials. Due to the increasingly globalized nature of the food supply chain, food safety needs to be addressed from a variety of perspectives.

Manufacturers are also facing stricter regulatory environments as governments aim to reduce foodborne illnesses and ensure the overall safety of the food supply. The manufacturing industry is finding innovative ways to protect the supply chain despite these challenges, using various solutions, such as advanced traceability systems and strict quality control measures (Ayan, Güner and Son-Turan, 2022). Despite this, there are still many obstacles to overcome to create a completely safe and transparent food supply chain. With Blockchain and the Internet of Things (IoT), we can achieve an unprecedented level of transparency, traceability, and accountability (Iftekhar *et al.*, 2020).

Lack of Transparency and Traceability

The food supply chain faces systemic issues including lack of transparency and traceability, quality and safety concerns, organized crime and forced labor influences, and poor communication across channels (Bonanni, 2010; Haji *et al.*, 2020). The food supply chain is being made more transparent using state-of-the-art technologies.

Such new technologies, including Blockchain and IoT, offer promising solutions to persistent problems often faced in the food sector. These technologies can enhance traceability and transparency, address safety risks, and mitigate criminal influence. Even though they pose implementation challenges, ensuring safety, sustainability, and trust makes them attractive.

Role of IoT

The Internet of Things (IoT) represents an exciting technological breakthrough that may offer solutions to the issues affecting the food supply chain. This cutting-edge technology is characterized as a system where hardware/devices possess the ability to identify, sense, connect, and process information to achieve particular objectives (Nižetić *et al.*, 2020). Such devices hold the promise of substantially enhancing supply chain transparency. In addition to allowing real-time monitoring and data gathering, the Internet of Things (IoT) offers an opportunity to improve decision-making processes within the supply chain through the linking of various devices.

Implementing IoT in the food supply chain includes the connection of IoT devices that can gather data throughout the food's journey, From the point of production to the point of consumption. Continuous monitoring can minimize errors, reduce waste, and enhance food safety. As per Abdul Zahra *et al.* (2022) n the supply chain, IoT has a profound impact

Traceability and IoT

Traceability, a procedure that leverages technology to monitor every raw material involved in a product and oversee its lifecycle, can be profoundly enhanced by IoT (Kaur *et al.*, 2022). Through IoT, real-time tracking of food products in the supply chain, from the source to the consumer, becomes a viable reality. This capability provides invaluable data that can be leveraged to improve food safety measures, manage inventory more efficiently, and ultimately reduce food waste.

Imagine an IoT-powered traceability system that can immediately alert stakeholders when a batch of products has been exposed to harmful conditions or contaminants. Such timely information can prevent the spread of foodborne diseases, saving lives and reducing costs. Furthermore, this data can also be used to optimize logistics, ensuring that perishable items are transported and stored efficiently, thus maintaining their freshness and quality (Zhang, Zhao and Qian, 2017; Cil, Abdurahman and Cil, 2022).

Traceability and Sustainability

Agricultural production's environmental, economic, and social impacts can be assessed easily through traceability by showing the inputs used in food production, their origins, and their environmental, economic, and social impacts. Food production's adverse environmental and societal impacts can be effectively understood and mitigated through this level of transparency. With traceability and transparency at the helm, we can move towards a more sustainable, safe, and efficient food supply chain. Providing detailed insights about a product's lifecycle and associated impacts, traceability provides the knowledge base, whereas transparency enables manufacturers and consumers to demonstrate these insights and make informed choices (Rejeb *et al.*, 2022).

Thus, these twin pillars of traceability and transparency, supported by technologies like IoT and blockchain, are critical to reshaping our food systems into a model of sustainable consumption and production.

Benefits of IoT in the Food Supply Chain

The integration of Internet of Things (IoT) technology in the food supply chain brings with it a slew of advantageous features that could drastically improve the state of food security and supply chain efficiency. One of the key benefits of IoT lies in its capacity to enhance product safety. IoT technology enables real-time monitoring of food items, facilitating early detection of contamination which, in turn, mitigates the risk of foodborne diseases (Osmólska, Stoma and Starek-Wójcicka, 2022). This early detection system can save invaluable time, resources, and most importantly, lives. Beyond safety, IoT aids in streamlining inventory tracking, providing insights into real-time stock levels, and enhancing demand forecasting, thus reducing overproduction and food waste. Additionally, it aids in ensuring responsible sourcing, creating an avenue for increased consumer satisfaction as people become more discerning about the provenance and sustainability of their food (Shelake *et al.*, 2020). Moreover, the role of IoT extends to the arena of environmental sustainability. By minimizing waste through more efficient logistics and enabling smarter farming techniques, IoT can significantly reduce the carbon footprint of the food supply chain. Consequently, the integration of IoT into the food supply chain serves the dual purpose of bolstering economic efficiency and promoting environmental sustainability.

Blockchain: An Emerging Solution

While the IoT has provided significant advancements in food supply chain management, the potential to further revolutionize this sector lies in another emerging technology - blockchain. This technology, a decentralized and distributed ledger system, has the capacity to provide an immutable, transparent record of all transactions within the food supply chain (Pandey, Pant and Snasel, 2022). The inherent nature of blockchain, with its tamper-proof records and transparency, enhances traceability beyond the capabilities of traditional systems (Gad *et al.*, 2022). Each product can be tracked from farm to fork, and each transaction along its journey is recorded and verified on the blockchain. This level of traceability can drastically reduce fraud and misrepresentation in the food industry, a problem of increasing concern in our globalized markets (Kshetri, 2018).

Blockchain technology can enhance supply chain performance by eliminating intermediaries, reducing costs and enabling efficient transactions, optimizing processes for a safer, more efficient and transparent food system (Tokkozhina, Martins and Ferreira, 2022). IoT and blockchain together present an opportunity to address pressing challenges, increasing efficiency, reducing waste, improving safety and promoting sustainability through synergistic effects. As we explore the intricacies of the food supply chain, IoT and blockchain emerge as transformative technologies capable of enhancing transparency, traceability and sustainability. IoT data collection and blockchain record-keeping can build trust and transparency in the global food system. These technologies can address the lack of accountability, improving food safety and reducing foodborne disease risk. Integrating IoT and blockchain also supports sustainability, allowing stakeholders to understand the environmental, economic and social impacts of food production, leading to sustainable consumption and production (Tsoukas *et al.*, 2022). IoT and blockchain have immense potential to revolutionize the food supply chain, making it more transparent, traceable and sustainable to meet escalating food demand and future challenges. IoT-enabled traceability specifically supports sustainability through responsible sourcing, tracking impacts and enabling sustainable consumer choices. It provides visibility of inputs and their origin, enabling comprehensive tracking of environmental, economic and social consequences of agricultural production. It empowers consumers with information to make sustainable choices, driving demand for responsibly and sustainably produced food (Yadav, Luthra and Garg, 2023).

Thus, blockchain and IoT are transformative technologies enhancing transparency, traceability and sustainability in the food supply chain. Blockchain improves supply chain performance by eliminating intermediaries and optimizing processes. Together, IoT and blockchain address pressing challenges by increasing efficiency, reducing waste, improving safety and promoting sustainability (Rana *et al.*, 2021). First, IoT-enabled traceability supports sustainability through responsible sourcing, tracking impacts and sustainable consumer choices. IoT data and blockchain records build trust and transparency in the global food system. Second, it allows for comprehensive tracking of environmental, economic health, and social consequences of agricultural production, thus aiding in the assessment and mitigation of the supply chain's environmental impact. However, the adoption of these technologies in the food supply chain is not without challenges. For instance, technological transformations need to be implemented in both developed countries, where giant retailer chains dominate the market, and in developing countries, where the sector is unorganized or less organized (Altenburg *et al.*, 2016; Jerath, Sajeesh and John Zhang, 2016). This requires a tailored approach, taking into account the unique characteristics and constraints of different markets. While challenges exist, the benefits that these technologies offer in terms of inventory management, food safety tracking, and sustainability make them a promising solution for the future of food safety.

Problem Statement

Disturbing instances of food fraud like the deliberate adulteration of food with harmful substances, as seen in the melamine scandal in China, underscore the dire need for improved traceability and transparency within the food supply chain (Rana *et al.*, 2021). It is still difficult to manage inventory, manage waste in the food supply chain, and communicate effectively between these channels despite the existence of regulatory bodies (Sharma *et al.*, 2020). In addition, many food products can't be traced from farm to fork, posing a considerable threat to the safety, quality, and sustainability of those products.

Furthermore, the impact of organized crime, forced labor, and pandemics compounds these challenges, revealing a pressing need for technological transformation within the food supply chain. This calls for the incorporation of advanced technologies like the Internet of Things (IoT) and blockchain to revolutionize transparency, traceability, and thereby ensuring food safety (Spink *et al.*, 2017; Momtaz, Bubli and Khan, 2023). Considering the substantial challenges in the global food supply chain, this research asks: *How can the integration of IoT and blockchain technologies revolutionize the supply chain's efficiency, sustainability, and transparency, specifically improving food safety, traceability, and inter-channel communication?*

This inquiry aims to guide future strategic planning and policy development in the food supply industry.

Approach

The research draws on literature related to food supply chain challenges, blockchain, IoT applications and case studies (Duan *et al.*, 2020). It evaluates how data collection with IoT and data management with blockchain can improve inventory tracking, food safety and sustainability. The potential benefits and challenges of the technologies are also discussed. The literature review was conducted using academic databases and keywords such as "food supply chain," "food safety," "IoT," "blockchain," "traceability," and "sustainability" will be used. The literature review included academic articles, books, and reports published in the last ten years. The review also encompassed case studies of the successful implementation of IoT and blockchain technologies in the food supply chain (van Hilten, Ongena and Ravesteijn, 2020).

Applications

Blockchain technology, specifically the Hyperledger framework, can be deployed to facilitate supply chain transparency in the food sector. Hyperledger, an open-source collaborative effort to advance cross-industry blockchain technologies, can be used to develop an application that records and verifies transactions in the food supply chain. Smart contracts, a key feature of blockchain technology, can automate the execution of transactions based on pre-set rules and conditions (van Hilten, Ongena and Ravesteijn, 2020). This can facilitate verifiability by providing an immutable and tamper-proof record of all transactions.

In accordance with Iftekhar *et al.* (2021) Hyperledger's modular architecture allows for the integration of different components like Certificate Authority (CA), Peers, Ordering Service, and Channels, each with their specific roles in blockchain deployment (refer to Figure 1).

Figure 1. Environment for distributed ledger Blockchain platform
(Credit: IBM 2022)

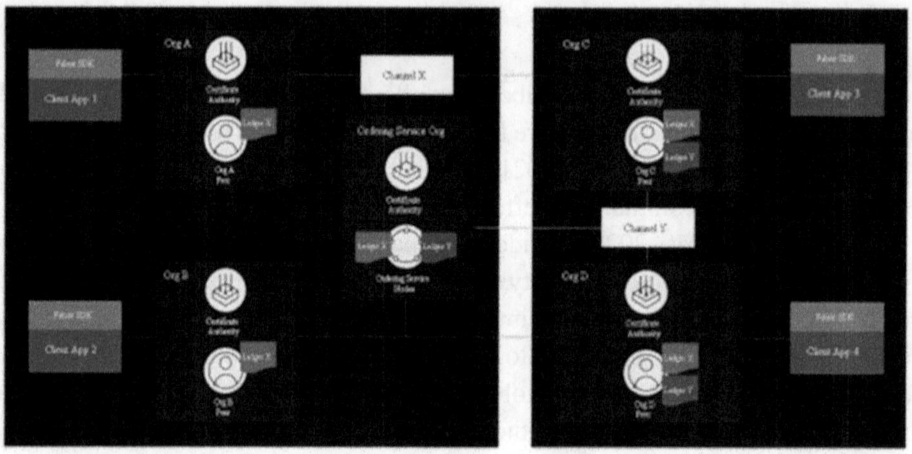

The Certificate Authority (CA) is responsible for issuing and managing digital identities, crucial for establishing trust among participants. Peers, nodes in the network, endorse transactions, maintain the ledger, and run chaincode (smart contracts). The Ordering Service batches transactions into blocks, creating a single chain of blocks, ensuring the consistency of transactions. Channels are private communication mechanisms that allow only specific network members to interact and share a private ledger. These components together ensure a secure, private, and efficient blockchain network that can handle the complex demands of a food supply chain (IBM, 2020, 2022).

By implementing blockchain using Fabric Components as per Figure 2, the food sector can address several critical issues. Product safety can be ensured with the ability to track and trace food items from their origin to the consumer. Inventory tracking can be optimized, reducing waste and inefficiencies.

Figure 2. Hyperledger fabric components
(Credit: IBM 2022)

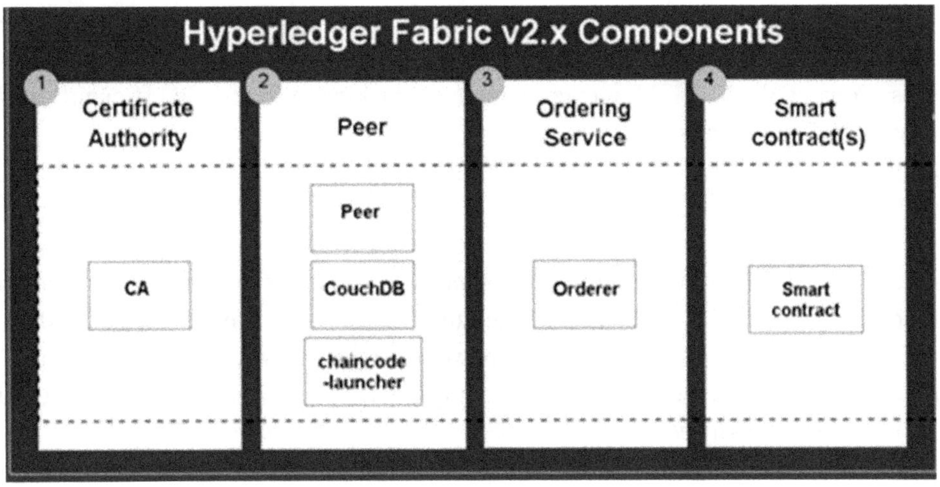

Responsible sourcing can be validated, promoting fair trade and ethical practices. Consumer satisfaction can be improved with transparency, fostering trust and confidence. Minimal waste and carbon footprint can be achieved by optimizing resources and eliminating unnecessary processes. Ultimately, the deployment of blockchain technology can revolutionize the food sector by addressing traceability and transparency in the food supply chain (Sekuloska and Erceg, 2022).

As the food supply chain becomes more transparent and traceable, Blockchain technology and the Internet of Things (IoT) offer transformative solutions (Kumar *et al.*, 2022). As the global population grows, these technologies will be critical to improving food safety and sustainability across the food supply chain. Integrating the Internet of Things and blockchain in the food supply chain addresses traceability, transparency and sustainability; IoT monitors production, processing and distribution, ensuring safety and quality, blockchain makes data immutable, transparent and verifiable, enhancing traceability, addressing fraud; IoT and blockchain promote responsible sourcing and sustainability, IoT monitors impacts, blockchain provides transparency for ethical and sustainable food; implementing these technologies faces adoption, regulatory and technical challenges.

IMPLEMENTATION

IoT for Data Collection

Agriculture and Harvesting

Smart farming or precision agriculture involves the integration of IoT technologies into farming practices. IoT sensors help farmers make informed decisions by monitoring crucial factors like soil moisture, temperature, and nutrient levels, which in turn maximize crop yield and reduce waste (Wolfert *et al.*, 2017). IoT can also enhance harvesting through automated harvesters that use real-time data to improve efficiency and minimize spoilage.

Processing and Packaging

IoT plays a significant role in food processing and packaging as well. Sensors can monitor temperature and humidity in real-time during food processing and storage, ensuring optimal conditions and mitigating spoilage risks (Popa *et al.*, 2019). Moreover, automated packing systems can efficiently sort, package, and label food items, reducing errors.

Transportation and Retail

Sensors on transport vehicles offer real-time data about the food items' condition and location, allowing for quick responses to potential issues like temperature fluctuations (Verdouw *et al.*, 2016; Verdouw, Wolfert and Tekinerdogan, 2016). In retail, IoT-powered smart shelves automatically track inventory, expiration dates, and even customer interactions, leading to effective inventory management and enhanced customer service (Mashayekhy *et al.*, 2022).

Blockchain for Data Management

Supply Chain Smart Contracts

As conveyed earlier programmable agreements with self-executing provisions, known as smart contracts, are crucial to the blockchain (Taherdoost, 2023). In the food supply chain, smart contracts can mechanize procedures such as payments, validating proper temperatures during transportation, and transferring ownership. These actions execute automatically when predetermined conditions are fulfilled, boosting productivity and confidence.

Food Item Traceability

Blockchain's decentralized and unchangeable ledger is ideal for traceability. Recording each transaction or event in the supply chain on the blockchain results in a transparent and unchangeable record (Agrawal *et al.*, 2021). This permits absolute traceability from farm to fork, allowing rapid, accurate responses to issues such as foodborne disease outbreaks and reducing recall ranges.

Case Studies: Blockchain and IoT in the Food Supply Chain

Illustrating the potential of these technologies, several organizations have launched projects that utilize the blockchain to trace produce from farms to stores, ensuring quality and sustainability for customers (Difrancesco, Meena and Kumar, 2022). These case studies offer practical insights into implementing these technologies and their impacts.

Case Study 1: AgriDigital

In the contemporary digital landscape, the implementation of advanced technologies like Blockchain and the Internet of Things (IoT) in various sectors has revolutionized traditional operational mechanisms, driving efficiency, transparency, and sustainability. Particularly within the food supply chain, these technologies have immense potential to offer transformative solutions to age-old challenges (Kamble, Gunasekaran and Arha, 2019). A compelling case that echoes this technological shift is that of AgriDigital, an agricultural commodity management platform that employs blockchain technology for digitizing farming procedures and agricultural goods.

The agricultural industry, despite being the linchpin of our survival, has been slower in adopting digital record-keeping compared to other sectors. Predominantly relying on paperwork records to track crucial data, such as pharmaceuticals used on animals, the sector has been grappling with issues related to inefficiency, errors, and lack of transparency. However, companies like AgriDigital are setting the stage for a disruptive transformation, pivoting the industry towards digital practices (Xu *et al.*, 2019). The pivot centers on blockchain technology - a decentralized, immutable ledger that holds the promise of solving critical supply chain issues by introducing a layer of trust, transparency, and traceability. The blockchain-powered platform developed by AgriDigital records each step in the agricultural supply chain, from planting seeds to the shipment of the final product. By doing so, it facilitates a seamless, reliable traceability system, which ensures that the provenance of the product is readily available to customers. Moreover, blockchain technology helps establish the verification of the organic status of the products.

In a world increasingly concerned about sustainability and health, the organic status of food products is of paramount importance to consumers. Herein, the decentralized ledger of blockchain can hold immutable records of a product's lifecycle, indicating the use (or non-use) of pesticides, fertilizers, and pharmaceuticals. Hence, AgriDigital's implementation of blockchain not only satisfies the consumer's desire for organic products but also strengthens their trust in the food supply chain (Pufahl *et al.*, 2021).

The incorporation of IoT further amplifies the potential of blockchain within this context. IoT can provide real-time data about agricultural practices, such as soil quality, irrigation, weather patterns, and pest activity. Such data, when recorded on the blockchain, can validate the quality and organic nature of the products at each stage (Javaid *et al.*, 2022). The amalgamation of blockchain and IoT, thus, enhances the transparency and accuracy of information, revolutionizing the transparency in the food supply chain. The intersection of blockchain and IoT represents a paradigm shift in the agricultural sector, bringing unprecedented levels of transparency, traceability, and efficiency. The AgriDigital case study substantiates the promise of these technologies in advancing a sustainable food supply chain, a feat that has significant implications for not just the farming community, but also the consumers and the planet at large. Therefore, as we trace the future, the proliferation of such technological interventions becomes a pivotal necessity for creating an equitable, sustainable, and transparent food supply chain (Tönnissen and Teuteberg, 2020).

Case Study 2: Techrock

A transformative development currently revolutionizing various sectors, including the food supply chain, is the integration of the Internet of Things (IoT) and blockchain technology. This potent fusion has created unprecedented opportunities to foster transparency, efficiency, and safety, thereby mitigating numerous entrenched issues (Kshetri, 2018). A case study that underscores this technological shift is that of Techrock, a Beijing-based firm committed to ensuring the authenticity of infant formula products through the innovative utilization of smart packaging and blockchain technology (Rogerson and Parry, 2020). Infant formula is a highly sensitive commodity, given its direct impact on infant health. Issues such as product counterfeiting and contamination with harmful substances like melamine pose serious threats to infant safety (Gossner *et al.*, 2009). Owing to the same Techrock has devised a unique solution that relies on Blockchain to counter these threats.

It employs Hyperledger-based blockchain technology, a permissioned and scalable distributed ledger system. It controls each infant formula product with a unique verification tag. This innovative use of blockchain enables secure, transparent, and verifiable imprints of each product's journey from origin to consumer. Counterfeiting of infant formula is made easier with traceability (Shi *et al.*, 2020; Ahmed and MacCarthy, 2021).

The concept of smart packaging is integrated as an IoT solution. Each product has a tag consumers can verify with a smartphone. This simple yet potent combination of blockchain and IoT allows consumers to check a product's authenticity and track its supply chain journey. Blockchain provides an immutable record of a product's path to consumers. IoT tags allow consumers to access a product's blockchain journey record with a smartphone scan (Uddin *et al.*, 2021). Together, they combat counterfeiting and build consumer trust through transparency. In addition to ensuring product authenticity, the system also safeguards consumers from potential health hazards associated with counterfeit or contaminated products.

Beyond assuring parents, Techrock's application of blockchain and IoT presents a model for the broader food supply chain. The paradigm demonstrated here has vast potential, with implications stretching across various food sectors, especially those dealing with sensitive products. By instilling a high degree of confidence, it significantly contributes to the promotion of food safety and sustainability. Thus, the fusion of blockchain and IoT, as embodied by Techrock, offers a sustainable and efficient solution to perennial issues plaguing the food supply chain. It illustrates the considerable strides that these technologies can facilitate toward the creation of a safer, more transparent, and trustworthy food ecosystem. Thus, as we trace the future of the food supply chain, it is evident that the integration of such technological solutions will become increasingly indispensable.

Potential Benefits

Reduction in Food Fraud

Food fraud, including mislabeling and adulteration, is a significant concern in the food industry. With the integration of blockchain and IoT, food fraud can be effectively mitigated. IoT devices can collect data about the condition and location of food items, and this data can be securely stored on the blockchain. The resulting transparency can help expose fraudulent activities and assure consumers of the authenticity and quality of their food (Ismail *et al.*, 2023).

Greater Efficiency and Sustainability

The convergence achieved with blockchain and IoT results in increased efficiency in the food supply chain. The automated nature of these technologies gets rid of manual record-keeping and hence eliminates associated errors (Javaid *et al.*, 2021). Additionally, smart contracts on the blockchain can automate transactions based using 'Smart contracts', accelerating processes and reducing inefficiencies. It is also a valuable tool for identifying waste and inefficiencies, which will contribute to more sustainable practices.

Empowering Consumers With Information

In a world where consumers are increasingly concerned about the source and safety of their food, the utility of the Internet of Things and blockchain together can play a critical role (Astill *et al.*, 2019). It can provide reliable information to consumers about the source and safety of food they are consuming.

Consumers can access a wealth of information via QR codes or similar identifiers linked to blockchains about their food, such as its origin, journey through the supply chain, and any certifications associated with it (like organic food). This transparency culminates in the strengthening of the consumer's relationship with food providers and Boosts their confidence in what they eat (Wu *et al.*, 2021; Pozelli Sabio and Spers, 2022).

Potential Challenges and Limitations

Technical Challenges

Implementing blockchain and IoT in the food supply chain is not without its technical hurdles. Blockchain's computational demands, particularly in Proof of Work algorithms, can lead to significant energy consumption (Awan *et al.*, 2021). Moreover, the integration of IoT devices requires robust, reliable connectivity, and potential bandwidth limitations can pose challenges to data transmission (R. L. Kumar *et al.*, 2022). Additionally, managing the massive data generated by IoT devices in a blockchain framework may necessitate advanced data-handling solutions (Maftei *et al.*, 2023).

Regulatory and Legal Considerations

The decentralization of blockchain and the widespread use of IoT devices presents unique legal and regulatory challenges (Bodó, Gervais and Quintais, 2018). Legal uncertainties exist around data ownership, smart contract validity, and jurisdiction for decentralized networks. Regulators may face difficulties in creating and enforcing regulations that can keep up with such rapidly evolving technology (Raskin, 2016).

Privacy and Security Concerns

Despite the inherent security of blockchain and encrypted data transmission in IoT, concerns remain. Blockchain's transparency could unintentionally reveal sensitive business information, while IoT devices could be vulnerable to cyber-attacks, potentially compromising the integrity of the data collected (Christidis and Devetsikiotis, 2016). Privacy is also a concern, as the detailed data collected by IoT devices must be carefully managed to prevent misuse (Peng *et al.*, 2021; Knijnenburg *et al.*, 2022).

RECOMMENDATIONS

As the world's population is projected to reach nearly 10 billion by 2050, ensuring food safety and security is increasingly critical. It is possible to address many challenges associated with food production, processing, and distribution by integrating Blockchain and the Internet of Things.

Blockchain can provide a perfectly immutable, secure, decentralized, and transparent platform for ongoing and completed food supply chain transactions. Food product information, such as their origin, processing, storage, and transportation conditions, can be obtained and verified by the stakeholders such as farmers, processors, distributors, retailers, and consumers. Blockchain, for example, could have prevented consumers from eating pet food contaminated with melamine. A second benefit of the Internet of Things is that it lets us monitor and control food safety in real time. IoT devices like sensors and RFID tags can monitor conditions like temperature and humidity, track food products, and detect pathogens or contaminants. This data informs decision-making to improve food safety, inventory management, and waste reduction.

For instance, IoT sensors can detect when perishable goods are nearly expired, prompting retailers to discount or donate them to minimize waste. Third, integrating Blockchain and IoT amplifies the benefits of each technology. IoT devices can collect and share data, such as a product's origin or transportation conditions, on the Blockchain. This combination makes the data transparent, traceable, and unchangeable. For example, IoT devices could monitor poultry health, storing the data on Blockchain. This would allow quick detection and containment of affected poultry, preventing disease spread and protecting consumers. Fourth, successful implementation requires collaboration among stakeholders, including governments, NGOs, tech providers, food producers, processors, and retailers. Standards for data collection, storage, and sharing must be developed to ensure Blockchain information is reliable and IoT devices are compatible. Regulatory agencies like the FDA could help set and oversee standards.

Finally, consumers drive demand for safe and sustainable food. Educating consumers about how Blockchain and IoT improve food safety and sustainability can encourage them to support businesses using these technologies.

FUTURE RESEARCH DIRECTIONS

While Blockchain and the Internet of Things show significant promise for transforming the food supply chain, more research is required to fully achieve this potential. Several key areas must be the focus of future research.

One key area is developing compatible standards and protocols for IoT devices and Blockchain systems in the food supply chain. As noted previously, the IoT allows everyday objects to communicate over the Internet, but they can only do so effectively if they can understand each other regardless of their origin or manufacturer. Currently, a lack of compatibility leads to missed opportunities and inefficiencies. Future research should focus on creating universal standards for IoT and Blockchain in the food supply chain, similar to web standards. Integrating IoT and Blockchain also presents unique security and privacy challenges. While Blockchain is inherently secure, IoT devices are often vulnerable to cyberattacks, which could introduce vulnerabilities. Research is needed to develop security measures that ensure data integrity and confidentiality in IoT-Blockchain systems.

Advanced analytics and artificial intelligence could provide useful insights from the data IoT devices generate in the food supply chain. For instance, they could help identify inefficiencies, predict supply chain disruptions, and improve decision-making. However, research is required to develop effective analytics and AI for the complex food supply chain. The societal and economic impacts of IoT and Blockchain in the food supply chain should also be studied, including effects on jobs, inequality, consumer trust, and perceptions of food safety.

Safe food is a basic human right, so technologies that improve food safety must be thoroughly researched and their benefits communicated. While IoT and Blockchain have the potential to revolutionize the food supply chain, research is needed to address key challenges and opportunities. By focusing on these areas, researchers can help ensure these technologies enhance not only food safety and traceability but also sustainability and social equity.

CONCLUSION

The integration of Blockchain and IoT technologies has immense potential to revolutionize the food supply chain by enhancing traceability, transparency, and sustainability. As discussed, these technologies can enable real-time monitoring, reduce food waste, and improve food safety and inventory management. However, there are substantial hurdles that must be overcome for their full benefits to materialize.

First, organizational, and behavioral challenges pose barriers to implementation. Food companies must adapt to new systems and processes, which can be difficult. Consumers must also trust these technologies, demanding education, and outreach. Second, legal and regulatory frameworks have been slow to adapt to Blockchain and IoT, creating uncertainty around data ownership, privacy, and liabilities. Governments will need to develop new policies that account for these technologies. Third, the technologies face economic hurdles, such as high costs for deployment, integration, and data management.

Going forward, research should focus on a few priority areas. Compatibility standards for IoT devices and Blockchain systems must be developed to facilitate integration. Security measures that ensure data integrity and privacy must also be implemented. Advanced analytics can provide insights from the vast data generated but requires further research. Finally, the societal impacts of these technologies, such as their effects on jobs and inequality, should be studied to ensure their benefits are widely shared.

While new technologies offer hope of a safer, sustainable, and transparent food system, overcoming hurdles to their adoption will be challenging. Addressing key issues through research, new policies and industry collaboration can help maximize their benefits for all stakeholders and the planet. As the need for a resilient food supply chain becomes more urgent, the innovation of Blockchain and IoT provides a pathway forward - but only if implemented responsibly and inclusively to ensure outcomes are truly sustainable and equitable.

REFERENCES

AbdulZahra, M. M. (2022). In Y. Zhang (Ed.), *'Internet of Things-Based Smart and Connected Supply Chain: A Review', International Journal of Antennas and Propagation* (p. 8182813)., doi:10.1155/2022/8182813

Abideen, A.Z. (2021). Food Supply Chain Transformation through Technology and Future Research Directions - A Systematic Review. *Logistics*. . doi:10.3390/logistics5040083

Agrawal, T. K., Kumar, V., Pal, R., Wang, L., & Chen, Y. (2021). Blockchain-based framework for supply chain traceability: A case example of textile and clothing industry. *Computers & Industrial Engineering*, *154*, 107130. doi:10.1016/j.cie.2021.107130

Ahmed, W. A. H., & MacCarthy, B. L. (2021). Blockchain-Enabled Supply Chain Traceability in the Textile and Apparel Supply Chain: A Case Study of the Fiber Producer, Lenzing. *Sustainability, 13*(19), 10496. doi:10.3390u131910496

Altenburg, T. (2016) *Making retail modernisation in developing countries inclusive: a development policy perspective.*

Astill, J., Dara, R. A., Campbell, M., Farber, J. M., Fraser, E. D. G., Sharif, S., & Yada, R. Y. (2019). Transparency in food supply chains: A review of enabling technology solutions. *Trends in Food Science & Technology, 91*, 240–247. doi:10.1016/j.tifs.2019.07.024

Awan, S. (2021). In S. Khan (Ed.). *IoT with BlockChain: A Futuristic Approach in Agriculture and Food Supply Chain. Wireless Communications and Mobile Computing* (p. 5580179)., doi:10.1155/2021/5580179

Ayan, B., Güner, E. & Son-Turan, S. (2022). Blockchain Technology and Sustainability in Supply Chains and a Closer Look at Different Industries: A Mixed Method Approach. *Logistics*. . doi:10.3390/logistics6040085

Bodó, B., Gervais, D., & Quintais, J. P. (2018). Blockchain and smart contracts: The missing link in copyright licensing? *International Journal of Law and Information Technology, 26*(4), 311–336. doi:10.1093/ijlit/eay014

Bonanni, L. (2010). *Beyond transparency : collective engagement in sustainable design.*

Christidis, K., & Devetsikiotis, M. (2016). Blockchains and Smart Contracts for the Internet of Things. *IEEE Access : Practical Innovations, Open Solutions, 4*, 2292–2303. doi:10.1109/ACCESS.2016.2566339

Cil, A. Y., Abdurahman, D., & Cil, I. (2022). Internet of Things enabled real time cold chain monitoring in a container port. *Journal of Shipping and Trade, 7*(1), 9. doi:10.118641072-022-00110-z

Damoska Sekuloska, J. & Erceg, A. (2022). Blockchain Technology toward Creating a Smart Local Food Supply Chain. *Computers*. . doi:10.3390/computers11060095

Difrancesco, R. M., Meena, P., & Kumar, G. (2022). How blockchain technology improves sustainable supply chain processes: A practical guide [Preprint]. *Operations Management Research*, (0123456789). doi:10.100712063-022-00343-y

Djekic, I., Jankovic, D., & Rajkovic, A. (2017). Analysis of foreign bodies present in European food using data from Rapid Alert System for Food and Feed (RASFF). *Food Control, 79*, 143–149. doi:10.1016/j.foodcont.2017.03.047

Duan, J., Zhang, C., Gong, Y., Brown, S., & Li, Z. (2020). A Content-Analysis Based Literature Review in Blockchain Adoption within Food Supply Chain. *International Journal of Environmental Research and Public Health, 17*(5), 1784. doi:10.3390/ijerph17051784 PMID:32182951

Dutta, P., Choi, T.-M., Somani, S., & Butala, R. (2020). Blockchain technology in supply chain operations: Applications, challenges and research opportunities. *Transportation Research Part E, Logistics and Transportation Review, 142*, 102067. doi:10.1016/j.tre.2020.102067 PMID:33013183

Gad, A. G. (2022). Emerging Trends in Blockchain Technology and Applications: A Review and Outlook. *Journal of King Saud University - Computer and Information Sciences, 34*(9), 6719–6742.

Gossner, C. M.-E., Schlundt, J., Ben Embarek, P., Hird, S., Lo-Fo-Wong, D., Beltran, J. J. O., Teoh, K. N., & Tritscher, A. (2009). The melamine incident: Implications for international food and feed safety. *Environmental Health Perspectives, 117*(12), 1803–1808. doi:10.1289/ehp.0900949 PMID:20049196

Haji, M. (2020). Roles of Technology in Improving Perishable Food Supply Chains. *Logistics*. . doi:10.3390/logistics4040033

IBM (2020) *Blockchain Component Overview, BM Blockchain Platform for IBM Cloud*. Available at: BM Blockchain Platform for IBM Cloud.

IBM. (2022) *Hyperledger Fabric Access Control System for Internet of Things Layer in Blockchain-Based Applications*. Available at: https://www.ibm.com/products/blockchain-platform-hyperledger-fabric

Iftekhar, A. (2020). In L. Patarata (Ed.), *'Application of Blockchain and Internet of Things to Ensure Tamper-Proof Data Availability for Food Safety', Journal of Food Quality* (p. 5385207). doi:10.1155/2020/5385207

Iftekhar, A., Cui, X., Tao, Q., & Zheng, C. (2021). Hyperledger Fabric Access Control System for Internet of Things Layer in Blockchain-Based Applications. *Entropy (Basel, Switzerland), 23*(8), 1054. doi:10.3390/e23081054 PMID:34441194

Ismail, S., Reza, H., Salameh, K., Kashani Zadeh, H., & Vasefi, F. (2023). Toward an Intelligent Blockchain IoT-Enabled Fish Supply Chain: A Review and Conceptual Framework. *Sensors (Basel), 23*(11), 5136. doi:10.339023115136 PMID:37299864

Javaid, M. (2021). Blockchain technology applications for Industry 4.0: A literature-based review. *Blockchain: Research and Applications, 2(*4), 100027.

Javaid, M., Haleem, A., Singh, R. P., & Suman, R. (2022). Enhancing smart farming through the applications of Agriculture 4.0 technologies. *International Journal of Intelligent Networks, 3*, 150–164. doi:10.1016/j.ijin.2022.09.004

Jerath, K., Sajeesh, S., & John Zhang, Z. (2016). A model of unorganized and organized retailing in emerging economies. *Marketing Science, 35*(5), 756–778. doi:10.1287/mksc.2015.0962

Kamble, S., Gunasekaran, A., & Arha, H. (2019). Understanding the Blockchain technology adoption in supply chains-Indian context. *International Journal of Production Research, 57*(7), 2009–2033. doi :10.1080/00207543.2018.1518610

Kaur, A., Singh, G., Kukreja, V., Sharma, S., Singh, S., & Yoon, B. (2022). Adaptation of IoT with Blockchain in Food Supply Chain Management: An Analysis-Based Review in Development, Benefits and Potential Applications. *Sensors (Basel), 22*(21), 8174. doi:10.339022218174 PMID:36365871

Knijnenburg, B. P. (2022). *Modern Socio-Technical Perspectives on Privacy*. Modern Socio-Technical Perspectives on Privacy. doi:10.1007/978-3-030-82786-1

Knorr, D., Augustin, M. A., & Tiwari, B. (2020). Advancing the Role of Food Processing for Improved Integration in Sustainable Food Chains. *Frontiers in Nutrition, 7*(April), 1–8. doi:10.3389/fnut.2020.00034 PMID:32309287

Kshetri, N. (2018). 1 Blockchain's roles in meeting key supply chain management objectives. *International Journal of Information Management*, *39*, 80–89. doi:10.1016/j.ijinfomgt.2017.12.005

Kumar, M., Choubey, V. K., Raut, R. D., & Jagtap, S. (2023). Enablers to achieve zero hunger through IoT and blockchain technology and transform the green food supply chain systems. *Journal of Cleaner Production*, *405*, 136894. doi:10.1016/j.jclepro.2023.136894

Kumar, R. L., Khan, F., Kadry, S., & Rho, S. (2022). A Survey on blockchain for industrial Internet of Things. *Alexandria Engineering Journal*, *61*(8), 6001–6022. doi:10.1016/j.aej.2021.11.023

Kumar, S., Raut, R. D., Agrawal, N., Cheikhrouhou, N., Sharma, M., & Daim, T. (2022). Integrated blockchain and internet of things in the food supply chain: Adoption barriers. *Technovation*, *118*, 102589. doi:10.1016/j.technovation.2022.102589

Maftei, A. A., Lavric, A., Petrariu, A. I., & Popa, V. (2023). Massive Data Storage Solution for IoT Devices Using Blockchain Technologies. *Sensors (Basel)*, *23*(3), 1570. doi:10.339023031570 PMID:36772609

Mashayekhy, Y. (2022). *Impact of Internet of Things (IoT) on Inventory Management: A Literature Survey*. Logistics. doi:10.3390/logistics6020033

Miller, K. B., Eckberg, J. O., Decker, E. A., & Marinangeli, C. P. F. (2021). Role of Food Industry in Promoting Healthy and Sustainable Diets. *Nutrients*, *13*(8), 2740. doi:10.3390/nu13082740 PMID:34444900

Momtaz, M., Bubli, S. Y., & Khan, M. S. (2023). Mechanisms and Health Aspects of Food Adulteration: A Comprehensive Review. *Foods (Basel, Switzerland)*, *12*(1), 199. doi:10.3390/foods12010199 PMID:36613416

Nižetić, S., Šolić, P., López-de-Ipiña González-de-Artaza, D., & Patrono, L. (2020). Internet of Things (IoT): Opportunities, issues and challenges towards a smart and sustainable future. *Journal of Cleaner Production*, *274*, 122877. doi:10.1016/j.jclepro.2020.122877 PMID:32834567

Nosratabadi, S., Mosavi, A., & Lakner, Z. (2020). Food Supply Chain and Business Model Innovation. *Foods*, *9*(2), 132. doi:10.3390/foods9020132 PMID:32012751

Osmólska, E., Stoma, M., & Starek-Wójcicka, A. (2022). Application of Biosensors, Sensors, and Tags in Intelligent Packaging Used for Food Products—A Review. *Sensors (Basel)*, *22*(24), 9956. Advance online publication. doi:10.339022249956 PMID:36560325

Pandey, V., Pant, M., & Snasel, V. (2022). Blockchain technology in food supply chains: Review and bibliometric analysis. *Technology in Society*, *69*, 101954. doi:10.1016/j.techsoc.2022.101954

Peng, L. (2021). Privacy preservation in permissionless blockchain: A survey. *Digital Communications and Networks, 7(*3), 295–307.

Pigłowski, M. (2020). Food hazards on the European Union market: The data analysis of the Rapid Alert System for Food and Feed. *Food Science & Nutrition*, *8*(3), 1603–1627. doi:10.1002/fsn3.1448 PMID:32180969

Popa, A., Hnatiuc, M., Paun, M., Geman, O., Hemanth, D., Dorcea, D., Son, L., & Ghita, S. (2019). An Intelligent IoT-Based Food Quality Monitoring Approach Using Low-Cost Sensors. *Symmetry*, *11*(3), 374. doi:10.3390ym11030374

Pozelli Sabio, R., & Spers, E. E. (2022). 'Consumers' Expectations on Transparency of Sustainable Food Chains'. *Frontiers in Sustainable Food Systems*, *6*(April), 1–8. doi:10.3389/fsufs.2022.853692

Pufahl, L. (2021). Enabling Financing in Agricultural Supply Chains Through Blockchain: Interorganizational Process Innovation Through Blockchain. Business Process Management Cases Vol. 2: Digital Transformation-Strategy, Processes and Execution. Springer. doi:10.1007/978-3-662-63047-1_4

Rana, S. (2021). Blockchain-Based Model to Improve the Performance of the Next-Generation Digital Supply Chain. *Sustainability*. . doi:10.3390/su131810008

Raskin, M. (2016). The law and legality of smart contracts. *Geo. L. Tech. Rev.*, *1*, 305.

Rejeb, A., Suhaiza, Z., Rejeb, K., Seuring, S., & Treiblmaier, H. (2022). The Internet of Things and the circular economy: A systematic literature review and research agenda. *Journal of Cleaner Production*, *350*, 131439. doi:10.1016/j.jclepro.2022.131439

Rogerson, M., & Parry, G. C. (2020). Blockchain: Case studies in food supply chain visibility. *Supply Chain Management*, *25*(5), 601–614. doi:10.1108/SCM-08-2019-0300

Sethi, P., & Sarangi, S. R. (2017). In R. Khanna (Ed.), *'Internet of Things: Architectures, Protocols, and Applications', Journal of Electrical and Computer Engineering* (p. 9324035)., doi:10.1155/2017/9324035

Sharma, H. B., Vanapalli, K. R., Cheela, V. R. S., Ranjan, V. P., Jaglan, A. K., Dubey, B., Goel, S., & Bhattacharya, J. (2020). Challenges, opportunities, and innovations for effective solid waste management during and post COVID-19 pandemic. *Resources, Conservation and Recycling*, *162*, 105052. doi:10.1016/j.resconrec.2020.105052 PMID:32834486

Shelake, A. S. (2020). IoT Based Food Inventory Tracking System for Domestic and Commercial Kitchens. *International Journal of Scientific Research and Engineering Development*, *3*(4). www.ijsred.com

Shi, S., He, D., Li, L., Kumar, N., Khan, M. K., & Choo, K.-K. R. (2020). Applications of blockchain in ensuring the security and privacy of electronic health record systems: A survey. *Computers & Security*, *97*, 101966. doi:10.1016/j.cose.2020.101966 PMID:32834254

Spink, J., Ortega, D. L., Chen, C., & Wu, F. (2017). Food fraud prevention shifts the food risk focus to vulnerability. *Trends in Food Science & Technology*, *62*, 215–220. doi:10.1016/j.tifs.2017.02.012

Statista. (2023) *Food - Worldwide*. Statista. https://www.statista.com/outlook/cmo/food/worldwide

Taherdoost, H. (2023). Smart Contracts in Blockchain Technology: A Critical Review. *Information (Basel)*, *14*(2), 117. doi:10.3390/info14020117

Tan, W. C., & Sidhu, M. S. (2022). Review of RFID and IoT integration in supply chain management. *Operations Research Perspectives*, *9*, 00229.

Tokkozhina, U., Martins, A. L., & Ferreira, J. C. (2022). Use of Blockchain Technology to Manage the Supply Chains: Comparison of Perspectives between Technology Providers and Early Industry Adopters. *Journal of Theoretical and Applied Electronic Commerce Research*, *17*(4), 1616–1632. doi:10.3390/jtaer17040082

Tönnissen, S., & Teuteberg, F. (2020). Analysing the impact of blockchain-technology for operations and supply chain management: An explanatory model drawn from multiple case studies. *International Journal of Information Management*, *52*, 101953. doi:10.1016/j.ijinfomgt.2019.05.009

Tsoukas, V., Gkogkidis, A., Kampa, A., Spathoulas, G., & Kakarountas, A. (2022). Enhancing Food Supply Chain Security through the Use of Blockchain and TinyML. *Information (Basel)*, *13*(5), 213. Advance online publication. doi:10.3390/info13050213

Uddin, M. A. (2021). A survey on the adoption of blockchain in IoT: challenges and solutions. *Blockchain: Research and Applications*, *2*(2), 100006.

United Nations. (2017) *World population projected to reach 9.8 billion in 2050, and 11.2 billion in 2100No Title, Department of Economic and Social Affairs*. UN. https://www.un.org/development/desa/en/news/population/world-population-prospects-2017.html (Accessed: 12 June 2023).

van Hilten, M., Ongena, G., & Ravesteijn, P. (2020). Blockchain for Organic Food Traceability: Case Studies on Drivers and Challenges. *Frontiers in Blockchain*, *3*(September), 1–13. doi:10.3389/fbloc.2020.567175

Verdouw, C., Wolfert, S., & Tekinerdogan, B. (2016). Internet of Things in agriculture. CABI Reviews, 1–12. doi:10.1079/PAVSNNR201611035

Verdouw, C. N., Wolfert, J., Beulens, A. J. M., & Rialland, A. (2016). Virtualization of food supply chains with the internet of things. *Journal of Food Engineering*, *176*, 128–136. doi:10.1016/j.jfoodeng.2015.11.009

Wolfert, S., Ge, L., Verdouw, C., & Bogaardt, M.-J. (2017). Big data in smart farming–a review. *Agricultural Systems*, *153*, 69–80. doi:10.1016/j.agsy.2017.01.023

Wu, W., Zhang, A., van Klinken, R. D., Schrobback, P., & Muller, J. M. (2021). Consumer Trust in Food and the Food System: A Critical Review. *Foods (Basel, Switzerland)*, *10*(10), 2490. doi:10.3390/foods10102490 PMID:34681539

Xu, X. (2019). Case study: AgriDigital: Blockchain technology in the trade and finance of agriculture supply chains. *Architecture for blockchain applications*, 239–255.

Yadav, N., Luthra, S., & Garg, D. (2023). Blockchain Technology for Sustainable Supply Chains : A Network Cluster Analysis for Future Research Propositions. *Environmental Science and Pollution Research International*, *30*(24), 64779–64799. doi:10.100711356-023-27049-3 PMID:37086323

Young, I., & Waddell, L. (2016). Barriers and Facilitators to Safe Food Handling among Consumers: A Systematic Review and Thematic Synthesis of Qualitative Research Studies. *PLoS One*, *11*(12), e0167695. doi:10.1371/journal.pone.0167695 PMID:27907161

Zhang, Y., Zhao, L., & Qian, C. (2017). Modeling of an IoT-enabled supply chain for perishable food with two-echelon supply hubs. *Industrial Management & Data Systems*, *117*(9), 1890–1905. doi:10.1108/IMDS-10-2016-0456

Chapter 11
The Use of 3D Printing in the Food Industry:
From Novelty to Necessity

Majid Shoeibi
Yazd University, Iran

ABSTRACT

The potential of 3D printing in the food industry is enormous, but its current limitations hinder widespread adoption. Despite its ability to create intricate designs, mass produce, and reduce waste and costs, large-scale implementation remains a challenge. One major advantage of 3D printing in food is its ability to customize and personalize products, allowing for the creation of unique and complex designs. This opens up new possibilities for culinary creativity and experimentation. Additionally, 3D printing improves efficiency and precision by automating tasks, reducing production time, and ensuring consistent quality. Overall, 3D printing is transforming the food sector by offering customization, efficiency, personalized nutrition, and waste reduction, revolutionizing the way food is prepared, consumed, and enjoyed.

INTRODUCTION

The food industry is undergoing a technological revolution, and one of the most intriguing advancements is the use of 3D printing. While 3D printing was initially regarded as a novelty in the culinary world, it has rapidly evolved into a necessity with the potential to transform the way food is produced, customized, and consumed. This innovative technology offers a range of possibilities, from creating intricate and customizable food designs to addressing food sustainability and meeting the diverse dietary needs of individuals. As 3D printing in the food industry continues to gain traction, it is poised to revolutionize not only the way we eat but also how food is manufactured and distributed.

Food 3D printing has emerged as the cutting-edge technology in the past years because of accuracy, customization, and breakthrough ability in food structure and texture. For the first time, FLM was used to prepare cake mix, in the form of the pulp of a mixture that consisted of starch, yeast, sugar, corn syrup, and cake topping by Lille, Nurmela, Nordlund, Metsa-Kortelainen, & Sozer (2018). Among various methods, the extrusion-based method is the most suitable for 3D food printing. The fascinating

DOI: 10.4018/978-1-6684-9094-5.ch011

applications of these printed edible items with custom-designed shapes have also been used for early childhood education to teach children about flavour, colour, and form (Sun, Zhou, Yan, Huang, & Lin, 2018). FLM consists of three-steps: (a) 3D modeling, (b) printing, and (c) post-processing. In the first step, a 3D model is designed by Computer-Aided Design (CAD) (Attarin & Attaran, 2020), then this model is converted into STL file or AMF file format. For customized and texture-modified food printing AMF file format is used because STL files express only the surface geometry of a 3D object and do not represent colour, texture, or other common CAD model attributes. After that, the model is sliced into layers to get each section's outline. Then the products are made by depositing the materials layer-by-layer. In the end, the printed products are post-processed such as cooked or baked as per the requirement (Yang, Zhang, & Bhandari, 2017).

The concept of 3D printing, also known as additive manufacturing, involves the layer-by-layer construction of objects based on digital designs. In recent years, this technology has been adapted to the food industry, enabling the creation of edible structures and products. By using food-grade materials and precise deposition techniques, 3D printers can produce intricate and visually stunning food items that were once thought to be beyond the realm of possibility. One area where 3D printing has made significant strides is in personalized nutrition. As individuals have distinct dietary requirements and preferences, customization has become a key driver in the food industry. 3D printing allows for the creation of tailored meals that cater to specific nutritional needs, allergens, and taste preferences. For example, researchers at Columbia University have developed a 3D food printer capable of creating meals with precise nutrient compositions, tailored to an individual's unique dietary requirements, such as low-sodium or high-fiber diets (Dang, 2017). Moreover, 3D printing has the potential to address food sustainability challenges.

Traditional food production methods often result in significant food waste, inefficient resource utilization, and high carbon emissions. In contrast, 3D printing offers the possibility of producing food with optimized ingredient usage and reduced waste. Researchers at Wageningen University in the Netherlands have explored the use of 3D printing to create plant-based protein-rich structures, which could potentially reduce the environmental impact of meat production (Bos, 2019). Furthermore, 3D printing technology enables chefs and culinary artists to push the boundaries of creativity and design in the culinary world. Intricate food sculptures, visually appealing desserts, and geometrically complex dishes can be crafted with precision, presenting a new dimension of culinary artistry. For instance, Chef Jan Smink, a Michelin-starred chef, has utilized 3D printing to create unique and visually stunning dishes that captivate diners through their intricate designs (Kane, 2020).

The use of 3D-printed food applications in the food service industry has been largely unnoticed by the consumer but is set to bring about drastic changes in the way our food is prepared and presented. This leads to the need to assess the impact of Novel Food Technology Neophobia (NFTN) on consumer willingness to use. The evidence suggests a direct negative effect of NFTN on willingness to use, which is to be expected. However, by looking at indirect effects, it is evident that NFTN has the potential to undermine an individual's perception of the relevance of 3D food printing technology to them personally. For marketing strategists to be successful in promoting 3D food printing, they must be aware of this indirect effect and focus on the personalisation factor in order to make the technology more relevant to their target audience. Introducing 3D food printing applications into the food service sector gives consumers the opportunity to try 3D-printed foods with minimal risk, which may help to reduce initial neophobic inclinations. Research has also shown that contexts, where consumers are in control of their exposure to hazards, are more accepted than those where exposure is not voluntary. Despite the fact that 3D food printing technologies are still relatively unknown among consumers, it is important to gain a deeper understanding of the thought processes and behaviours that could potentially hinder acceptance before attempting to extend the use of the technology into the mainstream marketplace.

The findings of this study indicate that there are multiple influences on consumer intention to use, with perceived relevance being a significant determinant. Therefore, a consumer-oriented mindset is essential for the design and development of new products/technologies. As the technology advances, 3D printing in the food industry is also making inroads in large-scale food production and distribution. Companies like Natural Machines and BeeHex are developing 3D printers capable of producing food items rapidly and efficiently, potentially revolutionizing the way food is manufactured and distributed, particularly in institutional settings such as hospitals and schools (Lambert, 2019). In conclusion, what started as a novelty in the culinary world, 3D printing has rapidly emerged as a game-changer in the food industry. Its potential to create customized nutrition, address sustainability challenges, and push the boundaries of culinary creativity makes it a necessity rather than a mere novelty. As the technology continues to advance and more research is conducted, 3D printing holds the promise of transforming not only our dining experiences but also the way food is produced, distributed, and consumed on a global scale.

BIOPRINTING

Bioprinting is the use of 3D printing technologies to construct biomedical parts that closely resemble natural tissue properties by amalgamating cells, growth factors, and biomaterials. Bioprinting is a process that uses a layer-by-layer method to deposit bio-inks to make tissue-like structures that can be used in medical and tissue engineering applications. Bioprinting can be used for a wide variety of biomaterials. To assist in the development of new medicines and pills, Bioprinting can be used to construct tissues and organs (Hinton et al., 2015). This technique has been used to grow tissues without using any biomaterial base by researchers. The method entails layer-by-layer deposition of biomaterials and living cell cultures. The first step is to choose a biocompatible platform. Then, layer-by-layer, biomaterials, and living cell cultures are deposited on this platform. This is then transferred to a bioreactor that provides an in-vitro environment to cells and the required biomechanical properties are developed during the maturation period. Bio-printing is a term that refers to printing with cellular material, and it is most commonly used to create tissue-based structures using extrusion, SLS, or inkjet printing (Serizawa et al., 2014).

MATERIAL SUPPLIES

Since the characteristics of ingredients directly influence the printing process and quality of printed food items, it is important to choose the ingredients precisely. Based on printability, material supplies can be divided into three categories: (a) natively printable materials, (b) non-printable traditional materials, and (c) alternative ingredients (Dankar, Haddarah, Omar, Sepulcre, & Pujola, 2018).

Non-printable traditional materials are those materials that cannot be printed in their natural form. To make them printable, some additional ingredients such as agar, gellan gum, methylcellulose may be added. Staple foods such as rice, meat, fruit, vegetables are some examples of non-printable materials. Due to overpopulation and the limitation of resources, now researchers are trying to use eco-friendly materials as alternative ingredients. These include algae, chia seeds, flower seeds, insect powders, etc. These ingredients are a rich source of various nutrients. Under the "Insects Au Gratin" project (Soares &

Forkers, 2014), researchers combined entomophagy with 3D food printing technologies. In this project, insects were dried and ground to make a fine powder. The flour was then combined with other ingredients like chocolate, cream cheese, spices, and icing butter to prepare edible foods.

Personalized Nutritious Foods

The need for nutrition for individuals varies, e.g., children, old-age people, athletes, women, etc. require different nutrition. By traditional food processing methods, it is not possible to provide them with the kind of food they need. 3D printing technology provides the solution of the same. Based on daily nutrient requirements, personalized food can be fabricated by incorporating those nutrients in the food items. Though everyone takes proper meals still lack some of the nutrients. Personalized foods can be the solution to overcome this problem. Personalized foods are like nectar for those who are suffering from swallowing problems or mastication problems. By using the FLM technique, texture can be modified in such a way that it can be easy to swallow or chew. Biozoon Food Innovations has produced cookies by using flours made from insects for people who have mastication issues (Pinna et al., 2016).

Researchers have also developed softer foods for old age people by using edible gel to overcome swallowing problems. Nutrient-rich personalized foods are developed by two methods: (a) regulating the quantity of food, and (b) adjusting the levels of dietary ingredients while designing. Fruits and vegetables contain various nutrients and bio-active compounds that are desired for us. Nowadays people are cautious about health and the demand for nutritious food has increased. Besides these obesity and other chronic diseases are afflicting people these days. FLM can be a great tool to fight these problems by creating personalized and nutritious foods using fruits, vegetables, and other natural resources. Hence it is an ecological technology (Tomasevic et al., 2021).

Customized Food Recipes

Food customization is the most attractive application of FLM. In general, commercial food manufacturers focus on mass production and not on the nutritional value and aesthetics of the products. FLM has made it possible to print the customized foods, along with customization of the shape, structure, texture, and flavour. Colourful, multi-structure, and multi-flavour foods can also be prepared. customized food recipes have become increasingly popular in recent years as people seek personalized and tailored approaches to their dietary needs and preferences.

This trend is fueled by advancements in technology and a growing awareness of the impact of food on individual health and well-being. One notable example of this is the rise of personalized meal delivery services like PlateJoy. PlateJoy offers a unique and innovative approach to customized food recipes by using an algorithm that takes into account an individual's dietary restrictions, food preferences, and health goals. Users simply input their information, and PlateJoy creates personalized meal plans and recipes that cater to their specific needs. By considering factors such as allergies, dietary restrictions (e.g., gluten-free, vegan), and calorie requirements, PlateJoy ensures that each recipe is tailored to the individual's unique needs. This approach not only saves time and effort for users who may struggle with meal planning but also promotes healthier eating habits by encouraging the consumption of nutrient-rich ingredients.

The customization aspect of these food recipes extends beyond dietary restrictions. PlateJoy also takes into account personal preferences, such as flavor profiles and ingredient preferences. Whether someone enjoys spicy foods or prefers a milder taste, PlateJoy incorporates these preferences into the recipe recommendations, ensuring an enjoyable dining experience. Furthermore, personalized food recipes promote sustainability by reducing food waste. PlateJoy's platform provides users with precise ingredient measurements, reducing the likelihood of excess food purchases and unused ingredients. This approach aligns with the growing concern for environmentally friendly practices in the food industry. the rise of customized food recipes exemplified by services like PlateJoy demonstrates a shift towards personalized and tailored approaches to nutrition. By leveraging technology and considering individual needs and preferences, these services provide a convenient and efficient way for people to enjoy healthy, delicious meals that align with their unique dietary requirements and goals.

CHALLENGES

Recently, researchers have made great efforts to apply 3D food printing to the food industry. However, there are still many difficulties with the widespread application of this technology in the food industry for several reasons: 1) printing precision and accuracy 2) process productivity 3) production of colourful, flavorful and textured products. Printing precision and accuracy are critical to the application of 3D printing technology in the food sector.

One of the advantages of 3D printing is to produce an exquisite and intriguing texture of edible products to increase consumer interest and appetite. However, at present, few works are published to address printing accuracy. To achieve precise and accurate printing, material properties (e.g., rheological properties, particle size, etc.), process parameters (e.g., nozzle diameter, printing speed, printing distance, etc.), and post-processing methods (e.g., baking, frying, cooking, etc.) must be considered. More efforts should be made to achieve precise and accurate printing. Improving production efficiency can reduce production costs. A common example of increasing process productivity is increasing printing speed and using a large nozzle or laser diameter. However, this often leads to a decrease in the precision and resolution of the printed objects, which puts 3D food printing in an unfavourable circumstance. We emphasise that under the premise of ensuring acceptable printing accuracy, a large nozzle diameter and high printing speed should be selected (Liu, Z., Zhang et.al. 2017).

Another possible way to improve print productivity is to use printers with multiple nozzles to produce multiple objects simultaneously. However, this will increase the complexity of the control system and the technical challenges, so extensive studies are needed to achieve both high print accuracy and high process productivity. Every technology has some advantages and disadvantages and the same is with FLM. Though it has revolutionized the food processing industry, it has several challenges. In traditional food processing, the quality depends on the skill of the operator but in FLM, the printing process depends on the operation, so process parameters need to be optimized to achieve good printability. These parameters include temperature, nozzle diameter, viscosity, etc. Food safety is a major concern. FLM produces food in a little while, which can cause food to be cooked at definite temperatures or may result in temperature fluctuations, allowing microbes growth that contaminates the food. So, while printing food, the best procedures and protocols must be followed to avoid contamination-related issues. It is difficult to mix all ingredients to prepare the ink because every ingredient has distinct cooking requirements, such as an ideal temperature, that must be taken care of. All the ingredients should not be put in the same container. Since digital recipes are created and stored in cloud-based storage, so there are chances of the piracy of digital recipes. The commercialization of this technique is also a big challenge because of its low printing speed. Mass production is another big challenge.

RESEARCH LIMITATIONS

The limitations of using 3D printers in the food industry can be stated:

- **Limited range of ingredients:** The availability of suitable ingredients for 3D printing can be a constraint. The range of edible materials that can be used in 3D printers is more limited compared to traditional cooking methods. The ingredients need to meet certain criteria such as being in a printable form, having appropriate texture and consistency, and being safe for consumption.
- **Texture and taste limitations:** Achieving the desired texture and taste in 3D-printed food can be challenging. The layer-by-layer printing process may result in a different texture compared to traditional food preparation methods. It can be difficult to replicate the complexity and mouthfeel of certain foods, such as crispy or flaky textures. Additionally, the taste of 3D-printed food might differ from conventionally prepared food due to variations in cooking techniques and ingredient interactions.
- **Equipment and cost considerations:** 3D printers suitable for food applications can be expensive and require specialized equipment. The cost of purchasing and maintaining 3D printers, as well as the associated software and materials, may pose a financial barrier for some food businesses. Moreover, the need for trained personnel to operate and maintain the equipment could add to the overall costs.
- **Food safety and hygiene:** Ensuring food safety and hygiene is crucial in the food industry. With 3D printing, there may be concerns about maintaining proper sanitation standards, preventing cross-contamination between different food materials, and ensuring the printer itself is clean and free from any potential contaminants. Research is necessary to establish best practices for maintaining hygiene during the printing process.
- **Regulatory considerations:** The use of 3D printers in the food industry may be subject to specific regulations and standards that need to be met. Depending on the jurisdiction, there might be requirements related to food labeling, ingredient sourcing, manufacturing practices, and quality control. Compliance with these regulations can pose challenges and may require additional resources and expertise.

CONCLUSION

Three-dimensional (3D) food printing is a revolutionary technology that has aroused the interest of different sectors in academia and industry. This promising technology allows the fabrication of intricate and personalized food designs involving since the texture, and appearance until the nutritional profile (Alain Le-Bail et.al. 2020). Currently, this technology is still in its early stages and has some limitations. A systematic way needs to be found to study digital recipes, platform architecture, printing technologies and their impact on food printing. while 3D printing in the food industry is currently more commonly used for prototyping and customized production, its potential to revolutionize large-scale food manufacturing should not be overlooked. The ongoing advancements in 3D printing technology and research efforts are addressing the existing limitations and paving the way for its widespread adoption in the food industry. One area of progress is the development of food-grade materials specifically designed for 3D printing. These materials need to meet strict safety and regulatory standards while also providing the necessary properties for successful printing and desirable food quality. Researchers and manufacturers are actively working on creating a wider range of food-safe materials that can be used with 3D printers, expanding the possibilities for food production. Additionally, the cost and availability of 3D printing equipment are expected to improve over time, making it more accessible to food manufacturers. As the technology becomes more mainstream and economies of scale come into play, the barriers to entry will likely decrease, encouraging wider adoption and integration of 3D printing into existing food production processes.

Standardization and quality control are also key areas of focus for the future of 3D printing in the food industry. Establishing standardized recipes, guidelines, and certification processes will ensure consistency in food production and guarantee that printed food items meet quality and safety standards. This will help build consumer trust in 3D-printed food and drive its acceptance in the market. Furthermore, ongoing research is exploring the potential of using alternative ingredients and additives in 3D printing to enhance the nutritional value and sensory experience of printed food. By incorporating functional ingredients, such as vitamins, minerals, and dietary fibers, 3D-printed food could become an innovative way to address specific dietary needs and improve overall nutrition. Although several technologies are available for food printing, there are many challenges in their commercialization. Process parameters, material properties, and post-processing methods are some of the main factors that affect the quality of printed food. Therefore, these factors should be considered and the parameters should be optimized to produce a delicate and complex edible structure. FLM can be used for a variety of food applications, such as personalized food for soldiers, space food for astronauts, food for the elderly, and customized food. In summary, while there are current limitations to the widespread adoption of 3D printing in the food industry, ongoing advancements in technology, materials, cost-effectiveness, standardization, and research are paving the way for its transformative potential. With continued progress, 3D printing has the capacity to revolutionize food manufacturing, providing consumers with a wider range of healthy, personalized, and sustainable food options.

REFERENCES

Attarin, S., & Attaran, M. (2020). Food Printing: Evolving technologies, challenges, opportunities, and best adoption strategies. *Journal of International Technology and Information Management*, *29*(1), 25–55. https://scholarworks.lib.csusb.edu/jitim/vol. doi:10.58729/1941-6679.1442

Bos, M. (2019). *3D printing meat: The potential of the food industry's new frontier*. WUR. https://www.wur.nl/en/newsarticle/3D-printing-meat-The-potential-of-the-food-industrys-new-frontier.htm

Dang, J. (2017). *How 3D printing could change the way we eat*. CNN. https://www.cnn.com/2017/06/22/health/3d-printing-food/index.html

Dankar, I., Haddarah, A., Omar, F. E. L., Sepulcre, F., & Pujola, M. (2018). 3D printing technology: The new era for food customization and elaboration. *Trends in Food Science & Technology, 75*, 231–242. doi:10.1016/j.tifs.2018.03.018

Hinton, T. J., Jallerat, Q., Palchesko, R. N., Park, J. H., Grodzicki, M. S., Shue, H. J., Ramadan, M. H., Hudson, A. R., & Feinberg, A. W. (2015). Three-dimensional printing of complex biological structures by freeform reversible embedding of suspended hydrogels. *Science Advances, 1*(9), e1500758. doi:10.1126ciadv.1500758 PMID:26601312

Kane, P. (2020). *Jan Smink: How 3D printing is transforming fine dining*. CNN. https://www.cnn.com/style/article/3d-printing-fine-dining-jan-smink/index.html

Lambert, F. (2019). *3D printed food: A comprehensive review.* Science Direct. https://www.sciencedirect.com/science/article/pii/S0924224419305242

Le-Bail, A., Maniglia, B. C., & Le-Bail, P. (2020). Recent advances and future perspective in additive manufacturing of foods based on 3D printing. *Current Opinion in Food Science, 35*, 54–64. doi:10.1016/j.cofs.2020.01.009

Lille, M., Nurmela, A., Nordlund, E., Metsa-Kortelainen, S., & Sozer, N. (2018). Applicability of protein and fiber-rich food materials in extrusion-based 3D printing. *Journal of Food Engineering, 220*, 20–27. . JFOODENG.2017.04.034 doi:10.1016/J

Liu, Z., Zhang, M., Bhandari, B., & Wang, Y. (2017). 3D printing: Printing precision and application in food sector. *Trends in Food Science & Technology, 69*, 83–94. doi:10.1016/j.tifs.2017.08.018

Pinna, C., Ramundo, L., Sisca, F. G., Angioletti, C. M., Taisch, M., & Terzi, S. (2016 September). Additive manufacturing applications within food industry: An actual overview and future opportunities. *XXI Summer School "Francesco Turco" - Industrial Systems Engineering*.

PlateJoy. (n.d.). *Personalized Meal Plans and Recipes*. PlateJoy. https://www.platejoy.com/

Rahul Soni, K. (2022). Ponappa, Puneet Tandon,A review on customized food fabrication process using Food Layered Manufacturing. *Lebensmittel-Wissenschaft + Technologie, 161*, 113411. doi:10.1016/j.lwt.2022.113411

Ross, M. M. (2022). *3D printing of dairy-based ingredients and investigation into Irish consumer acceptance of 3D food printing applications.* [PhD Thesis, University College Cork].

Serizawa, R., Shitara, M., Gong, J., Makino, M., Kabir, M. H., & Furukawa, H. (2014). 3D jet printer of edible gels for food creation, 2014 Behavior and Mechanics of Multifunctional Materials and Composites, 9058. *Article, 90580A*, 90580A. doi:10.1117/12.2045082

Soares, S., & Forkers, A. (2014). Insects Au Gratin - an investigation into the experiences of developing a 3D printer that uses insect protein-based flour as a building medium for the production of sustainable food. In E. Bohemia, A. Eger, W. Eggink, A. Kovacevic, B. Parkinson, & W. Wits (Eds.), *DS 78: Proceedings of the E&PDE 2014 16th international Conference on Engineering and product design,* (pp. 426–431). The Design Society & Institution of Engineering Designers. https://www.designsociety.org/publication/35727/ds_78_proceedings_of_the_e_pde_2014_16th_international_conference_on_engineering_and_product_design_university_of_twente_the_netherlands.

Sun, J., Zhou, W., Yan, L., Huang, D., & Lin, L. (2018). Extrusion-based food printing for digitalized food design and nutrition control. *Journal of Food Engineering*, *220*, 1–11. doi:10.1016/j.jfoodeng.2017.02.028

Tomasevic, I., Putnik, P., Valjak, F., Pavlic, B., Sojic, B., Markovinovic, A. B., & Kovacevic, D. B. (2021). 3D printing as novel tool for fruit-based functional food production. *Current Opinion in Food Science, 41,* 138–145. . COFS.2021.03.015 doi:10.1016/J

Yang, F., Zhang, M., & Bhandari, B. (2017). Recent development in 3D food printing. *Critical Reviews in Food Science and Nutrition*, *57*(14), 3145–3153. doi:10.1080/10408398.2015.1094732 PMID:26479080

Chapter 12
From Kitchen to Table:
The Transformative Journey of 3D Printing in the Food Industry

Ümit Can Kaya
https://orcid.org/0000-0001-6451-9766
Ankara Hacı Bayram Veli University, Turkey

Aybüke Elif Ceyhun Sezgin
Ankara Hacı Bayram Veli University, Turkey

ABSTRACT

3D food printers, which technology has brought to life, have been produced in order to facilitate the production of three-dimensional objects with different and complex structures, to give flexibility to original qualities and changes, and to find solutions for people who need a different nutrition system. This technology, which is developing rapidly in the food sector but still needs to be developed further, has brought with it many conveniences. It is thought that the 3D food printer technology, which entered the food sector in 2001 and has developed considerably today, will be a necessity in the future. In this context, the studies conducted on 3D printer technology and 3D food printer technologies were evaluated. Within the framework of the studies conducted, it is thought that 3D food printers will help people with special nutritional needs, people who need to be fed under a certain diet, and people who are fed according to their personal preferences.

INTRODUCTION

The rapid increase in the world population, climate change, environmental pollution and famine levels, the decrease in arable agricultural areas on Earth and negative changes in the global world order have made it difficult to access food resources. With these increasing challenges, digitalization is developing rapidly and changing the world order. With the increasing difficulty of accessing the food sources that people need to meet their basic physiological needs, a number of alternative ways have emerged. Digitalization and the resulting technological developments have been effective in the field of gastronomy

DOI: 10.4018/978-1-6684-9094-5.ch012

as well as in all areas of life. The use of three-dimensional printers (3D) in the food field is one of the important reflections of technology on the food field.

The first robotic 3D printer was developed by Charles Hull in 1984. The developed 3D printer technology was commercialized in 1989 and has been widely used in many fields. Emphasis has been placed on 3D printer technology in areas such as healthcare, textile, education, automotive, construction, defense industry and jewelry. First periods of the 21st century, the food industry has also started to take part in these areas. 3D printers work on the principle that a predetermined shape is processed by layering on top of each other until it is obtained. For this reason, these printers are also described as layered production. Since additive manufacturing is carried out in 3D printers, the amount of waste and increased material is much less compared to traditional production (Horvath and Cameron, 2020:5; Meral and Demirdöven, 2022: 92; Aldemir, 2022: 153-154).

3D food printers have provided the opportunity to create changes in sensory properties such as taste, texture and color, as well as shape in foods. For this reason, 3D printer technology offers a more sensitive, original, structurally innovative and open to development process to the food industry (Aday and Aday, 2020:444).

The advantages of 3D food printers are that they create less food waste compared to traditional production, speed up the sample product production process, save time and energy and provide a long shelf life (Şahin and Turan, 2018: 99-100). The advantages provided by 3D food printers are considered as an important step towards ensuring sustainability in the food industry. The necessity of modeling the food in advance at the beginning of the production process provides the opportunity to give the product originality both in terms of nutritional content and shape. In this process, which includes a digital design and production, affordability is ensured by minimizing the margin of error caused by humans. 3D food printers, which bring standardization with them, reduce the workload in recipe preparation and preservation and speed up the process (Zoran and Coelho, 2011:427).

In this part of the book, a general assessment is made about 3D printers and 3D food printers and the innovations brought by these applications to the food sector are detailed within the scope of scientific studies. In addition, within the scope of the study, the applications that should be made in this field in the future and the possible difficulties were discussed. The points to which the applications made will contribute have been evaluated.

Types of 3D Food Printing

Typical 3D printers consist of (1) a control circuit integrating computer and printer, (2) motor filament and driver to guide motors, (3) mixing compartment for mixing and storing raw materials, (4) raw material spray cylinders, (5) flow sensors, (6) pressure regulators, (7) hoses and (8) a printing platform on which food is printed (Figure 1).

In 3D food printers, the process consists of three stages: creating a model of the food item, printing the product, and post-printing the product. Printing techniques vary according to the characteristics of the food to be printed. There are four kinds of printing techniques which are commonly sintering technology (hot air/laser), direct printing by extrusion, binder (adhesive) spraying and inkjet printing techniques (Değerli and El, 2017:594; Seker and Karaman, 2020:193).

Figure 1. Schematic flow diagram of a typical 3d food printing process
Source: Nachal, Moses, Karthik and Anandharamakrishan (2019:125)

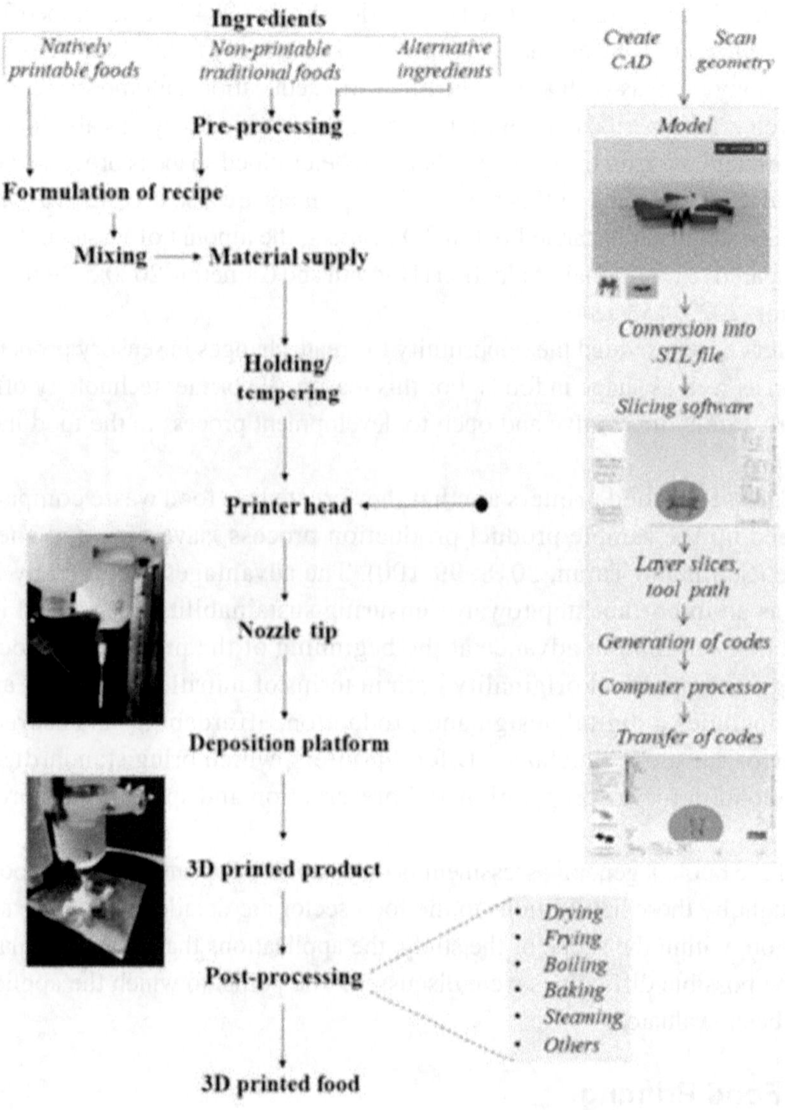

Sintering Technology (Hot Air/Laser)

Sintering is the process of increasing the density and gaining consistency as a result of applying temperature to a powdered substance suitable for 3D food printing. In this technique, the powder raw material is heated by laser and a 3D layer is formed. While the platform where the printing surface is located is moving, a header prepares the powder raw material to be sintered and shapes the printed product. Katmanlar oluşturuldukça platform aşağı eksene doğru iner ve 3D şeklin korunması sağlanmaktadır (Godoi, Prakash ve Bhandari, 2016:48).

Figure 2. Sintering technology (a)
Source: Sun et al.(2015)

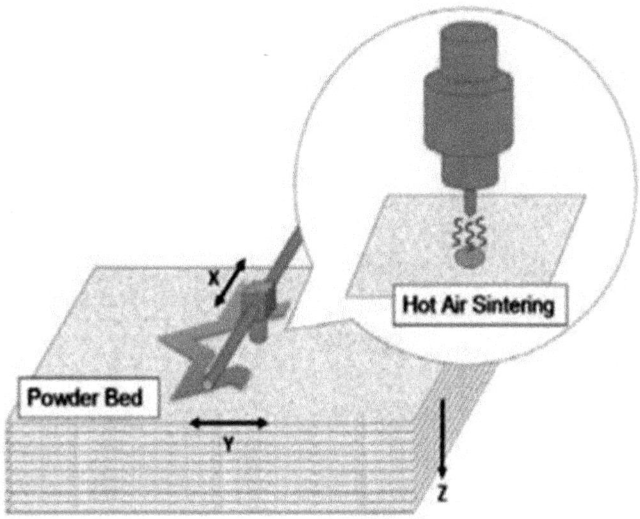

Figure 3. Sintering technique (b)
Source: Sun et al.(2015)

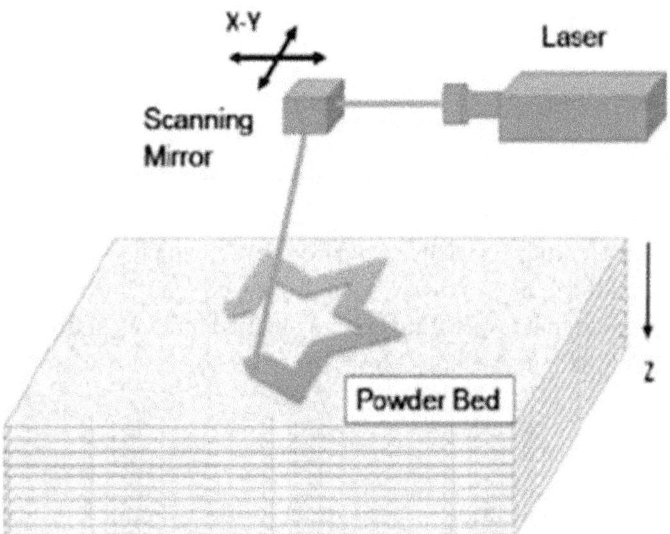

Direct Printing by Extrusion

This technique, developed for the purpose of modeling plastic materials, has been integrated into the food field with modifications made to the printer heads. In this technique, the food material located just above the melting point is transferred to the platform with the help of the header. The important thing in this technique is that the ambient temperature at which the food material is transferred is lower than the temperature of the food material. In this way, the food material will be printed on another layer by ensuring rapid solidification (Lipton, Arnold, Nigl, Lopez, Cohen, Noren and Lipson, 2015: 116-117).

Figure 4. Printing technique by extrusion
Source: Sun et al.(2015)

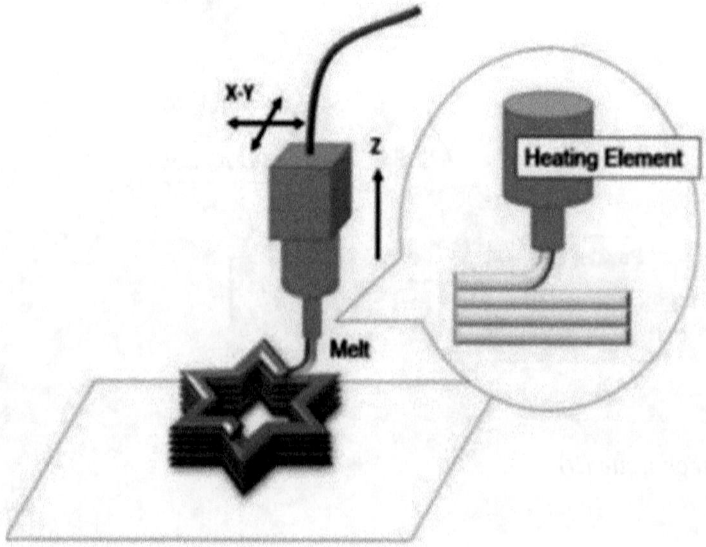

Binder (Adhesive) Spraying

It shows similarities with the sintering method in terms of the platform and material form on which the product is printed. However, temperature is not used as a power source in this technique. Layers are formed by spraying liquid on products such as sugar or starch, which are in powder form. As the layers are created, the printing platform moves towards the downward axis. It is especially known that sugar is used as a raw material. Although it is fast compared to other techniques, it is high in terms of installation cost (Sun, Peng, Zhou, Fuh, Hong, & Chiu, 2015:314).

Inkjet Printing

The working principle of classical printers used in daily life is applied in inkjet printing technique. The product to be printed is sprayed onto the platform in the form of small droplets with the help of a header. It can spray December or continuous according to the shape of the desired product. The drops can create a two-and-a-half-dimensional digital image as a decoration or surface filling. It is the preferred printing technique for cookies, cakes and pastries. It gives good results especially in the completion of pre-prepared products (Değerli and El, 2017:595).

Figure 5. Binder spraying technique
Source: Sun et al.(2015)

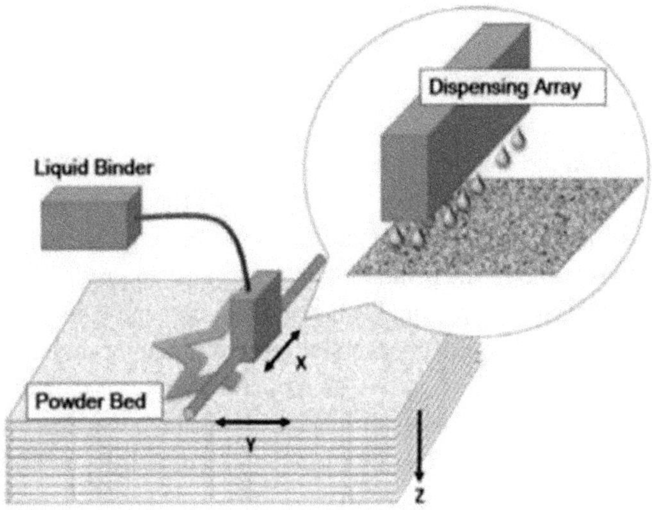

Figure 6. Inkjet printing technique
Source: Sun et al.(2015)

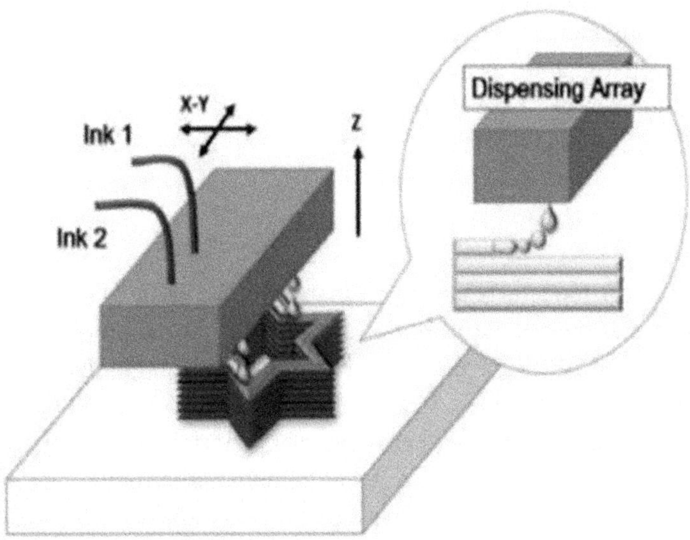

Types of Raw Materials in 3D Food Printers

In 3D food printers, it is required that the raw material be in powder or liquid form that can flow to the printing area. It is necessary that the raw material should not be distorted during the processing process to be performed after printing. In terms of these properties, chocolate is the most suitable food item. On the other hand, the raw materials used are also divided into three categories: printable, non-printable traditional and alternative raw materials according to their printability characteristics (Seker and Karaman 2020:195). The printing technique to be used plays a role in determining the type of raw material.

Printable Raw Materials

The most important features that the raw material used in 3D food printers should carry are that the raw material flows smoothly from the cartridge part of the printer to the printing platform and retains the solid shape it has formed. Substances that can form hydrogels together with water, such as starch and protein, and foods such as cheese, chocolate and hummus have these properties. Sugar Decoction, starch and mashed potatoes are also among the printable raw materials. In products where natural foodstuffs are used as printing materials, taste, nutritional value and texture can be kept completely under control. It is difficult for the products that require cooking process to maintain their shapes after they are produced. The most successful product obtained by using natural food ingredients is pasta dough. When looking at the products produced from these substances, it is noteworthy that there are foods that are in the sweet or snack group (Cohen, Jeffrey, Cutler, Coulter, Vesco, & Lipson, 2009:808; Lipton, Arnold, Nigl, Lopez, Cohen, Noren, & Lipson, 2010: 810).

Non-Printable Traditional Raw Materials

Foods such as dried legumes, rice, meat, fruits and vegetables consumed by humans every day are not printable by their nature. These foods need to be subjected to a number of preliminary processes such as boiling, juicing, grinding and pureeing. Then, the added xanthan is brought into a printable form by adding gelling agents such as gamut, agar and pectin. Printing of turkey meat with the addition of transglutaminase enzyme and agar, plant-based beef production can be given as an example of this category (Cohen et al., 2009: 809; Sun et al., 2015: 312; Kim, Bae and Park, 2017:24).

Alternative Raw Materials

Alternative raw materials are foodstuffs that have emerged in the name of rapidly increasing the world's population and meeting the nutritional needs of the growing population. Alternative raw materials, which are rich in protein and fiber, are satisfying but printable. Fungi, algae, seaweed and insects are used as alternative raw materials (Dankar, Haddarah, Omar, Sepulcre and Pujola, 2018:234). The first product printed in three dimensions using alternative raw materials was produced by Susana Soares in 2011 under the name "Insects Au Gratin". Soares obtained an alternative raw material by mixing the insects he ground into powder and the cheese he brought into molten form (Soares, 2011). Apart from the mentioned alternative raw material sources, it is possible that residues resulting from agricultural practices and food processing can be used as a three-dimensional food printing agent by converting them into biologically active metabolites, enzymes and food flavor components (Yang, Zhang and Bhandari, 2017:3153; Sun et al., 2015: 315).

Comparison of 3D Food Printer Technologies

When the characteristics of the technologies used in 3D food printers are examined, it is understood that each of them is similar to each other and there are points where they differ. Soft substances such as chocolate, dough and cheese are used in the printing made by extrusion. Powdery substances such as sugar, chocolate and oil are suitable for printing with sintering technology. Powder substances such as starch, sugar and protein are used together with liquid binder in the prints made by binder spraying, while substances with low viscosity such as pizza sauce are preferred in inkjet prints (Table 1).

Table 1. Comparison of different 3D food technologies

		Extrusion Based Printing	Selective Laser Sintering	Binder Jetting	Inkjet Printing
	Available material	Chocolate, soft-material such as dough, cheese, meatpuree	Powdered materials such as sugar, chocolate, fat	Liquid binder and powdered materials such as starch, sugar, protein	Low viscosity material such as pizza sauce
Factors affecting printing precision	*Material properties*	Rheological properties, mechanical strength	Melting temperature, flowability, particle size, wettability	Flowability, particle size, wettability and binder's viscosity and surface tension	Compatibility, ink rheological properties, surface properties
	Processing factors	Printing height, nozzle diameter, printing rate, nozzle movement rate	Laser types, laser power, laser energy density, scanning speed, laser spot	Head types, printing rate, nozzle diameter, layer thickness	Temperature, printing rate, nozzle diameter, printing height
	Post processing	Additive, recipe control	Removal of excess parts	Heating, baking, surface coating, removal of excess parts	No
	Advantages	More material choices, simple device	Complex 3D food fabrication, varying textures	Complex 3D food fabrication, full color potential, varying flavors and textures	More material choices, better printing quality, fast fabrication
	Limitations	Incapable of fabricating of complex food designs, difficult to hold 3D structures in post-processing	Limited materials, less nutritious products	Limited material, less nutritious products	Simple food design, only for surface filling or image decoration

Source: Liu, Zhang, Bhandari and Wang, 2017: 85

When the advantages and limitations of printing technologies are examined, it attracts attention as an advantage that it has more printing raw material options and a simple working principle in printing by extrusion than others. The inability to produce fabricated complex foods and the fact that it is almost impossible to hold them after 3D printing draw the limitations of the technique. The fact that complex foods can be written with 3D technology as a fabrication in sintering technique increases their preferability, keeping the tissue diversity more. Limited raw materials and low nutrient substances are the limitations of the technique. Advantages of binder spray printing include printing of complex fabricated 3D products and providing the exact color of the product. Decoupling is one of the advantages of binder spray printing. A limited number of raw materials and low nutrient substances constitute the limitations of the technique. Having more raw material options in inkjet food printers, a higher quality printing and fast production are among the advantages. Simple food designs constitute the limitations that only surface filling or shape filling can be made (Table 1).

When comparing 3D food printing technologies, it is known that the most Decently used method among them is the direct printing technique by extrusion. The main reason for this is that the application methods of the technique are suitable for raw materials of different density (He, Zhang and Fang, 2020:2380).

3D Printer Applications in the Food Industry

The first 3D printing system produced for food was produced by two researchers working at Cornell University in 2007 and introduced with the name "Fab @ Home Model 1". The results of the study have attracted the interest of researchers who want to conduct research in the field. Therefore, various companies have produced new printer models. RIG company has released FoodformTM 3D printer, 3D Systems company ChefjetTM and CocoJetTM printers, Natural Machines company FoodiniTM printer, Choc Edge company Choc CreatorTM printer, Essential Dynamics company Imagine3DprinterTM printer and Makerbot company ReplicatorTM printer (Nachal, Moses, Karthik and Anandharamakrishnan, 2019: 123-124).

In the food industry, 3D printers were used for the first time by "Nanotek Instruments" for the production of a cake in 2001 (Sun et al., 2015: 309).

In 2014, within the scope of the "3D Pasta Printing" project carried out jointly by Barilla and Dutch Applied Research Centre institutions, 3D pasta production was carried out using only durum wheat and water. The pasta takes 5 minutes to print, cook and serve. At the same time, the shape, color and taste of the pasta can be adjusted at the request of the customer (Savastano, Amendola and D'Ascenzo, 2018: 138-139).

In various studies conducted by the Netherlands Organization for Applied Scientific Research (TNO), various foods have been produced using non-traditional materials such as algae, insects, lupines and mushrooms instead of traditional materials (Daniel, 2015).

In the Netherlands, the company Upprinting Food has produced snacks using stale bread, crushed fruits and vegetables in order to evaluate the increasing number of foods. Thus, the aim of minimizing the amount of waste targeted by 3D food printers has been tried to be realized (Meral and Demirdöven, 2022:93).

As a result of the increase in living standards, people's demand for healthy and functional foods has increased. 3D food printers have become a technology that plays an important role at this point. The elements that have functional properties in human nutrition such as protein, fat, dietary fiber, vitamins and minerals are balanced in accordance with the demand in 3D food printing technology. Therefore, it can respond to the expectations of consumer groups with special nutritional needs (He, Zhang and Fang, 2020:2379).

Within the scope of the "Performance" project conducted by the Netherlands Organization for Applied Scientific Research (TNO) in 2012, a pureed food design was carried out for groups of elderly people who have difficulty chewing and swallowing. In the project, the use of functional food items and modified food prints for pregnant women and athletes has been proposed (Liu et al., 2017:91; Noort, Bommel and Renetti, 2017: 273).

These developments in food production have also paved the way for the opening of restaurants where food produced by 3D printing is served. The "FoodInk" restaurant, established in London in 2016, was the first restaurant to produce food with 3D printing. opened in Tokyo in 2020, the restaurant named "Sushi Singularity" has announced its name as the first restaurant that prepares personalized meals by analyzing the nutritional needs of its customers and brings the concept of "Hyper Personalized Menu" to the industry (Aldemir, 2022:160).

CONCLUSION

3D printing technology has shown a rapid development since 2001, when it was first used in the food field. This technology, which is a part and result of the intense impact of technology on today's living standards, is trying to create solutions to many problems in the field of food. The increasing world population and the expected food crisis on a global scale are among these problems. Studies are being carried out with raw materials that are both used in human nutrition in the current situation and made suitable for its use and are being tried to be brought. When we look at these studies, factors such as the 3D food printing system's optimal use of resources compared to the current food production system, minimizing the waste generated during the production process, and being able to produce personalized and instant meals indicate that this technology will become a necessity in human life in the future.

3D food printing technology aims to help produce personalized and nutritious foods that every individual in society can buy by adding a new dimension to processed foods. 3D food printers will also play an important role in the emerging Sunday of digital food. However, 3D foods are still in the early stages of development and need multiple optimization trials before they are introduced to the market. Regarding the safety of printed foods, different food materials should be tested by printing.

The idea of "Downloadable Food" offered by Open Meals company to the sector is quite remarkable. As a result of the analysis of blood samples taken from the consumer, personalized food designs and prints are made by determining the nutritional elements needed by the individual. The "Foodbase" platform, which came into being with the combination of these two ideas, aims to transfer the food printing Decodes prepared specially for the individual to the environment. The codes of those who want to eat these dishes created in this way will be downloaded and printed in 3D (Aldemir, 2022:162).

In the future, it is likely that 3D food printer prototypes will be developed and made available for use by citizens in the public domain. Like other technological devices in human life, this device is one of the expected results of suitable models for home use. Situations such as saving time and raw materials, making production for the immediate needs of a person, making personalized changes in the sensory characteristics of food, preventing food waste by ensuring the desired amount of production, visually highlighting the plates make these devices preferable.

It is thought that when 3D food printing technology is effective in human life, the need to go to a restaurant will decrease. This system gives a person the chance to eat what he wants whenever he wants. Therefore, it is likely to eliminate dependence on a business or person in the food sector. In addition to the money and time savings it will provide to people, it is likely to negatively affect employment, as employees in the kitchen system have the ability to perform many of their job descriptions.

The limited raw materials used in 3D food printing technology and, accordingly, new product development pose another problem. It is very important to diversify sustainable raw materials other than the raw materials currently consumed and to make them suitable for human consumption. In addition, it is necessary to work on issues related to printers such as increasing the ink flow rate, diversifying the print size, and inkjet tip distance.

REFERENCES

Aday, S., & Aday, M.S. (2020). *Üç Boyutlu Yazıcıların Gıda Endüstrisinde Kullanımı*. Türkiye 13. Gıda Kongresi, 21-23 Ekim 2020, Çanakkale, Türkiye, 444 s.

Cohen, D. L., Lipton, J. I., Cutler, M., Coulter, D., Vesco, A., & Lipson, H. (2009, September). Hydrocolloid Printing: A Novel Platform for Customized Food Production. In *2009 International Solid Freeform Fabrication Symposium*. University of Texas at Austin.

Daniel, V. L. (2015). *3D Food Printing Creating Shapes and Texture*. Elesvier. http://refhub.elsevier.com/S0924-2244(17)30082-1/sref8

Dankar, I., Haddarah, A., Omar, F. E., Sepulcre, F., & Pujolà, M. (2018). 3D Printing Technology: The New Era for Food Customization and Elaboration. *Trends in Food Science & Technology*, *75*, 231–242. doi:10.1016/j.tifs.2018.03.018

Değerli, C., & El, S. N. (2017). Üç boyutlu (3D) Yazıcı Teknolojisi ile Gıda Üretimine Genel Bakış. *Turkish Journal of Agriculture-Food Science and Technology*, *5*(6), 593–599. doi:10.24925/turjaf.v5i6.593-599.1062

Godoi, F. C., Prakash, S., & Bhandari, B. R. (2016). 3D Printing Technologies Applied for Food Design: Status and Prospects. *Journal of Food Engineering*, *179*, 44–54. doi:10.1016/j.jfoodeng.2016.01.025

Horvath, J., & Cameron, R. (2020). *Mastering 3D Printing: A Guide to Modeling, Printing and Prototyping*. Apress Media. doi:10.1007/978-1-4842-5842-2

Kim, H. W., Bae, H., & Park, H. J. (2017). Classification of The Printability of Selected Food for 3D Printing: Development of An Assessment Method Using Hydrocolloids As Reference Material. *Journal of Food Engineering*, *215*, 23–32. doi:10.1016/j.jfoodeng.2017.07.017

Lipton, J., Arnold, D., Nigl, F., Lopez, N., Cohen, D., Norén, N., & Lipson, H. (2010, September). Multi-material Food Printing with Complex Internal Structure Suitable for Conventional Post-Processing. In *2010 International Solid Freeform Fabrication Symposium*. University of Texas at Austin.

Lipton, J. I., Cutler, M., Nigl, F., Cohen, D., & Lipson, H. (2015). Additive Manufacturing for The Food Industry. *Trends in Food Science & Technology*, *43*(1), 114–123. doi:10.1016/j.tifs.2015.02.004

Liu, Z., Zhang, M., Bhandari, B., & Wang, Y. (2017). 3D Printing: Printing Precision and Application in Food Sector. *Trends in Food Science & Technology*, *69*, 83–94. doi:10.1016/j.tifs.2017.08.018

Meral, H., & Demirdöven, A. (2022). Gıdanın Özelleştirilmesi ve Detaylandırılmasında Üç Boyutlu (3D) Yazıcının Kullanımı. *Gida*, *47*(1), 91–106. doi:10.15237/gida.GD21112

Nachal, N., Moses, J. A., Karthik, P., & Anandharamakrishnan, C. (2019). Applications of 3D Printing in Food Processing. *Food Engineering Reviews*, *11*(3), 123–141. doi:10.100712393-019-09199-8

Noort, M., Van Bommel, K., & Renzetti, S. (2017). 3D-printed Cereal Foods. *Cereal Foods World*, *62*(6), 272–277. doi:10.1094/CFW-62-6-0272

Şahin, K., & Turan, B. O. (2018). Üç Boyutlu Yazıcı Teknolojilerinin Karşılaştırmalı Analizi. *Stratejik ve Sosyal Araştırmalar Dergisi*, *2*(2), 97–116. doi:10.30692isad.441648

Savastano, M., Amendola, C., & D'Ascenzo, F. (2018). How Digital Transformation is Reshaping The Manufacturing Industry Value Chain: The New Digital Manufacturing Ecosystem Applied to A Case Study From The Food Industry. In *Network, Smart and Open: Three Keywords for Information Systems Innovation* (pp. 127–142). Springer International Publishing. doi:10.1007/978-3-319-62636-9_9

Şeker, İ. T., & Karaman, E. E. (2020). Dijital Gastronomi. Abdullah Karaman ve Kürşad Şahin (Ed.) Dijital Turizm: Sektörün Yeni Geleceği içinde (189-196). İstanbul: Eğitim Yayınevi.

Soares, S. (2011). *Insects au Gratin*. Sunsan Soares. http://www. susanasoares.com/index.php?id=79

Sun, J., Peng, Z., Zhou, W., Fuh, J. Y., Hong, G. S., & Chiu, A. (2015). A Review on 3D Printing for Customized Food Fabrication. *Procedia Manufacturing*, *1*, 308–319. doi:10.1016/j.promfg.2015.09.057

Yang, F., Zhang, M., & Bhandari, B. (2017). Recent Development in 3D Food Printing. *Critical Reviews in Food Science and Nutrition*, *57*(14), 3145–3153. doi:10.1080/10408398.2015.1094732 PMID:26479080

Zoran, A., & Coelho, M. (2011). Cornucopia: The Concept of Digital Gastronomy. *Leonardo*, *44*(5), 425–431. doi:10.1162/LEON_a_00243

ADDITIONAL READING

Aldemir, T. (2022). Üç Boyutlu (3D) Gıda Yazıcıları. Tuba Aldemir (Ed.) Güncel Gastronomi Trendleri içinde (151-168). Ankara: Nobel Akademik Yayıncılık.

Dankar, I., Haddarah, A., Omar, F. E., Sepulcre, F., & Pujolà, M. (2018). 3D Printing Technology: The New Era for Food Customization and Elaboration. *Trends in Food Science & Technology*, *75*, 231–242. doi:10.1016/j.tifs.2018.03.018

Godoi, F. C., Prakash, S., & Bhandari, B. R. (2016). 3D Printing Technologies Applied for Food Design: Status and Prospects. *Journal of Food Engineering*, *179*, 44–54. doi:10.1016/j.jfoodeng.2016.01.025

Kim, H. W., Bae, H., & Park, H. J. (2017). Classification of The Printability of Selected Food for 3D Printing: Development of An Assessment Method Using Hydrocolloids As Reference Material. *Journal of Food Engineering*, *215*, 23–32. doi:10.1016/j.jfoodeng.2017.07.017

Yang, F., Zhang, M., & Bhandari, B. (2017). Recent Development in 3D Food Printing. *Critical Reviews in Food Science and Nutrition*, *57*(14), 3145–3153. doi:10.1080/10408398.2015.1094732 PMID:26479080

Chapter 13
Assessing Food Safety and Food Waste Relatively Among Food Handlers' Practices in Developing Countries

Birsen Bulut Solak
https://orcid.org/0000-0003-1583-1504
Selcuk University, Turkey

Feridun Aydinli
https://orcid.org/0000-0002-0230-5129
Selcuk University, Turkey

ABSTRACT

Food waste is among the most important natural problems that threaten the future. In the last decade, food waste has received increased attention on both academic and societal levels. Food waste (FW) and food loss (FL) are also ecological, economic, and social problems. In developing countries, food handlers are one of the biggest sources of food waste due to lack food safety knowledge. It is figured out that people working in food production companies, such as restaurants and hotels, may cause food waste due to incomplete or incorrect information about food safety. It is necessary to have a clear understanding of the factors influencing food waste-related food handlers' perceptions, knowledge, and behaviours. Examining the hygiene and sanitation habits of food handlers and the level of knowledge of food handlers and their practice in the field are other reasons to carry out this book chapter linking with food waste.

INTRODUCTION

The continuity of life on the earth depends on the balance of the environment living in and the continuation of the cycle's especially ecological cycle, of substances such as water, oxygen, CO_2, nitrogen, sulphur, magnesium, which are indispensable components for human beings' existence. All living things survive their lives in ecosystems, which are the habitats of the earth. Pollution of substances necessary

DOI: 10.4018/978-1-6684-9094-5.ch013

for life such as air, water, soil or a disruption in oxygen, water, CO_2, nitrogen cycles makes the life and habitats for living things difficult (Akın, 2014; Kılınç Şahin & Bekar, 2018: p, 1040). Food is defined as liquid, solid, frozen or dried nutrients consumed by living organisms for their survival and development (Yüzüncüyıl & Coşkun, 2020).

Food waste and rapid world population growth have become a global problem, with people facing hunger in many countries and negative socio-economic and environmental impacts (Gilland, 2002: p. 48; Parfitt et al., 2010: pp. 3078-3079; Yılmaz, 2022: p. 1). Food waste also ranks first among global problems such as malnutrition of individuals, hunger and famine (Ademoğlu, 2021: p. 64). For this reason, food waste is one of the important issues that have recently come to the fore all over the world (Marra, 2013; Kılınç Şahin & Bekar, 2018: p. 1040).

Food waste is a complex issue that seriously affects natural resources, economy, agriculture, soil fertility, environment and natural balance, health, global warming and climate, international relations and policies, food security, development efforts, social equality and income distribution and many other areas (T.R. Ministry of Trade, 2018; Tekiner et al., 2021: p. 124). So that, food waste directly threatens environmental, social and economic sustainability. Therefore, in order to understand food waste, the concepts of food waste and food loss need to be properly understood. FAO defines food waste as the loss of food products suitable for consumption within the food supply chain or waste due to behavioural reasons such as poor stock management, negligence; while the US Environmental Protection Agency defines it as unconsumed food and beverages or their residues that occur in residences, commercial and institutional establishments. Based on these definitions, food waste is generally defined as edible materials that are not ultimately consumed or left over for a number of sensory and physiological reasons (Yiğitoğlu & Cerrah, 2021). The concept of food waste is defined as unnecessary expenditure of money, time and labour wastefulness (TDK, 2020; Demir, 2022: p. 6). As a cause of negative economic, environmental and social effects, food waste is considered to be one of the sustainability issues that need to be addressed.

Food loss, which is another concept, refers to the losses that occur in the food production chain due to administrative and health-related factors. In this context, Ceylan et al. (2017) defines the concept of food loss as the inability to use non-consumable parts such as shells, stems and seeds despite having the necessary controls, product processing skills and knowledge in the process of processing food products (Çetin, 2022: p. 8). In other words, food losses are the reduction in the amount of edible food available for human consumption at various stages of the supply chain. Food waste, which briefly refers to spoilt and discarded foods, is called the event that occurs for many reasons in the process from the field, which is the first beginning of production, until it reaches the houses.

Food and Agriculture Organization (FAO) states that approximately one third of all edible food produced for human consumption is wasted or lost after harvest (Karakas, 2019: 467). In order to minimise these problems, food resources should be used and consumed efficiently. From this point of view, it is of critical importance to determine the extent of food waste generation and the reasons for the generation of these wastes, to take preventive measures for food waste and to realise successful applications for recycling, especially in terms of human health, environment and economy. Reducing food waste and food safety issues are key points defined in the nutrition strategy of 9 billion people in the future (Godfray et al., 2010; Parfitt et al., 2010). While over 840 million people in the world lose their health due to malnutrition, from an ethical point of view, it is unacceptable to throw away food (Betz et al., 2015; Pirani and Arafat, 2014: p. 328; Altınel, 2011: p. 175). Although more than 800 million people in the world struggle with hunger, one third of the food produced for human consumption turns into waste before reaching the consumer. If this picture prevailing in the world must not be prevented, it is thought that the diminishing resources will not be able to maintain the supply-demand balance in the future (Drewitt, 2013; Cirişoğlu & Akoğlu, 2020: p. 39). Along with the innovations brought by globalisation, changing consumption habits have brought along the waste of resources.

This chapter of the book aims to find and investigate food safety deficiencies in a coordinated manner during food processing. In addition, the authors will investigate the level of awareness of food processors on food safety and food waste by analysing scientific articles and books. It is also aimed to collect some information about food waste and food safety by paying attention to food handlers' evaluations about food waste and food safety. The examination of food handlers' hygiene and sanitation habits, knowledge levels and practices in the field is another reason to conduct this research on food waste. Additionally the authors will research the awareness levels of the food handlers about food safety and food waste according to reviewing the scientific articles and books.

CONCEPTUAL FRAMEWORK

According to Bontoux and Leone (1997), the concept of waste is substances or objects which are disposed of, or planned to be disposed of, or intended to be disposed of in accordance with the conditions of national law. In order to understand the concept of food waste, it is first necessary to examine the concept of food. Focusing on food waste is important because there are also many definitions in the literature about food waste. Each case of food waste can be explained as a 'rational' economic decision based on comparing the cost of time and materials (infrastructure, proper storage, etc.) required to prevent food waste with the value of the food at risk of waste. To simplify, food wastage occurs in many areas starting from agricultural production to the production, distribution and consumption of food (FAO, 2013: pp. 7-8; Yılmaz, 2022: p. 20). However, the decision-making process of actors in the supply chain is further complicated by various considerations (Bajželj et al., 2020).

BACKGROUND

Food Waste

As it is mentioned before, food waste consists of all edible materials in the food supplies that are intended for human consumption but are ultimately not consumed, lost, spoilt or contaminated. Food waste is into three different types: avoidable, partially avoidable and unavoidable food waste (Papargyropoulou et al., 2014). Preventable food waste is when food is still usable and consumable by people before it is thrown away. For example, whole bread that is not taken and eaten, food that is not consumed after it is put on the plate, sliced bread, fruits, packaged foods that are not expired (Ilyasov, 2017). Most preventable food losses occur at the household, processing and agricultural production stages of the food value chain. Households are responsible for almost half of total avoidable losses (in terms of calorific content) (Durmuş, 2022). Venkat (2011) found that avoidable food waste in the US exceeds 55 million metric tonnes per year, about 29% of annual production, and that these wastes produce at least 113 million metric tonnes of CO_2 per year, equivalent to 2% of national emissions and life-cycle greenhouse gas emissions, costing \$198 billion. Avoidable wastes are partially foods that some of people consume and the others do not consume (bread crumbs, potato peelings) (Akar, 2022). Beretta et al. (2013) determined that 23% of the energy of the purchased food is wasted and 5% of the wasted energy is partially avoidable, considering the Swiss consumer basket and the average calorie content of each food category. Unavoidable wastes are considered to be parts of manufactured products that cannot be consumed by humans. Elements such as bones, egg shells, nerve and fat parts of meat products, fruit and vegetable residues, nut shells can be shown as examples of unavoidable food waste (Lebersorger and Schneider, 2014; Riesenegger and Hübner, 2022; Çetin, 2022). In the UK, 8.3 million tonnes of food is annually wasted in average households. 1.5 million tonnes of these losses are unavoidable food waste (Quested et al., 2011: 461).

Causes of Food Losses

The problem of food waste is increasing, covering all sectors of waste management from collection to disposal; identifying sustainable solutions involves all parties contributing to food supply chains, agriculture and industry sectors as well as retailers and final consumers. Identifying the causes of food losses is the first step towards reducing them and identifying prioritised actions. Losses and wastage along the food supply chain are often due to interrelated causes. As with a production line, actions at any stage in the chain can affect the whole chain, some of which can be traced back to the harvest or even the pre-harvest period. In doing so, it is seen that not all causes are exactly "on the same level". There are various causes of food losses and wastage, including inadequate planning, buying more food than necessary, errors during industrial processing, managerial, financial and technical difficulties (Rohini et al., 2020, pp. 691-692; Aday and Aday, 2021: p. 293) and at all stages of the food supply chain, including the consumption stage (Martin-Rios et al., 2018).

Losses in agricultural production vary between 15-40%, and the distribution of this rate on the basis of supply chain processes is between 4-12% during harvesting, 2-8% during the transport of products to the market or wholesales market, 5-15% during the market preparation stage, 3-10% during the storage process and 1-5% at the consumer stage (Özdal, 2019: p. 17). Some articles have analysed the behaviour of companies and the population at different levels (home, restaurant, retail) to assess the factors affecting the waste of food products (Glanz and Schneider, 2009; Schneider and Lebersorger, 2009; Silvennoinen et al., 2012; Quested et al., 2013; Katajajuuri et al., 2014; Garrone et al., 2014; Graham-Rowe et al., 2014; Mena et al., 2014). In this direction, the factors affecting food losses based on the literature are listed below in general terms.

Pre-Harvest Factors and Unharvested Part

Environmental conditions as important factors appear in the first stage of food loss and waste (Akar, 2022: p. 28). Food loss during the process is caused by several factors; natural factors (weather), biological factors (insects, pests and diseases), economic factors (food prices in the market, cost of harvesting) and technological factors (nature and type of harvesting equipment and harvesting process) (Tirkey, 2011; Abdul- Kareem, 2021: p, 99). Damage occurs to crops, resulting in loss of market value and quality or complete destruction (ACF, 2014; Abdul Kareem, 2021: p, 99). This is occured because of differences in production and agricultural practices which can result in differences in quality at harvest, differences in suitability for transport, differences in storage durability and differences in post-harvest shelf life. In addition, failure of crops to meet certain quality standards in terms of shape, size and weight as prescribed by retailers, processors or target markets results in crops being left in the fields unharvested (Stuart, 2009; Abdul Kareem, 2021: p, 99). 17.7 million tonnes of agri-food products in Italy, accounting for 3.25% of total production, could not be harvested in 2009 (Segre et al. 2014). In the USA, it is stated that an average of 7% of cultivated areas are not harvested every year (Segre and Falasconi, 2011; Demirbaş, 2019: p. 290). Failure to follow meteorological forecasts, not having soil analysis, not following appropriate pesticide applications and not planning the timing of harvest, damages to the field before harvest due to biological and biotic factors such as weeds, insect pests and diseases are among the other causes of food losses in this area. Although the level of these damages is significant, they are not included in the scope of "food loss and waste" (FAO and Ministry of Agriculture and Forestry, 2020). Sometimes, failure in the harvest period is ableto be due to meso-level reasons, such as economic reasons such as low market price and high labour cost during the same period.

Harvest Process Losses

This food loss occurs during the period from the beginning to the end of the harvest. Improper planning of harvest time, careless handling of the product and improper packaging are able to lead to food loss. Incorrect harvesting techniques are also effective in increasing food losses. Perishable foods such as fruits and vegetables, exposure to more than one treatment increases the damage. Temperature and humidity management play a key role in maintaining the safety and quality of all foodstuffs. For example, failure to maintain the product at the appropriate temperature and humidity level immediately after harvesting significantly contributes to product spoilage at later stages of the food value chain (FAO and Ministry of Agriculture and Forestry, 2020).

Transport, Logistics and Warehousing

The most serious reduction in the nutritional value and quality of fresh food produce occurs immediately after harvest. Therefore, transport is considered to be the most important in preventing losses after harvest. Weight loss of harvested products is able to be very important to cause food deterioration during transport and storage (GTHB, 2018; Demirbaş, 2019: p. 291). Since there is a certain time between transport, production and consumption, the risk of physical, chemical, biological hazards and the risk of cross-contamination are able to be one of the main causes of food loss, especially for fresh products. The reasons for food losses due to transport and logistics are summarised as inappropriate vehicle loading and unloading methods, poor packaging, long transport processes, poor road structures, inadequate cooling and humidity adjustment systems during the transport process, uncontrolled movement of products in the vehicle, lack of cooling before loading, transporting products that are not suitable to be transported together, food losses caused by the vehicle driver (FAO and Republic of Turkey Ministry of Agriculture and Forestry, 2020). The transport of food products in developed countries is provided by refrigerated vehicles. Malfunctions, accidents and loading and unloading delays that may be occured in the vehicles cause food loss. On the other hand, some factors in developing countries are such as bad roads, lack of refrigerated vehicles, poor logistics management which prevent the proper preservation of food (Rolle, 2006: p. 29; Alıntaş, 2021: p. 19). At the same time, it causes various damages to the food products due to the unqualified employees carrying the products during the loading and unloading phase in developing countries. Since the products are transported by compressing for less transport cost, food losses are in question (Kader, 2002: p. 535; Alıntaş, 2021: p. 18).

Food products are stored in various places according to their intended use and time (Kurt, 2019: p. 24). The storage system is very important in terms of food losses. With the storage of products, time management is able to be done properly and time can be saved in marketing and consumption (FAO and Republic of Turkey Ministry of Agriculture and Forestry, 2020). However, post-harvest storage problems are one of the most important causes of food loss in countries without adequate storage facilities (Alıntaş, 2021: p. 18). Inappropriate storage conditions lead to undesirable changes in the taste, odour and appearance of foodstuffs. Products such as root crops may face unfavourable conditions such as sprouting and deterioration due to inappropriate storage conditions (Stuart, 2009).

Processing and Packaging

Food processing is defined in a broad sense, from simple packaging of fresh products to extensive cooking or preservation processes (Akkerman, Farahani and Grunow, 2010: p. 865; Demir, 2022: p. 36). In other words, food processing is the transformation of raw foods into food products for suitability for consumption, cooking or storage (European Food Information Council 2016). However, the processing losses are losses resulting from industrial and local processing (Kummu et al., 2014: 479). Factors that cause food losses during food processing include inadequate infrastructure, machinery, inefficient system design, damage during production, inaccurate supply and demand forecasting in production, contamination, trimming and lifting, supply/demand issues, inconsistent/confusing date labels, inconsistency in the quality of ingredients, food safety issues, production line changes, cold chain deficiencies, plant employee behaviour (CEC, 2017).

The food processing and manufacturing industry generates food losses and food waste throughout the entire production phase due to damage during transport or inappropriate transport systems, problems during storage, losses or contamination during processing, inappropriate packaging. Food losses during the processing stage are generally caused by technical errors and inefficiencies caused by personnel or production. Failures made during processing lead to incorrect size, weight, shape, appearance or damaged packaging in the final product. Although all these defects do not have a negative impact on the quality and safety of food, food that does not have the standards may have to be discarded. Furthermore, the lack of standards to ensure food safety and quality through appropriate processing management result in some processed products being unsafe and nutritionally poor (FAO and Ministry of Agriculture and Forestry, 2020).

The main reason for the loss of foods during processing and handling is contamination. The contamination is the contamination of food with harmful substances or microorganisms. This situation occurs when hygiene is not provided. In addition, contaminated foods are thrown away because they are harmful to human health. Some processed foods are not trusted due to the lack of food safety, inadequate processing management, lack of certain standards and quality practices in foods. These foods with low nutritional value are not purchased by consumers. This situation causes food waste (Oral, 2015: p. 25; Akan, 2022: p. 32).

Moreover, there are some researches on the use of food waste and food by-products that occur during food processing in various ways, both to provide economic gain and to eliminate the damage of these wastes to the environment. Regarding the situation, Redlingshöfer et al. (2017) stated that the use of food wastes released during primary production and processing as animal feed is a method used to reduce food waste. Mete (2012) stated that average 85 kilos of every 100 kilos of milk in cheese processing is converted into whey. Then, whey is used in areas such as whey butter, soups, hydrolysed protein, cheese, cream cheese, animal food, bread, cake, additive, sugar, children's food, lactose syrup, pill coating, penicillin, acrylic-plastic production, especially in developed countries. The author also emphasises that the whey released in cheese production should be utilised in bread making and in this case, both economic and environmental benefit will be obtained from whey and the amount of water needed for bread production will be saved.

Semolina flour released during the processing of durum wheat is used in bakery products to increase the amount of gluten in cereals such as rye and oats. The use of apple pulp as a source of fibre in bakery products has also been investigated. There are some research studies' stating that tomato pulp added at the rate of 10% improves the protein quality of bread and gives good sensory properties to bread (Yağcı et al., 2006: pp. 499-501). Ethanol, baker's yeast, single cell protein, lactic acid and ethyl alcohol are produced from molasses left as waste in sugar factories (Şener and Ünal, 2008: p. 1036). The scales, skin, bony structures of fish and wastes obtained from other parts, gelatin, collagen and many products are used in many areas such as health, biomedical, personal care and food (Bozkurt & Yüksel, 2019: p. 10; Demir, 2022: p. 37-38).

Packaging is an important element to extend shelf life and prevent water. While reducing packaging is an important element of waste policies, increasing the amount of wasted food in unpackaged products causes an undesirable result. Since packaging is relied upon by all actors in the food chain to maintain food safety and quality from the moment of harvest to the final consumer, packaging optimisation should be considered and proper packaging planning should be carried out (FAO and Ministry of Agriculture and Forestry, 2020).

Food Safety and Food Waste

Food losses (95%) in developing countries are involuntary losses in the early stages of the food supply chain (FAO, 2018). Nutrient losses can also be explained by factors such as decreased nutritional value, undesirable changes in taste, texture and colour, or quantity measured by reduced weight or volume (Bagherzadeh et al., 2014: 6). Food safety is the discipline that explains how food is handled, processed and stored in a way to prevent food diseases (Bulut Solak, Aydınlı, & Bin Amin, 2023). Food safety, that is, the assurance that food will not harm the consumer when prepared and consumed according to its intended use, continues to be a global problem affecting the health of societies in both industrialised and developing countries (Guillier et al., 2016). At the same time, food safety, which is one of the driving forces of global problems, brings along food waste. Misperception of food safety frequently leads individuals to dispose of otherwise edible food.

Food safety refers to the whole of the measures taken to eliminate physical, chemical, biological and all kinds of damage that may occur in foods (Koc and Uzmay, 2015). It is known that food production goes through many stages from the field to the table. During food supply stages, there is an increase in food waste at some food production stages. The sensitivity of consumers on food consumption in recent years has also increased because of food-borne diseases and food safety and quality standards at the food production (Mutlu and Yurdakul, 2008). When the lack of knowledge of producers and consumers about food safety is eliminated, reductions in food waste can be probably experienced. Carrying out studies to increase food safety in food and beverage businesses also prevents some of the food waste.

Many foods turn into waste before they are ready for consumption, especially due to incomplete and inadequate practices regarding food safety in food and beverage establishments, cross-contamination of food processors, and lack of knowledge of pathogen microorganisms. However, misperceptions of food safety often lead individuals to discard edible food (Neff, Kanter, & Vandevijvere, 2015). For example, individuals often use date labels (date of expiry, date of last freezing) as an anchor when assessing product freshness and safety (Aschemann-Witzel, 2018). However, the dating of most products is related to product quality and dates are not intended to be used as safety indicators (USDA 2019). Evidence suggests that the closer the food is to the date label, the more unsafe or risky the food is perceived to be. This occurs despite the absence of any visual or sensory evidence to indicate spoilage, demonstrating that the dates act as a heuristic that overrides evidence based-examination or an overreliance on the anchor provided by the date on the package and an inability to use food-related cues (e.g., appearance, smell) to make appropriate adjustments (Wilson et al., 2017). Consumers are often influenced by non-evidential prejudices (date labels, product and packaging defects) and this leads to a distorted perception of food safety. Experts emphasise that the decision to destroy food should be based on evidence-based observation (sensory changes in visual appearance, colour, smell or texture) (FDA, 2019). Still, other studies demonstrate that imperfect, blemished, or misshapen foods can lead to misattributions and greater food waste not through contagion, but through automatic categorization of these foods as "bad". Cooremans and Geuens rely on categorization theory-more specifically, the premise that "abnormal is bad" and "ugly is bad" to demonstrate that people infer lower quality, poorer taste, and lower intentions to purchase produce that is abnormally shaped (2019).

Retail

The definition of retail, which is a set of actions related to the supply of goods or services directly to the final consumer. It is defined in the Turkish Income Tax Law as follows: it is the sale of the goods and materials sold to those other than those who sell them as they are or after they have been processed (Öztürk, 2006: p. 69). Retailers have a significant impact on the supply chain as they offer food directly to the end consumer. Conditions at the presentation stage (temperature, humidity, light, ventilation, etc.), appropriate display conditions that minimise contact with consumers, and practices in areas such as loading or transport also affect the quality of the product, its shelf life and customers' decisions to purchase the product (FAO and Republic of Turkey Ministry of Agriculture and Forestry, 2020). Fruits, vegetables, bakery products and unpackaged foods are highly wasted in retail locations. The reason why they are wasted the most is that these foods are more affected by ambient temperature, humidity and microbial flora compared to other foods. The retail system generates food waste at various stages of food distribution and purchasing: damaging during transport or inappropriate transport systems, problems during intermediate storage, losses during processing or contamination, inappropriate packaging, preservation or handling problems, a lack of refrigeration/chill storage (Aiello et al., 2014). The retail sector is responsible for 5% of global food waste and loss (TISVA, 2020). Retailers sell foods with a distant expiry date to consumers. Therefore, retailers do not desire to buy food products that are close to their expiry date from producers. Although there is no harm in consuming a product with an approaching expiry date, such an attitude of retailers causes food waste (Oral, 2015: p. 30).

Consumption

Consumption is defined as the purchase of goods and services in the market, the spending of income and the expenditure of households for goods and services (Kocacık, 1998: p.18; Durmuş, 2022: p. 7). Moreover, consumers make a greater contribution to food waste than food producers, distribution function, market retailing and accommodation sectors (Griffin et al., 2009; Quested et al., 2011; Graham-Rowe et al., 2014: 15). According to one research study conducted by Özdemir and Güçer (2018) in hotel businesses, it was stated that although all institutional measures have been taken, a high rate of food waste occurs due to the lack of food awareness of customers.

Increasing income levels and demographic changes over the last 20 years have led to differences in eating habits. Different dietary preferences such as a massive increase in processed food consumption increased per capita consumption of meat, poultry, dairy and seafood products, and the emergence of obesity problems even in the poorest segments of the population are examples of these changes. However, as the wealth level of households increases, consumers' food waste may also increase (FAO and Republic of Turkey Ministry of Agriculture and Forestry, 2020). In addition, some factors cause food waste at the consumer level including inadequate planning, expiry of "use-by" dates, sometimes careless behaviour of consumers (Gustavsson et al, 2011), inappropriate storage of food, throwing away the product before its shelf life expires, not using the leftover food in a different meal, buying more than the amount needed, cooking more than the amount consumed and not eating it (WRAP, 2009; HISPACOOP, 2012; Baptista, Campos, Pires, & Sofia, 2012; Alıntaş, 2021: p. 20). This inevitably means that a large amount of resources used in food production are unnecessarily utilised, resulting in increased greenhouse gas emissions (Gustavsson et al., 2011). Overbuying food is also common consumer behaviour (Evans, 2012), and preparing and serving larger meals results in more uneaten and wasted food (Roe et. al., 2020). In short, consumers buy too much food and over-prepare food, which can lead to waste when these food products are not consumed (Block, Vallen, & Austin, 2022).

FOOD LOSSES AND FOOD WASTES IN DEVELOPING COUNTRIES

In the latest report on the State of Food Security and Nutrition in the World (SOFI), FAO reported that the number of people affected by hunger in the world reached 828 million in 2021, an increase of nearly 46 million since 2020 and 150 million since 2019. However, 3.1 billion people do not have access to a healthy diet. Meanwhile, according to FAO's report (2019), about 14% of the world's food (worth $400 billion annually) continues to be lost after it is harvested and before it reaches stores (FAO, 2022). UNEP's Food Waste Index Report shows that 17% of our food is wasted by consumers at retail and especially in households. According to FAO estimates, lost and wasted food could feed 1.26 billion hungry people every year (UNEP, 2021).

Depending on the population growth rate in the world, food wastage is also increasing, creating a favourable environment for other environmental problems such as environmental pollution, health risks and reduction of empty spaces (Tekiner et al., 2021: p. 124). Food loss and food waste occur at different stages of the production chain and are caused by different driving forces (Demir, 2022: p. 11). Therefore, the causes and sources of food waste vary. The causes of food waste differ from country to country (Heikkilä et al., 2016: 452; Joardder and Masud, 2019: 237), and the causes of food waste generation vary greatly when developed and developing countries are compared (Chalak et al., 2016: 418; Öcal, 2021: p. 14).

Developing economies and the inhabitants in the developing countries are heavily dependent on agriculture and livestock. According to the World Bank, the contributions of the food sector to gross domestic product (GDP) in developing countries are up to 30%. In South Asia, Africa and East Asia, up to 70% of the population lives in rural areas. In monetary terms, global food loss and waste is estimated at US$ one trillion, of which one third is lost in developing countries. However, many smallholder farmers in developing countries live on the margins of food insecurity and reducing food losses can have an immediate impact and has a significant impact on their livelihoods. Therefore, reducing food wastes in developing countries will not only increase food and nutrition security, but also increase their income levels and minimise the negative impact on the environment (Demirbaş, 2018: p. 522).

It is known that about half of the food waste in the world is experienced in developing countries (İlkbahar & Köksel, 2016; Özdal, 2019: p. 20). While 56% of food loss and waste occur in developed countries, the remaining 44% occur in developing countries (Gustavsson et al., 2011). 40% of food waste in developing countries occurs during pre-harvest, harvest and processing, while it occurs during retail chains and consumption in developed countries (Sputnik, 2019). In addition to these, the rate of food loss in developing countries is 14% in production, transport and storage processes, 2% in processing and packaging, 7% in distribution and marketing, and 7% in consumption (Lipinski, 2013: p. 8). This is due to inadequate infrastructure in developing countries, low education level of farmers and low investments in food production systems (Arah et al., 2016; Ali et al., 2021).

In Mexico, a developing country, 34.50% of the food reserved for human consumption is lost along the food supply chain. In addition to costing $ 23,380 million for Mexico alone, this rate generates more than 36.9 Gt CO_2 and causes the loss of more than 39.8 billion m^3 of fresh water (Aguilar, 2017; Lopez-Sanchez, 2021: p. 2). In the Republic of South Africa, the main sources of food waste and loss are domestic, fruit and vegetable and meat processing and distribution processes, with total wastage in the entire food supply chain reaching $ 7.7 billion (Termeer et al., 2018; Adenuga et al., 2020; Tekiner et al., 2021: p. 125). According to studies conducted by the Federation of Food and Beverage Industry Associations of Turkey, the amount of food wasted annually in Turkey is 26 million tonnes. This amount corresponds to 214 billion Turkish Liras. Of the 49 million tonnes of fruit and vegetables grown in Turkey in a year, 25-40% is lost or wasted. In addition, 4.9 million loaves of bread are wasted daily and 1.7 billion loaves of bread are wasted annually in Turkey (TISVA, 2019). According to the 2021 Food Waste Report presented by the UN; more than 12.5 million tonnes of food is wasted every year in Brazil, which is among the 10 countries with the highest food waste in the world, approximately 3.5 million tonnes in Argentina, which is in the Latin America region and in the category of developing countries, 92 million tonnes in China, where the most food waste is experienced, more than 2 million tonnes in Poland, approximately 21 million tonnes in Indonesia, approximately 500 thousand tonnes in Bulgaria and approximately 5 million tonnes in Russia. In addition, Russia wastes 33 kg of food per capita (the minimum amount of waste) (UNEP, 2021).

Economic Losses

Around 4 billion tonnes of food is produced in the world every year. Of this, 1.3 billion tonnes (1/3) of the food produced is thrown into the garbage for various reasons. According to the United Nations Food and Agriculture Organisation data, the monetary value of this amount is $ 990 billion (Gemeç & Gürbüz, 2022: p. 166). In addition to quantitative losses, food products can also suffer from quality deterioration, resulting in economic and nutritional losses (FAO, 2019). Moreover, economic impacts arise from the costs associated with food waste and their impact on farmers' and consumers' incomes (Lipinski et al., 2013; Papargyropoulou et al., 2014). Losses and wastes in food products cause great damage to national economies because when it is considered all the stages of a food from production to the consumer as a whole, the resources used in the preparation of agricultural lands (labour, water, fertiliser, pesticides, diesel oil, etc.), the time spent in the production and supply chain (the time required to procure, transport and store the product) are also wasted along with the wasted food (Vilariño et al., 2017: p. 3; M. Aday & S. Aday, 2021: p. 297). The loss of any edible food by consumers due to their own preferences negatively affects all individuals, especially the person himself/herself, the organisations in the food supply chain and the national economy (Thyberg, 2015: p. 16; Ilyasov, 2017: p. 43). It is also reported that households in the United States throw away 25% of the food they buy. This waste figure also corresponds to more than 25% of drinking water and the loss of 300 million barrels of oil. According to another study conducted in the USA, the rate of food loss during harvest is 31%. It is known that one third of the food loss is in the grocery and retail chain ($ 165.5 billion) (Ceryes et al., 2021; Tekiner et al., 2021: p. 124). In Finland, the amount of food waste in households is the same as the CO_2 emitted by 100 thousand vehicles. The sources of this wastage are households, restaurants, the food sector and the retail sector, and the financial equivalent is € 550 million (Katajajuuri et al., 2014). Food loss in New Zealand is estimated at $ 872 million per year, $ 26.6 billion in Canada and £ 15 billion in the UK (Reynolds et al., 2016; Tekiner et al., 2021: p. 125). It is stated that the approximate value of edible food wasted in the United Kingdom (UK) is € 12 billion (Quested et al., 2011: 460). It is estimated that Spanish citizens cause food waste of approximately € 180 per person per year (Garcia-Herrero et al., 2018: 24). In Germany, it has been determined that it causes a loss of $ 331 per person at the retail and consumer level (Bräutigam, Jörissen, & Priefer, 2014: p. 684; Öcal, 2021: p. 21).

Ecological Losses

Food waste has negative effects on the environment. Foods interact with the environment from their production to their consumption. Energy, water and soil losses during this period damage nature (Tekiner, 2023: p. 6). Food waste weakens the long-term resilience of the global food system by increasing ecosystem damage which is related to food security (Bajželj et al., 2020). Each food loss and waste leads to unnecessary greenhouse gas emissions, water and land degradation (Lipinski et al., 2013: p. 9; Ilyasov, 2017: p. 45). Food loss and waste (FLW) accounts for 8-10% of global greenhouse gas emissions (GHGs) associated with the food system in the futurein scenarios for 2050, contributing to an unstable climate and extreme weather events such as droughts and floods (FAO, 2022). In addition, changes in ecological balance and loss of biodiversity due to the impact of food waste cause 250 km^3 of water and 30% of the world's agricultural land to be wasted (Vilariño, et al., 2017: p. 21; Rohini et al., 2020: p. 690; Öcal, 2021: p. 18). Food waste in the landfill decomposes without oxygen and releases a strong methane gas into the environment. This means that for every tonne of food sent to landfill, one third of the greenhouse gas equivalent to one tonne of CO_2 gas will be released. Europe, North America, Oceania and industrialised Asia have the highest per capita carbon footprint of food waste (around 700- 900 kg CO_2 equivalent per person per year), while Sub-Saharan Africa has the lowest per capita carbon footprint (around 180 kg CO_2 equivalent). It is emphasised that the blue water footprint (i.e. consumption of surface and groundwater resources) caused by food waste is approximately 250 km^3 and that 1.4 billion hectares of land and 28% of the world's agricultural area are used annually to produce food that is lost or wasted every year (Racz, Marchesi, & Crnković, 2018: p. 34; Demir, 2022: p. 56). Beyond the numerical impacts, the natural environment and the ecosystem services provided by it are also negatively affected by the resources used in the production of lost and discarded food (Songür & Çakıroğlu, 2016: p. 22).

CONCLUSION

It was determined that there is limited number of studies on waste and food safety. When the food waste generated on both household and business basis is totalled, it is seen that the amount of food waste generated constitutes a significant amount for the economies of the develeoping countries. The authors could be able to identify actions that governments, societal stakeholders and retailers can undertake to reduce food handlers' related food waste, highlighting that synergistic actions among all food preparation stages with food safety knowledge. Some results show that food handlers' true knowledge to avoid food waste, their management skills of food provisioning and food handling and their trade-offs between priorities have an extensive influence on their food waste behaviours. It is thought that the results obtained in this chapter will contribute to the gap in the literature partly. It is also thought that the evaluation of surplus food will contribute to businesses in terms of sustainability and is very important in terms of reducing food waste. This contribution will also contribute to the economy positively. In addition, the importance of the issue will become even more evident when it is considered that each food loss harms animals, plants and the entire ecosystem as well as humans (Gjerris & Gaiani, 2013: p. 16; İlyasov, 2017: p. 43). Consequently, next researches should focus on exploring specific food waste contexts and interactions more in depth in the future. For these reasons, better production and consumption statistics would be needed.

REFERENCES

Abdul- Kareem, M. M. (2021). Gana'da Tarım İşletmelerinde Gıda Kaybının Ekonomik Ölçümü ve Etkileri. Ankara: Ankara Üniversitesi Fen Bilimleri Enstitüsü.

ACF. (2014). *Postharvest Losses and Strategies to Reduce Them.* ACF International.

Aday, M. S., & Aday, S. (2021). Gıda Kayıp ve İsrafının Azaltılmasında Gıda Bankacılığı. *IBAD Sosyal Bilimler Dergisi*, *9*(9), 291–310. doi:10.21733/ibad.844971

Ademoğlu, A. (2021). Atık Gıda Yönetimi ve Sürdürülebilirlik. In Y. Oğan (Ed.), *Gastronomi Araştırmaları* (pp. 64–73). Çizgi Kitabevi Yayınları.

Adenuga, O. T., Mpofu, K., & Modise, R. K. (2020). An Approach for Enhancing Optimal Resource Recovery From Different Classes of Waste in South Africa: Selection of Appropriate Waste to Energy Technology. *Sustainable Futures, 2*(2), 100033.

Aguilar, G. (2017). *Pérdida y Desperdicio de Alimentos en México.* México, Cd.MX: Caso de estudio para el Banco Mundial.

Akan, S. (2022). *Gıda Kaybı ve İsraf Sorunu.* Gaziantep Üniversitesi Sosyal Bilimler Enstitüsü.

Akar, E. (2022). *Yalın Üretim Anlayışının Otel İşletmelerinde Uygulanabilirliği: Gıda İsrafı ve Kaybı Üzerine Bir Çalışma.* Mersin Üniversitesi Sosyal Bilimler Enstitüsü.

Akın, G. (2014). İnsan Sağlığı ve Çevre Etkileşimi. *Ankara Üniversitesi Dil ve Tarih-Coğrafya Fakültesi Dergisi*, *54*(1), 105–116.

Akkerman, R., Farahani, P., & Grunow, M. (2010). Quality, Safety and Sustainability in Food Distribution: A Review of Quantitative Operations Management Approaches and Challenges. *OR-Spektrum, 32*(4), 863–904. doi:10.100700291-010-0223-2

Ali, A., Xia, C., Ouattara, N., Mahmood, İ., & Faisal, M. (2021). Economic and Environmental Consequences' of Postharvest Loss Across Food Supply Chain in the Developing Countries. *Journal of Cleaner Production, 323*, 129–146. doi:10.1016/j.jclepro.2021.129146

Alıntaş, O. (2021). *Gastronomi ve Mutfak Sanatları Öğrencilerinin Gıda İsrafı ve Gıda Güvenliği Hakkında Görüşleri: İstanbul İli Örneği*. İstanbul Gelişim Üniversitesi Lisansüstü Eğitim Enstitüsü.

Altınel, H. (2014). *Menü Yönetimi ve Menü Planlama*. Detay Yayıncılık.

Arah, I. K., Ahorbo, G. K., Anku, E. K., Kumah, E. K., & Amaglo, H. (2016). Postharvest Handling Practices and Treatment Methods for Tomato Handlers in Developing Countries: A Mini Review. *Advances in Agriculture, 2016*, 1–9. doi:10.1155/2016/6436945

Aschemann-Witzel, J., Giménez, A., & Ares, G. (2018). Consumer in-store choice of suboptimal food to avoid food waste: The role of food category, communication and perception of quality dimensions. *Food Quality and Preference, 68*, 29–39. doi:10.1016/j.foodqual.2018.01.020

Baptista, P., Campos, I., Pires, I., & Sofia, V. (2012). *Do Campo ao Garfo*. Desperdício Alimentar em Portugal.

Betz, A., Buchli, J., Göbel, C., & Müller, C. (2015). Food Waste İn The Swiss Food Service İndustry – Magnitude And Potential For Reduction. *Waste Management (New York, N.Y.), 35*, 218–226. doi:10.1016/j.wasman.2014.09.015 PMID:25305683

Bontoux, L., & Leone, F. (1997). *The Legal Definition of Waste and its Impact on Waste Management in Europe*. European Commission - Joint Research Centre Institute for Prospective Technological Studies WTC.

Bozkurt, R., & Yüksel, A. Y. (2019). Günümüzde Balık Atıklarının Helal Sektörde Kullanım Alanları. *Türk Doğa ve Fen Dergisi, 8*(2), 10–16.

Bräutigam, K.-R., Jörissen, J., & Priefer, C. (2014). The Extent of Food Waste Generation Across EU-27: Different Calculation Methods and the Reliability of Their Results. *Waste Management & Research, 32*(8), 683–694. doi:10.1177/0734242X14545374 PMID:25161274

CEC. (2017). *Characterization and Management of Food Loss and Waste in North America*. Commission for Environmental Cooperation. https://rescuefood.ca/wp-content/uploads/2021/04/11772-characterization-and-management-food-loss-and-waste-in-north-america-en.pdf

Ceryes, C. A., Antonacci, C. C., Harvey, S. A., Spiker, M. L., Bickers, A., & Neff, R. A. (2021). "Maybe It's Still Good?" A Qualitative Study of Factors Influencing Food Waste and Application of the E.P.A. Food Recovery Hierarchy in U.S. Supermarkets. *Appetite, 161*, 105–111. doi:10.1016/j.appet.2021.105111 PMID:33482300

Çetin, H. C. (2022). *Zincir Restoran İşletmelerinde Gıda Atığı Yönetimi: İstanbul Örneği*. Akdeniz Üniversitesi Sosyal Bilimler Enstitüsü.

Ceylan, Z., Bulkan, S., & Tozan, H. (2017). Tek ve Çok Dönemli Envanter Kontrol Modelleri. *Mühendislik Bilimleri ve Tasarım Dergisi*, *5*(2), 441–455.

Chalak, A., Abou-Daher, C., Chaaban, J., & Abiad, M. G. (2016). The Global Economic and Regulatory Determinants of Household Food Waste Generation: A Cross-Country Analysis. *Waste Management (New York, N.Y.)*, *48*, 418–422. doi:10.1016/j.wasman.2015.11.040 PMID:26680687

Cirişoğlu, E., & Akoğlu, A. (2021). Restoranlarda Oluşan Gıda Atıkları ve Yönetimi: İstanbul İli Örneği. *Akademik Gıda*, *19*(1), 38–48.

Cooremans, K., & Geuens, M. (2019). Same but different: Using anthropomorphism in the battle against food waste. *J Publ Pol Market*, *38*(2), 232–245. doi:10.1177/0743915619827941

Demir, Y. (2022). *Eğitim Mutfaklarında Gıda İsrafını Belirlemeye Yönelik Bir Araştırma*. İstanbul Üniversitesi Sosyal Bilimler Enstitüsü.

Demirbaş, N. (2018). Dünyada ve Türkiye'de Gıda Israfını Önleme Çalışmalarının Değerlendirilmesi. *VIII*. [Plovdiv.]. *IBANESS Kongreler Serisi*, *21*(22), 521–526.

Demirbaş, N. (2019). yi Tarım Uygulamaları ile Meyve Bahçelerinde Ortaya Çıkan Üretim, Hasat ve Hasat Sonrası Kayıpları Azaltılabilir mi? *I. IBANESS Kongreler Serisi, Tekirdağ. Türkiye*, *9*(10), 289–296.

Drewitt, T. (2013). *Food Waste Prevention in Quick Service Restaurants*. Lund University Master of Science in Environmental Management and Policy.

Durmuş, H. (2022). *Gıda İsrafının Boyutu ve Ekonomik Etkilerinin Tespiti Üzerine Bir Araştırma: Sakarya İli Örneği.* Sakarya: Sakarya Üniversitesi Sosyal Bilimler Enstitüsü.

Evans, D. (2012). Beyond the throwaway society: Ordinary domestic practice and a sociological approach to household food waste. *Sociology*, *2012*(46), 41–56. doi:10.1177/0038038511416150

FAO. (2013). *Food Wastage Footprint Impacts on Natural Resources.* FAO. https://www.fao.org/3/i3347e/i3347e.pdf

FAO. (2018). *The future of food and agriculture Alternative pathways to 2050.* Licence: CC BY-NC-SA 3.0 IGO. FAO.

FAO. (2022). *Tackling food loss and waste: A triple win opportunity.* FAO. https://www.fao.org/newsroom/detail/FAO-UNEP-agriculture-environment-food-loss-waste-day-2022/en.

FDA. (2019). How to cut food waste and maintain food safety. FDA. https://www.fda.gov/food/consumers/how-cut-food-waste-andmaintain- food-safety.

Garcia-Herrero, I., Hoehn, D., Margallo, M., Laso, J., Bala, A., Batlle-Bayer, L., Fullana, P., Vazquez-Rowe, I., Gonzalez, M. J., Durá, M. J., Sarabia, C., Abajas, R., Amo-Setien, F. J., Quiñones, A., Irabien, A., & Aldaco, R. (2018). On the Estimation of Potential Food Waste Reduction to Support Sustainable Production and Consumption Policies. *Food Policy*, *80*, 24–38. doi:10.1016/j.foodpol.2018.08.007

Gemeç, E., & Gürbüz, İ. B. (2022). Gıda İsrafının Önlenmesine Yönelik Tutum ve Davranışların Belirlenmesi; Üniversite Öğrencileri Üzerine Bir İnceleme. *Balkan ve Yakın Doğu Sosyal Bilimler Dergisi*, *8*, 164–175.

Gilland, B. (2002). World Population and Food Supply: Can Food Production Keep Pace with Population Growth in the Next Half-century? *Food Policy*, *27*(1), 47–63. doi:10.1016/S0306-9192(02)00002-7

Gjerris, M., & Gaiani, S. (2013). Household Food Waste in Nordic Countries: Estimations and Ethical İmplications. *Nordic Journal of Applied Ethics*, *7*(1), 6–23. doi:10.5324/eip.v7i1.1786

Godfray, H. J., Crute, I. R., Haddad, L., Lawrence, D., Muir, J. F., Nisbett, N., Pretty, J., Robinson, S., Toulmin, C., & Whiteley, R. (2010). The Future of the Global Food System. *Philosophical Transactions of the Royal Society of London. Series B, Biological Sciences*, *365*(1554), 2769–2777. doi:10.1098/rstb.2010.0180 PMID:20713383

Graham-Rowe, E., Jessop, D. C., & Sparks, P. (2014). Identifying Motivations and Barriers to Minimising Household Food Waste. *Resources, Conservation and Recycling*, *84*, 15–23. doi:10.1016/j.resconrec.2013.12.005

Griffin, M., Sobal, J., & Lyson, T. A. (2009). An Analysis of a Community Food Waste Stream. *Agriculture and Human Values*, *26*(1-2), 67–81. doi:10.100710460-008-9178-1

GTHB. (2018). Meyve ve Sebzelerin Depolama ve Nakliye Uygulamaları Eğitim Materyali. In Gıda Zincirindeki Hasat Sonrası Kayıpları Azaltmak için Yenilikçi Yaklaşımlar.

Gustavsson, J., Cederberg, C., & Sonesson, U. (2011). *Food Losses and Food Waste: Extent, Causes and Prevention.* FAO. https://www.fao.org/3/i2697e/i2697e.pdf:

Heikkila, L., Reinikainen, A., Katajajuuri, J.-M., Silvennoinen, K., & Hartikainen, H. (2016). Elements Affecting Food Waste in the Food Service Sector. *Waste Management (New York, N.Y.)*, *56*, 446–453. doi:10.1016/j.wasman.2016.06.019 PMID:27373724

HISPACOOP. (2012). *Estudio Sobre el Desperdicio de Alimentos en Los Hogares.* HISPACOOP. http://www.hispacoop.es/home/index.php?option=com_docman&task=doc_view&gi: onfederación Española de Cooperativas de Consumidores y Usuarios.

İlkbahar, A., & Köksel, H. (2016). *Ekmek.* Ankara Halk Ekmek ve Un Fabrikası AŞ.

İlyasov, A. (2017). *Türk Misafirlerin Gıda İsrafı Tutumları Üzerine Beş Yıldızlı Otel İşletmelerinde Bir Araştırma: Alanya Örneği.* Süleyman Demirel Üniversitesi Sosyal Bilimler Enstitüsü.

Joardder, M. U., & Masud, M. H. (2019). *Food Preservation in Developing Countries: Challenges and Solutions.* Cham.

Kader, A. A. (2002). *Postharvest Technology of Horticultural Crops.* University of California Agriculture and Natural Resources.

Katajajuuri, J.-M., Silvennoinen, K., Hartikainen, H., Heikkilä, L., & Reinikainen, A. (2014). Food Waste in the Finnish Food Chain. *Journal of Cleaner Production*, *73*, 322–329. doi:10.1016/j.jclepro.2013.12.057

Kılınç Şahin, S., & Bekar, A. (2018). Küresel Bir Sorun "Gıda Atıkları": Otel İşletmelerindeki Boyutları. *Journal of Tourism and Gastronomy Studies*, *6*(4), 1039–1061.

Kocacık, F. (1998). *Tüketim Eğilimleri Ve Sorunları.* Cumhuriyet Üniversitesi Yayınları.

Kummu, M., de Moel, H., Porkka, M., Siebert, S., Varis, O., & Ward, P. J. (2012). Lost Food, Wasted Resources: Global Food Supply Chain Losses and Their İmpacts on Freshwater, Cropland, and Fertiliser Use. *The Science of the Total Environment, 438*, 477–488. doi:10.1016/j.scitotenv.2012.08.092 PMID:23032564

Kurt, Y. (2019). *Yiyecek İçecek Personelinin Gıda İsrafı Hakkında Bilgi, Görüş ve Davranışları: Eskişehir İli Örneği*. Gazi Üniversitesi Sosyal Bilimler Enstitüsü.

Lipinski, B., Hanson, C., Waite, R., & Searchinger, T. (2013). *Reducing Food Loss and Waste.* (World Resources Institute Working Paper, 1-40). World Resources Institute Working Paper. https://files.wri.org/d8/s3fs-public/reducing_food_loss_and_waste.pdf

Lopez-Sanchez, A., Luque-Badillo, A. C., Orozco-Nunnely, D., Alencastro-Larios, N. S., Ruiz-Gomez, J. A., Garcia-Cayuela, T., & Gradilla-Hernandez, M. S. (2021). Food Loss in the Agricultural Sector of a Developing Country: Transitioning to a More Sustainable Approach. The Case of Jalisco, Mexico. *Environmental Challenges, 5*, 1–16. doi:10.1016/j.envc.2021.100327

Marra, F. (2013). *Fighting Food Loss and Food Waste in Japan.* Leiden University: M. A. in Japanese Studies –. *Asian Studies (Quezon City, Philippines)*, 2011–2013.

Mena, C., Terry, L. A., Williams, A., & Ellram, L. (2014). Causes of waste across multi-tier supply networks: Cases in the UK food sector. *International Journal of Production Economics, 152*, 144–158. doi:10.1016/j.ijpe.2014.03.012

Mete, H. (2012). Peynir Altı Suyunun Ekmekçilikte Değerlendirilmesi ve Ekonomik Önemi. *Tekirdağ S.M.M.M Odası Sosyal Bilimler Dergisi, 1*, 1-10.

Öcal, D. (2021). *Yiyecek İçecek İşletmelerinde Gıda İsrafı Üzerine Bir Araştırma: Kırıkkale Örneği.* Sakarya: Sakarya Üniversitesi Lisansüstü Eğitim Enstitüsü.

Oral, Z. (2015). *Dünya'da ve Türkiye'de Gıda İsrafı ve Önlenmesine Yönelik Uygulamalar.* Ankara: AB Uzmanlık Tezi, TC Gıda Tarım ve Hayvancılık Bakanlığı Avrupa Birliği ve Dış İlişkiler Müdürlüğü.

Özdal, Z. N. (2019). *Türkiye'de Tüketicilerin Gıda İsrafını Engellemeye Yönelik Tutumlarının Analizi ve Mobil Uygulama Tasarımı*. Çukurova Üniversitesi Fen Bilimleri Enstitüsü.

Özdemir, G., & Güçer, E. (2018). Food Waste Management within Sustainability Perspective: A Study on Five Star Chain Hotels. *Journal of Tourism and Gastronomy Studies, 6*(1), 280–299. doi:10.21325/jotags.2018.187

Öztürk, İ. (2006). Türkiye'de Perakende Sektörü. *Çağ Üniversitesi Sosyal Bilimler Dergisi, 3*(1), 69-81.

Parfitt, J., Barthel, M., & Macnaughton, S. (2010). Food Waste Within Food Supply Chains: Quantification and Potential for Change to 2050. *Phil. Trans. Biol. Sci., 365*(1554), 3065–3081. doi:10.1098/rstb.2010.0126 PMID:20713403

Pirani, S. I., & Arafat, H. A. (2014). Solid Waste Management in the Hospitality Industry: A Review. *Journal of Environmental Management, 146*, 320–336. doi:10.1016/j.jenvman.2014.07.038 PMID:25194519

Quested, T. E., Parry, A. D., Easteal, S., & Swannell, R. (2011). Food and Drink Waste From Households in the UK. *Nutrition Bulletin*, *36*(4), 460–467. doi:10.1111/j.1467-3010.2011.01924.x

Racz, A., Marchesi, V. V., & Crnković, I. (2018). Economical, Environmental and Ethical Impact of Food Wastage in Hospitality and Other Global Industries. *Jahr: Europski časopis za bioetiku, 9(1)*, 25-42.

Redlingshöfer, B., Coudurier, B., & Georget, M. (2017). Quantifying Food Loss During Primary Production and Processing in France. *Journal of Cleaner Production*, *164*, 703–714. doi:10.1016/j.jclepro.2017.06.173

Reynolds, C. J., Mirosa, M., & Clothier, B. (2016). New Zealand's Food Waste: Estimating the Tonnes, Value, Calories and Resources Wasted. *Agriculture*, *6*(1), 1–15. doi:10.3390/agriculture6010009

Roe, B. E., Qi, D., Apolzan, J. W., & Martin, C. K. (2020). Selection, intake, and plate waste patterns of leftover food items among U.S. consumers: A pilot study. *PLoS One*, *2020*(15), e0238050. doi:10.1371/journal.pone.0238050 PMID:32903260

Rohini, C., Geetha, P. S., Vijayalakshmi, R., Mini, M. L., & Pasupathi, E. (2020). Global Effects of Food Waste. *Journal of Pharmacognosy and Phytochemistry*, *9*(2), 690–699.

Rolle, R. S. (2006). *Improving Postharvest Management and Marketing in the Asia-Pacific Region: Issues and Challenges Trends.* Semantic Scholar.

Segre, A., & Falasconi, L. (2011). *Il Libro Nero Dello Spreco in Italia: Il Cibo.* Edizioni: Ambiente.

Segre, A., Falasconi, L., Politano, A., & Vittuari, M. (2014). *SAVE FOOD: Global Initiative on Food Loss and Waste Reduction Background Paper on the Economics of Food Loss and Waste.* Food and Agriculture Organization of the United Nations.

Şener, A., & Ünal, Ü. M. (2008). Gıda Sanayii Atıklarının Biyoteknolojik Yöntemlerle Değerlendirilmesi. Türkiye 10. Gıda Kongresi; 21-23 Mayıs 2008, (pp. 21-23). Erzurum.

Songür, A. N., & Çakıroğlu, F. P. (2016). Gıda Kayıpları ve Atık Yönetimi. *Türkiye Klinikleri Nutrition And Dietetics-Special Topics*, *2*(3), 21–26.

Sputnik. (2019). *Türkiye Yılda 214 Milyar Liralık Gıda İsraf Ediyor*. Sputnik. https://sputniknews.com.tr/20190929/turkiye-yilda-214-milyar-liralik-gida-israf-ediyor-1040278150.html

Stuart, T. (2009). *Waste: Uncovering the Global Food Scandal.* WW Norton & Company.

T.C. Tarım ve Orman Bakanlığı. (2020). *Türkiye'nin Gıda Kayıpları ve İsrafının Önlenmesi, Azaltılması ve Yönetimine İlişkin Ulusal Strateji Belgesi ve Eylem Planı.* Birleşmiş Milletler Gıda ve Tarım Örgütü.

TDK. (2020). *İsraf.* TDK. https://sozluk.gov.tr/

Tekiner, E. (2023). *Gıda İsrafı ve Çevre Koruma Bilinci: Kocaeli İli'nde Bir Uygulama.* Uludağ Üniversitesi Fen Bilimleri Enstitüsü.

Tekiner, İ. H., Mercan, N. N., Kahraman, A., & Özel, M. (2021). Dünya ve Türkiye'de Gıda İsrafı ve Kaybına Genel Bir Bakış. *İstanbul Sabahattin Zaim Üniversitesi Fen Bilimleri Enstitüsü Dergisi, 3*(2), 123-128.

Termeer, C. J., Drimie, S., Ingram, J., Pereira, L., & Whittingham, M. J. (2018). A Diagnostic Framework for Food System Governance Arrangements: The Case of South Africa. *NJAS Wageningen Journal of Life Sciences*, *84*(1), 85–93. doi:10.1016/j.njas.2017.08.001

Thyberg, K. L. (2015). *Food Waste and Sustainability: Quantifying Food Waste Disposal and Evaluating the Environmental Impacts of Technologies and Policies.* [Thesis, Stony Brook University].

Ticaret Bakanlığı, T. C. (2018). *Türkiye İsraf Raporu.* Tüketicinin Korunması ve Piyasa Gözetimi Genel Müdürlüğü.

Tirkey, K. (2011). *Effect of Harvest Time and Postharvest Operations on the Quality and Quantity of Sesame (Sesamum indicum L.).* [MSc Agriculture Thesis: Jawaharlal Nehru Krishi Vishwa Vidyalaya, Jabalpur, College of Agriculture].

UNEP. (2021). *UNEP Food Waste Index Report 2021.* UNEP. https://www.unep.org/resources/report/unep-food-waste-index-report-2021

Vilarino, M. V., Franco, C., & Quarrington, C. (2017). Food Loss and Waste Reduction as an Integral Part of a Circular Economy. *Frontiers in Environmental Science*, *5*, 1–5. doi:10.3389/fenvs.2017.00021

WRAP. (2009). *Household Food and Drink Waste in the UK.* https://wrap.org.uk/sites/default/files/2020-12/Household-Food-and-Drink-Waste-in-the-UK-2009.pdf:

Yağcı, S., Altan, A., Göğüs, F., & Maskan, M. (2006). Gıda Atıklarının Alternatif Kullanım Alanları. Türkiye 9. Gıda Kongresi: 24-26 Mayıs 2006, (pp. 24-26). Bolu.

Yiğitoğlu, V., & Cerrah, S. (2021). Yiyecek Artıklarıyla İlgili Yasal Düzenlemelerin Değerlendirilmesi. In G. S. Ergün (Ed.), *Multidisipliner Turizm Çalışmaları* (pp. 105–115). Çizgi Kitabevi Yayınları.

Yılmaz, M. (2022). *Konaklama İşletmelerinde Helal Konsept Uygulamalarının Gıda Atığı ve İsrafı Üzerindeki Etkisi.* Ankara Hacı Bayram Veli Üniversitesi Lisansüstü Eğitim Enstitüsü.

Chapter 14
Which Barriers Need More Attention in Food Waste Management and Blockchain Integration?

Salih Aka
Erzincan Binali Yıldırım University, Turkey

ABSTRACT

Blockchain is a digital innovative technology that is decentralized, open to the use of all stakeholders, does not allow data manipulation, and has many more features. Blockchain, which has the potential to solve problems in food safety, is attracting more and more attention from researchers in this field. In this study, a conceptual model for the integration of blockchain and food waste management is proposed. The combination of blockchain and food waste management has economic, social, and environmental dimensions that will serve the purpose of reducing food waste. However, some blockchain barriers make it difficult to integrate with food waste management. Blockchain barriers that affect the success of this integration are examined with the BEST-Worst Method (BWM). As a result of the analysis, infrastructure setup and adaptation challenges stand out as the most important barrier. The study would be a starting point for practitioners working on food waste management and the blockchain system, and government officials planning regulations in this area.

1. INTRODUCTION

The COVID19 pandemic was a time when economic sensitivities about food waste preceded environmental sensitivities. During this period, the disruption of supply chains and the deterioration of the demand structure brought along difficulties in accessing food (Heck et al., 2020). Although the negative effects of the pandemic have begun to disappear in global economies, the risks of product expiration dates have increased in perishable food products, especially with the loss of labor on the food production side and the slowdown in logistics activities (Kayikci et al., 2021). Due to its perishable, moist and smelly struc-

DOI: 10.4018/978-1-6684-9094-5.ch014

ture, food is an environmental threat, an important cost item and a serious harm to human health even during its disposal (Jin et al., 2021). The global identity of food production and the excessive increase in consumption trigger food waste.

Many events today make it difficult to reach food. For this reason, effective food waste management initiatives gain importance for national security (Ananno et al., 2021). Masud et al. (2020) states that a significant portion of food waste occurs during the production phase, especially in developing countries, and that the evaluation of this waste is an economic issue. Cooperation of stakeholders in food supply chains on issues such as timing, information sharing and coordination gains importance in terms of food waste management. Suppliers, manufacturers, retailers, customers, social initiatives and government authorities, which are stakeholders of food supply chains, have roles in reducing food waste (Aschemann-Witzel et al., 2017; Ghinoi et al., 2020). Raut et al. (2019) states that new generation technologies integrated into food supply chains offer effective results in reducing food waste and ensuring food safety. FAO (2020) reports a 14% reduction in food waste thanks to the application of modern technologies in food supply chains (Kumar et al., 2022).

Systems based on IOT and blockchain integration among smart technologies have become useful tools for food production and distribution systems with their innovative features such as product monitoring, transparency, data security, securing agreements and many more. The decentralized blockchain system provides transparency between stakeholders about authenticated records (Tillemann et al., 2019; Crosby et al., 2016). Blockchain has the potential to strengthen food security, reduce spoilage, and reduce waste and CO2 emissions from operations (Friedman and Ormiston, 2022).

Blockchain promises significant advantages for many systems due to its product-tracking feature. However, it takes time and experience to adapt new technologies and identify potential problems. Blockchain is one of the trending study topics of recent times, but most studies on its application areas are theoretical. Actually, there are various barriers to the use of blockchain in harmony with other systems. The energy problem caused by continuous data entry and computer processing is one of these barriers. The high-energy requirement has both economic and environmental effects due to the carbon footprint formation (Kouhizadeh et al., 2021). The high-energy requirement is an example of the mechanism of action that barriers to integration can reveal. Correctly identifying the factors affecting blockchain integration will facilitate the adoption and efficient use of blockchain (AlShamsi et al., 2022). This study aims to identify the barriers to the use of blockchain technology in food waste management systems and applications and to determine their degree of importance. Since there is no detailed data set about barriers, the opinions of experts on blockchain, food waste and food supply chains were taken. The Best-Worst method is a multi-criteria decision-making method that makes linear programming-based calculations in line with expert opinions. The Best-Worst method provides consistent results, so it was preferred as the analysis method in the study. This study is one of the first studies to examine the blockchain compliance in food waste management in line with the theoretical framework and expert opinions. The following questions are sought to be answered within the scope of the study:

1. What conceptual system model can be proposed for the integration of blockchain and food waste management?
2. What types of blockchain barriers might complicate this model?
3. Which of these barriers should be considered primarily?

In the remainder of this article, Section 2 compiles the literature on food waste management, uses of blockchain applications in food-related studies, and potential barriers to integration. Section 3 presents the Best-Worst method and analysis steps. Section 4 includes application and analysis results. In Section 5, the managerial and theoretical contributions of the study according to the findings are discussed.

2. LITERATURE SURVEY

2.1 Bibliometric Analysis

Systematic literature review on the subject is preferred in terms of revealing the original and theoretical contribution of the study. Keywords related to the primary research area should be chosen correctly. Bamakan et al. (2022) follow a useful systematic approach to bibliometric analysis. Figure 1 shows the systematic steps of literature review in our study.

Figure 1. The systematic steps of literature review

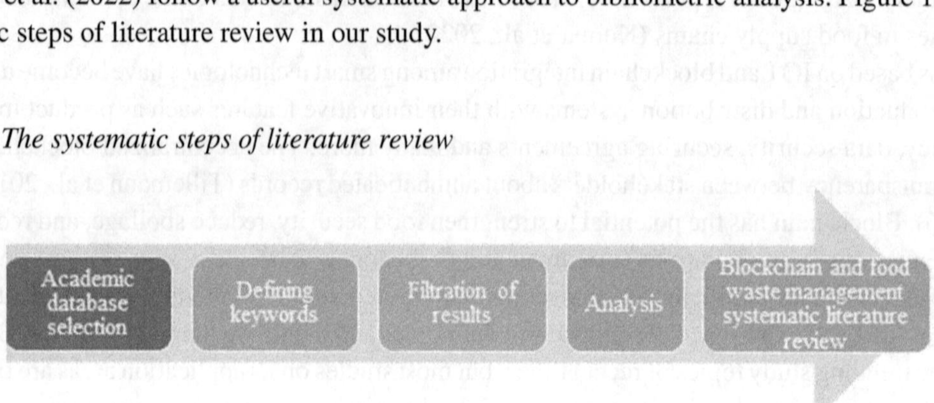

In the first step, the database to be researched was determined as "Web of Science". In the next step, the keywords "blockchain" and "food" were used for a wide range of searches. There are 778 studies using these two keywords in the database. In order to examine successful studies in the field, filtering was done based on only articles and English language. Because of these two processes, the number of studies decreased to 496. A total of 306 articles are in open access status. Blockchain attracts attention as a trend topic. Figure 2 shows the interest in blockchain and food-related studies over the years.

Figure 2. The interest in blockchain and food-related studies over the years

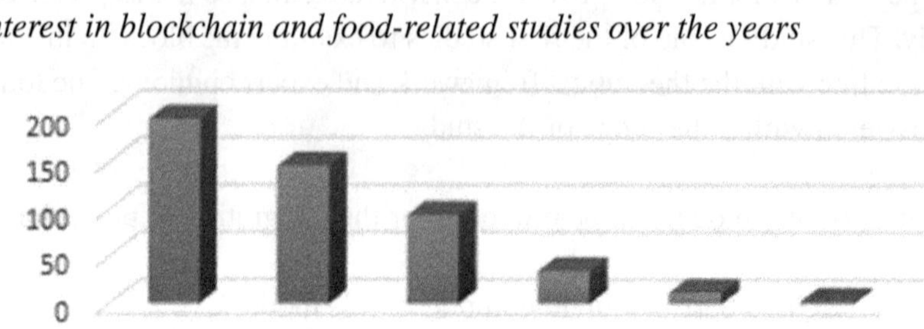

The first research in this area was made in 2017, and over the years, more studies have emerged each year than the previous year. Figure 3. shows in which areas the studies on blockchain and food are generally clustered after the analysis of the results. Keywords are grouped under six clusters. It is noteworthy that there is no cluster on food waste or food waste management within these main headings. From this point of view, the relationship between food waste or food waste management and blockchain emerges as a virgin area worth working on. Figure 4 shows the countries where the relevant studies are predominantly conducted. Researchers from 24 countries focus on this issue. However, studies of researchers in 16 countries are connected with each other. Figure 4 illustrates that eastern researchers show a strong interest in the food and blockchain association.

Figure 3. Areas the studies on blockchain and food are generally clustered after the analysis of the results

Figure 4. The countries where the relevant studies are predominantly conducted

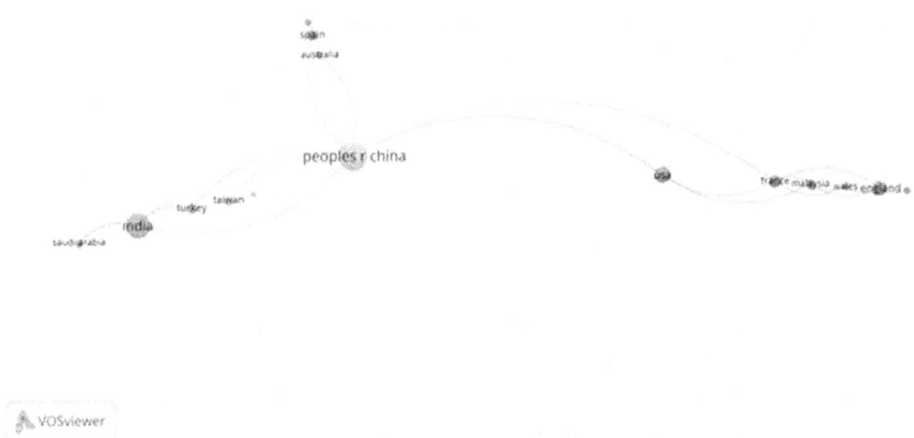

2.2 Blockchain for Food

Blockchain is a kind of shared database that can be managed by all stakeholders, is decentralized and protects transactional information with its cryptographic structure in a highly secure manner (Swan, 2015). Each stakeholder in the network can access the information recorded in the blockchain platform simultaneously. Transactional database has a security guarantee provided by a common mechanism for data collection, information distribution and security (Beck and Müller-Bloch, 2017). The transactional database is based on smart contracts that protect the interests of the parties. Blockchain is especially useful in studies in the food field, as it has a positive effect on cost reduction, data set consistency, transparency, accountability and efficiency (Zhang et al., 2022). Considering the perishable nature of the food product and the food safety sensitivity of consumers, it is important for the consumer to monitor all stages of the product. Blockchain systems record every process of the food product with its secure transaction mechanism and the food product becomes traceable (Bumblauskas et al., 2020). All stakeholders control the system, information cannot be changed, and all processes are transparent as the transaction mechanism is operated with smart contracts. The transparent blockchain system is a kind of precautionary tool for fraudsters who deliberately change the content of the raw materials in the composition of the food and the packaging or add some substances to the food afterwards. (Kohler and Pizzol, 2020).

The system provides very fast sharing of information. As soon as the nutritional content of the food product and any deterioration or disease information about the food product are entered into the system, all stakeholders receive information about the status of the food in a very short time. This functionality contributes to food safety (Creydt and Fischer, 2019). The system records the temperature, climate, transportation distance, locations and many more information of the product during the shipment of the food. Blockchain affects the quality of food products thanks to its data-oriented structure. (Ali et al., 2021).

2.3 Food Waste Management

FAO (2011) defines food loss as the loss of quantity, nutritional value, economic value or quality of food that is affected by food safety. While some of the food loss is wastage that cannot be economically recovered during harvest, another important part is food waste. Food waste arises because of the food being thrown away by spoiling or not being preferred for consumption even though it is suitable for use (McAdams et al., 2019; Munir, 2022). Various negligence causes the food to end its useful life and start to deteriorate. In developing countries, lack of organization mainly in storage, transportation, processing, air conditioning and handling activities increases food waste. In developed countries, food waste generation has spread to all production and consumption processes. Regulations and methods for food safety and quality standards can make it possible to re-evaluate many foods that are not suitable for consumption but are still consumable. Foods whose useful life is exceeded due to inadequate planning and misuse of food products at the purchasing stage also turn into undesirable food waste. Attempts to reduce food waste will result in higher efficiency in the food supply, less storage need, shorter transport distances and less energy loss. The value of food wasted or lost on a global scale is estimated at $1 trillion (FAO, 2015; Govindan, 2018).

It is considered that approximately one third of carbon emissions in developed countries is food waste created by the processing, distribution and consumption stages of the food supply chain (Garnett, 2011; Aschemann-Witzel et al., 2015). Food waste can be reduced by intervening waste at its source. The collected food waste can be reused as pet feed or used as compost to provide recycling material such as manure. Food waste contributes to both cost reduction and environmental awareness with such methods. In Figure 5, there are steps to create food waste and solutions to prevent waste (Balaji and Arshinder, 2016).

Figure 5. Steps to create food waste and solutions to prevent waste

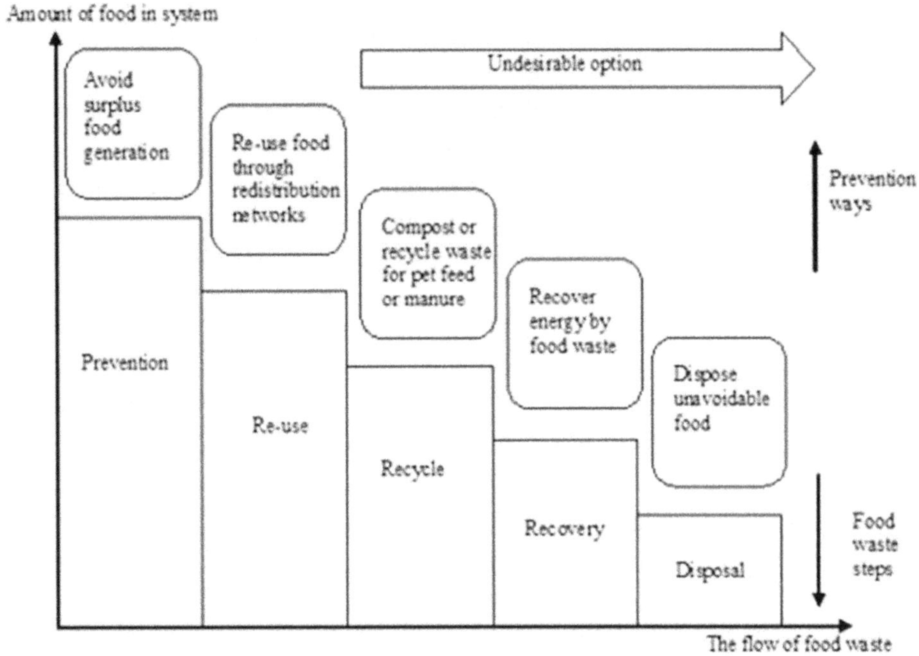

Food waste is an issue that should be given utmost importance by national and international actors; however, academic studies in this area are not sufficient (Munir, 2022). Recently, researchers try to interpret food waste through consumers and try to explain this relationship with behavioral models such as Theory of Planned Behaviour (Aka and Buyukdag. 2021; Coşkun and Özbük, 2020; Chen, 2023; Wang et al., 2023; Teoh et al., 2022).

Food waste is a topic in the field of sustainability due to its environmental, economic and social dimensions (Aschemann-Witzel et al., 2015). Separating existing energy into processing, active consumption and recycling operations will increase food efficiency (Kosseva, 2013). Food waste management aims to reduce costs by improving production and distribution operations, as well as to create a sustainable environment suitable for human health by reducing waste (Lo and Woon, 2016). Researchers have recently argued that the lack of sustainability and efficiency-based practices in food production may cause uncertainty and excessive expensiveness in food prices with the deterioration of environmental balance in the future (Kosseva, 2013). Although economic performance is the most important goal for businesses, activities carried out in line with sustainability provide positive contributions on company quality, customer satisfaction and operational efficiency (Li et al., 2014). Food management has a complex network of relations and the factors that can trigger deterioration in this network create economic and environmental costs that cause food waste. Food waste risks can be reduced with effective information management among stakeholders (Ghinoi et al., 2020). Innovative methods will also support food waste management organizations in decision-making and process management.

2.4 Blockchain and Food Waste Management Integration

Intelligent information sharing platforms such as blockchain and IOT are effective auxiliary platforms for waste management (Gopalakrishnan et al. 2021; França et al., 2020; Ahmad et al., 2021). Taylor et al. (2020) states that blockchain applications used in the field of waste management attract attention in terms of the functioning of the reward and punishment mechanism, the monitoring and transportation of waste. In this study, the role of blockchain applications for an effective food waste management system is evaluated through a conceptual model. The focus of the assessments is on the difficulties that blockchain-specific barriers will pose to this unity during the integration of blockchain into waste management. Vishkaei and De Giovanni (2022) proposed a mathematical model based on a scenario for blockchain and food waste. Apart from this study, there are not enough studies on the integration of blockchain and food waste management directly in the literature. The model proposed in our study is a pioneer in its field. Figure 6 shows the food waste management and blockchain integration model. The stakeholders involved in food waste management are connected by a blockchain application. Blockchain application is fed with data by smart contracts and data reporting system. During the food supply between Phase 1 and Phase 2, instantly changing information such as product quality, product quantity, air conditioning, product transportation conditions, status of vehicles and routes are recorded with the data reporting system. Conditions on subjects such as supply and distribution facilities, road data and waste data are secured with smart contracts. In Phase 2, food products shipped to manufacturers, retailers, markets and various service points are recorded by the system.

Food that spoils or cannot be sold during food production or at sales and service points has completed its economic life. Even if these food products have no resale value, they may still be consumable. Food banks are institutions that collect such products and organize food product donations to people with limited financial power. Those in need have access to foods that continue to be consumable in Phase 3. Foods that are not suitable for consumption reach the animal food and compost companies in Phase 4. Foods that cannot be evaluated are disposed in the final Phase.

Before and after each stage, data records for the quality and quantity of foods are created. All stakeholders in the chain have the right to access the data that cannot be changed. In this way, all information is transmitted instantly to the stakeholders in the system. In order to ensure the effectiveness of food management, each stakeholder has responsibilities in line with their own competencies. The system ensures that the operation of the process is kept under control. The system takes into account the predetermined process policies and the quality and quantity data between the phases of the food, and keeps the compliance of each stakeholder's responsibilities under control with the reward and penalties mechanism. It also processes the recorded big data and makes predictions about future problems or opportunities.

Data reporting system takes advantage of IOT, sensors and clouds. Air-conditioning and road information of vehicles during transportation, shelf and storage conditions of foods during sale and processing, points reached by foods at each stage and many other data are obtained and stored through these tools. An AI-powered blockchain system has the ability to understand food loss and waste. The system provides efficient evaluation of excess food. In addition, such a system detects food waste generation points thanks to its data and gains the ability to reduce the generation of waste (Kayikci et al., 2022).

Figure 6. Food waste management and blockchain integration model

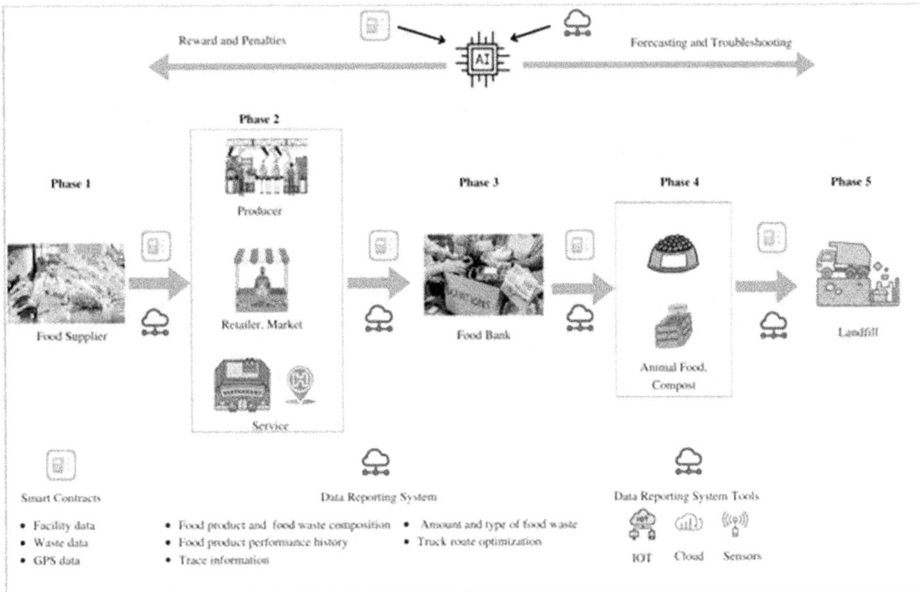

2.5 Barriers

Food waste is a phenomenon that is easy to occur and difficult to manage. Insufficient information on the origin, production and transportation process and quality of food raises concerns about food safety (Kumar et al., 2022). Despite the promised solution to this problem, blockchain has several disadvantages in terms of integration into different systems, as it is a new technology (Kouhizadeh et al., 2021). The food industry is very suitable for the use of blockchain. Actually, there are many studies investigating food supply chain and blockchain compatibility (Bumblauskas et al., 2020; Creydt and Fischer, 2019; Friedman and Ormiston, 2022; Kayikci et al. 2021; Kayikci et al. 2022; Köhler and Pizzol, 2020). Most studies are modeled conceptually. In these studies, the negative barriers during integration as well as the opportunities that blockchain integration promises are discussed. All the barriers that will arise during food waste management and blockchain integration are explained in the following subsections.

2.5.1 Infrastructure Setup and Adaptation Challenges

Blockchain is a new generation digital technology with complex components. For the establishment of the system, expensive digital cloud technologies, computer hardware, sensors and IOT tools are needed (Okorie et al., 2022). Educational platforms that will teach the operation of the system are not sufficient (Kamilaris et al., 2019). In addition, difficulties in accessing and using such an infrastructure by all stakeholders appear as barriers (Bumblauskas et al. 2020; Müßigmann et al. 2020; Kumar et al., 2022; Kouhizadeh et al., 2021).

2.5.2 Inadequate Government Regulations

There are no regulations or policies of government authorities regarding the use and supervision of blockchain. This situation creates pressure in the food industry circles who want to use the blockchain application and do not want to face any uncertainty in the future (Vu et al., 2021; Chen et al., 2021; Kumar et al., 2020; Howson, 2020; Biswas and Gupta, 2019; Kouhizadeh et al., 2021).

2.5.3 Cyber Risks

The existence of accounts, information and expenditures in the digital environment and the possibility that these elements may be affected by hacker operations increase prejudices against the system (Antonucci et al. 2019; Kayıkcı et al., 2021; Reyna et al., 2018; Okorie et al., 2022 Yadav et al., 2020).

2.5.4 High Energy Requirement

Uninterrupted use of the system by all stakeholders, continuous data entry and the operation of cloud, sensor and IOT tools connected to the system with energy increase the energy need (Khan et al., 2022). This may have a negative impact on carbon footprint emissions from energy production. The high dependence on energy is one of the important problem items of the blockchain system (Baralla et al., 2018; Okorie et al., 2022; Li et al., 2021).

2.5.5 The Ever-Increasing Need for Data Storage

The blockchain platform continuously records big data from all sensors and devices during activities such as waste monitoring, collection and transportation. Increasing memory needs arise after each new operation (Reyna et al., 2018; Kumar et al., 2022; Ahmad et al., 2021; Yadav et al., 2020; Kouhizadeh et al., 2021; Biswas and Gupta, 2019; Li et al., 2021).

2.5.6 Integration Challenges

Blockchain has its own characteristics originating from communication and data management. Consensus and transaction mechanism, data-enabled infrastructure and its own internal algorithms work differently from other systems. There are some problems regarding the integration of systems between blockchain and other digital management systems (Okorie et al., 2022; Vu et al., 2021).

2.5.7 Lack of Expertise

The development of expertise is a complex process in which knowledge, skills, abilities and the human factor are at the center (Kumar et al., 2022). It takes time to learn new technologies and to train experts with sufficient technical knowledge and equipment. As a matter of fact, the lack of sufficient number of experts in the use and development of blockchain technology limits the effective use of these systems (Kamilaris et al., 2019; Kouhizadeh et al., 2021; Okorie et al., 2022; Khan et al., 2022; Vu et al., 2021).

2.5.8 Low Investment Value

The fact that it is not seen as a profitable investment element for sector investors, on the contrary, the high investment cost negatively affects the choice of blockchain technology (Okorie et al., 2022, Mangla et al., 2022; Andoni et al., 2018).

2.5.9 Anonymous Data Ownership

The lack of data centralization can be seen as skeptical by the society. Although the blockchain system has an infrastructure that will not allow any violation in this regard, individuals may be dissatisfied with such platforms due to the discomfort they feel from illegal cyber environments such as the dark web (Kouhizadeh et al., 2021). On the other hand, concerns such as metadata being open to everyone and violation of privacy are also evaluated within this framework (Okorie et al., 2022; Kumar et al., 2022; Astill et al., 2019).

2.5.10 Insufficient Public Knowledge

The insensitivity to innovative technologies in society raises questions about the use of blockchain. Since there are very few applications on blockchain, time is needed for the adoption of the results of the applications and the development of awareness throughout the society (Kumar et al., 2022; Astill et al., 2019).

2.5.11 Scaling and System Speed Slowdown

Increasing transaction volume with continuous data entry causes the speed of the system to slow down and delays the transaction and data recording processes (Khan et al., 2022; Yadav et al., 2020; Rana et al., 2021).

2.5.12 Sectoral Indifference

The lack of incentives in the sectors for the adoption of blockchain technology is a negative reference for the system's experimenters. The government as a rule maker and industry stakeholders expected to support developments have a role in the formation of sectoral indifference (Kouhizadeh et al., 2021; Hughes et al., 2019).

In the study, after 12 barriers to be considered during the integration of blockchain and food waste management were determined, the Best-Worst method, which is one of the multi-criteria decision-making methods, was preferred for analysis.

3. BEST-WORST METHOD

The Best-Worst method (BWM) is a multi-criteria decision-making technique that makes two-stage comparisons over the most preferred and least preferred criteria. The decision maker evaluates the other criteria in terms of these two basic criteria after determining the most important criterion and the least important criterion in the problem. Rezaei (2016) determined criteria importance degrees with a non-linear minimax model that minimizes the maximum absolute difference between pairwise comparisons. Unlike other multi-criteria decision making methods, BWM offers very consistent results with much less comparison input. The method has alternative solutions when any inconsistency is encountered (Gupta & Barua, 2016). In multiple optimization situations, BWM using interval analysis can offer different solutions. Its robust results make the method especially preferable for barrier analysis (Gupta et al., 2017; Mostafaeipour et al., 2021; Malek and Desai, 2019; Govindan et al., 2022; Rejeb et al., 2022).

Below is the BMW calculation methodology: (Rezaei, 2015):

Step 1: Here, n criteria to be compared with each other are determined.

Step 2: Considering the problem subject, the most important criterion and the least important criterion are selected among the criteria.

Step 3: The decision maker evaluates the importance of the best criterion for all other criteria, taking into account the criteria range scaled between 1 and 9. a_{ij} represents the comparison value. Create the Best to Others vector as:

$$A_B = (a_{B1}, a_{B2}, \ldots a_{Bn})$$

Here a_{B_j} is the preference value of the best criterion (*B*) over criterion *j*. $a_{B_B} = 1$.

Step 4: In the second stage, the decision maker evaluates the importance of the other criteria in terms of the worst criterion, based on the same criteria scale. Create the Others to Worst vector as:

$$A_w = (a_{1W}, a_{2W}, \ldots a_{nW})^T$$

Here a_{jw} is the preference value of the best criterion *j* over the worst criterion (*W*). $a_{W_W} = 1$.

Step 5: Optimal weights ($w_1^*, w_2^*, \ldots, w_n^*$) are calculated in line with the evaluations. Optimal criteria weights are calculated, so that using the minimax model below the maximum absolute differences $\left|\frac{w_B}{w_j} - a_{Bj}\right|$ and $\left|\frac{w_j}{w_w} - a_{jW}\right|$ for all j is minimized:

$$\min max_j \left\{ \left|\frac{w_B}{w_j} - a_{Bj}\right|, \left|\frac{w_j}{w_w} - a_{jW}\right| \right\}$$

$$\text{s.t.} \sum_j w_j = 1$$

$$w_j \geq 0 \text{ for all } j \tag{1}$$

Model (1) is same as the following model:

$\min \xi$

s.t. $\left|\frac{w_B}{w_j} - a_{Bj}\right| \leq \xi$, *for all j*

$\left|\frac{w_j}{w_w} - a_{jW}\right| \leq$ ¾ *for all j*

$\sum_j w_j = 1$, $w_j \geq 0$, *for all j* (2)

Model (2) calculates optimal weights of (w_1*, w_2*, …., w_n*) and ξ*. Problems with more than three criteria have more than one optimal solution possibility. Center value of optimal criteria weights (Equation 5) can be used as the criteria weight fort his case. Model (3) and model (4) can be used to calculate the lower bound and upper bound weights of each criterion. Model (2) is solved to find ξ* (Rezaei et al. 2015).

$\min w_j$

s.t. $\left|\frac{w_B}{w_j} - a_{Bj}\right| \leq \xi^*$, for all j

$\left|\frac{w_j}{w_w} - a_{jW}\right| \leq$ ¾*, for all j

$\sum_j w_j = 1$

$w_j \geq 0$ for all j (3)

$\max w_j$

s.t. $\left|\frac{w_B}{w_j} - a_{Bj}\right| \leq \xi^*$, for all j

$$\left| \frac{w_j}{w_w} - a_{jW} \right| \leq \tfrac{3}{4}^{*}, \quad \text{for all } j$$

$$\sum_j w_j = 1,$$

$$w_j \geq 0 \text{ for all } j \tag{4}$$

$$w_j^* = \frac{\min w_j + \max w_j}{2} \tag{5}$$

Step 6: Consistency ratio check the consistency of the comparisons. When $a_{Bj}\ x\ a_{jW} = a_{BW}$ for all j, it is said that all comparisons are fully consistent. However, full consistency cannot be achieved in special cases. Table 1 contains consistency index (Rezaei et al. 2015).

Table 1. Consistency index (CI) table

a_{BW}	1	2	3	4	5	6	7	8	9
Consistency Index (max ξ)	0.00	0.44	1.00	1.63	2.30	3.00	3.73	4.47	5.23

For consistency ratio (CR), the method need to calculate ξ^* and the related consistency index value as given below:

$$\text{Consistency Ratio (CR)} = (\xi^*) / (\text{Consistency Index}) \tag{6}$$

The interpretation of the calculated consistency ratio is as follows: Values close to 0 indicate high consistency, values close to 1 indicate low consistency (Rezaei, 2016).

Step 7: While $\xi^*=0$ in problems with more than three criteria, the problem is fully consistent. However, the number of optimal solutions may increase as the value moves away from 0. Although Consistency Ratio is useful to solve this problem, Rezaei (2016) offers a new contribution to the solution of the problem with interval analysis. It is possible to arrive at a unique solution when Model (2) is formulated as a linear minmax problem. When the sets $\left\{ \left| \frac{w_B}{w_j} - a_{Bj} \right|, \left| \frac{w_j}{w_w} - a_{jw} \right| \right\}$ belonging to the minimization of the maximum value are converted to $\left\{ \left| w_B - a_{Bj} w_j \right|, \left| w_j - a_{Bj} w_w \right| \right\}$ by editing, the following notation appears:

$$\min max_j \left\{ \left| w_B - a_{Bj} w_j \right|, \left| w_j - a_{Bj} w_w \right| \right\}$$

s.t. $\sum_j w_j = 1$

$w_j \geq 0$ for all j (7)

Model (7) turns into the following formulation as a linear programming model:

min ξL

s.t. $\left| w_B - a_{Bj} w_j \right| \leq \xi^L$, *for all j*

$\left| w_j - a_{Bj} w_w \right| \leq \xi^L$, *for all j*

$\sum_j w_j = 1$

$w_j \geq 0$ for all j (8)

High consistency ξL^* and optimum weights $\left(w_1^*, w_2^*, \ldots, w_n^* \right)$ for criteria are obtained without the need for consistency index with Model (8) (Rezaei, 2016).

4. ANALYSIS AND RESULTS

4.1 Results

Multi-criteria decision-making methods serve different purposes as weighting the criteria, ranking the alternatives, and methods that can fulfill both functions together. The Best-Worst method was used to find the weights of the criteria in the problem. Before the analysis, the decision maker group consisting of experts in the field was determined. There is no lower limit or calculation process on the number of experts needed for multi-criteria decision-making methods in the literature. In one of the studies close to the subject, Okorie et al. (2022) reached 6 experts in his study and emphasized that 3 experts (Yildizbasi, 2021) and 7 experts (Singh et al., 2020) were sufficient in similar studies. Table 2 shows the profiles of decision makers experienced in food waste management and blockchain.

Experts operating in the private and public sectors gave their opinions on food waste management and barriers to blockchain integration. Although the expert group brought up different barriers, they agreed on 12 barriers in total. The identified 12 barriers are also seen in various other integration studies in the blockchain and food field. Relevant barriers are listed in Table 3.

Table 2. Decision maker group

No.	Years of Work Experience	Age	Position	Academic/ Practitioner
DM_1	3	25	Product Manager	Practitioner
DM_2	18	39	Expert	Practitioner
DM_3	15	35	Executive	Practitioner
DM_4	12	39	Supervisor	Practitioner
DM_5	15	42	Expert	Practitioner
DM_6	5	27	PhD	Academic
DM_7	13	38	Associate Professor	Academic
DM_8	12	36	Assistant Professor	Academic

Table 3. Blockchain barriers in food waste management

Code	Barrier	Reference
C_1	Infrastructure setup and adaptation challenges	Okorie et al., 2022; Kamilaris et al., 2019; Bumblauskas et al. 2020; Müßigmann et al. 2020; Kumar et al., 2022; Kouhizadeh et al., 2021
C_2	Inadequate government regulations	Vu et al., 2021; Chen et al., 2021; Kumar et al., 2020; Howson, 2020; Biswas and Gupta, 2019; Kouhizadeh et al., 2021
C_3	Cyber risks	Antonucci et al. 2019; Kayıkcı et al., 2021; Reyna et al., 2018; Okorie et al., 2022 Yadav et al., 2020
C_4	High energy requirement	Khan et al., 2022; (Baralla et al., 2018; Okorie et al., 2022; Li et al., 2021
C_5	The ever-increasing need for data storage	Reyna et al., 2018; Kumar et al., 2022; Ahmad et al., 2021; Yadav et al., 2020; Kouhizadeh et al., 2021; Biswas and Gupta, 2019; Li et al., 2021
C_6	Integration challenges	Okorie et al., 2022; Vu et al., 2021
C_7	Lack of expertise	Kumar et al., 2022; Kamilaris et al., 2019; Kouhizadeh et al., 2021; Okorie et al., 2022; Khan et al., 2022; Vu et al., 2021
C_8	Low investment value	Okorie et al., 2022, Mangla et al., 2022; Andoni et al., 2018
C_9	Anonymous data ownership	Kouhizadeh et al., 2021; Okorie et al., 2022; Kumar et al., 2022; Astill et al., 2019
C_{10}	Insufficient public knowledge	Kumar vd., 2022; Astill vd., 2019
C_{11}	Scaling and system speed slowdown	Khan et al., 2022; Yadav et al., 2020; Rana et al., 2021
C_{12}	Sectoral indifference	Kouhizadeh vd., 2021; Hughes vd., 2019

After the barriers took their final form, a questionnaire prepared in accordance with the technique was presented to the experts. Experts use a number between 1 and 9 to show the preference of most important barrier over the other barriers. In the second stage, they again use a number between 1 and 9 to show the preference of the barriers over the least important barrier. Table 4 and Table 5 show the pairwise comparisons of decision makers in terms of best and worst preferences.

Eq. (8) was modeled using the GAMS program to process the data and solve the BWM formulation with the help of a solvent. A unique solution can be reached without the need for consistency ratio with Eq. (8) (Rezaei, 2016). The values obtained the analysis are shown in Table 6.

Table 4. Pairwise comparison vector for the best criterion

DM	Best barrier	Barriers											
		C_1	C_2	C_3	C_4	C_5	C_6	C_7	C_8	C_9	C_{10}	C_{11}	C_{12}
DM_1	C_1	1	7	3	3	2	2	7	4	2	8	2	8
DM_2	C_1	1	6	7	6	7	5	4	5	5	4	5	5
DM_3	C_2	1	1	1	2	2	2	2	2	2	1	1	2
DM_4	C_{12}	4	9	2	3	4	7	6	2	2	8	2	1
DM_5	C_7	6	8	7	9	7	6	1	8	8	8	6	9
DM_6	C_{10}	9	7	9	9	4	6	9	5	4	1	5	2
DM_7	C_8	2	3	9	4	2	2	9	1	9	9	8	2
DM_8	C_1	1	3	5	4	7	7	3	7	3	5	3	6

Table 5. Pairwise comparison vector for the worst criterion

Barriers	DM_1	DM_2	DM_3	DM_4	DM_5	DM_6	DM_7	DM_8
	Worst barrier							
	C_8	C_{10}	C_9	C_{11}	C_8	C_3	C_5	C_8
C_1	3	2	8	2	9	8	9	8
C_2	5	3	9	5	5	8	2	8
C_3	2	4	9	3	6	1	2	3
C_4	3	3	8	4	8	9	9	4
C_5	3	4	8	5	8	6	1	3
C_6	3	4	8	6	7	8	5	5
C_7	8	5	8	3	9	9	6	6
C_8	1	4	7	7	1	3	9	1
C_9	3	4	1	2	4	6	2	3
C_{10}	8	1	9	8	5	9	3	3
C_{11}	3	5	8	1	9	9	3	6
C_{12}	9	4	8	2	3	1	5	2

4.2 Sensitivity Analysis

Sensitivity analysis is frequently used in multi-criteria decision-making problems to control the bias in the results and to get more robust results. In this way, the uncertainty arising from the subjective opinions of the decision makers is reduced. In the analysis, the value of the C_1 barrier with the highest value was changed by 0.1 in each step in the range of 0.1-0.9. The values of other barriers also change proportionally according to the change in the highest barrier (Gupta et al., 2017; Khan et al., 2022). Table 7 shows the results of the sensitivity analysis obtained after 15 iterations.

The sensitivity analysis shows that the most and least important barriers are consistent across all scenarios. Actually, Figure 7 indicates that the same results are achieved at the end of all scenarios.

Table 6. Importance weight and rank of barriers

DM	C_1	C_2	C_3	C_4	C_5	C_6	C_7	C_8	C_9	C_{10}	C_{11}	C_{12}	ξ^L
DM_1	0,153	0,037	0,087	0,087	0,13	0,13	0,037	0,015	0,13	0,032	0,13	0,032	0,107
DM_2	0,237	0,064	0,055	0,064	0,055	0,077	0,096	0,077	0,077	0,045	0,077	0,077	0,147
DM_3	0,129	0,079	0,129	0,065	0,065	0,065	0,065	0,065	0,014	0,129	0,129	0,065	0,050
DM_4	0,071	0,032	0,142	0,095	0,071	0,041	0,047	0,142	0,142	0,036	0,02	0,162	0,122
DM_5	0,078	0,058	0,067	0,052	0,067	0,078	0,329	0,024	0,058	0,058	0,078	0,052	0,139
DM_6	0,042	0,054	0,017	0,042	0,095	0,063	0,042	0,076	0,095	0,267	0,076	0,13	0,113
DM_7	0,151	0,101	0,034	0,076	0,023	0,151	0,034	0,174	0,034	0,034	0,038	0,151	0,129
DM8	0,231	0,102	0,061	0,077	0,044	0,044	0,102	0,022	0,102	0,061	0,102	0,051	0,076
Weight	0,136	0,065	0,074	0,069	0,068	0,081	0,094	0,074	0,081	0,082	0,081	0,090	
Rank	1	12	9	10	11	7	2	8	5	4	6	3	

Table 7. Weights of barriers in different runs

	R#1	Normal	R#2	R#3	R#4	R#5	R#6	R#7	R#8	R#9
C_1	0,100	0,137	0,2	0,3	0,4	0,5	0,6	0,7	0,8	0,9
C_7	0,098	0,094	0,087	0,076	0,065	0,054	0,044	0,033	0,022	0,013
C_{12}	0,094	0,090	0,083	0,073	0,063	0,052	0,042	0,031	0,021	0,012
C_{10}	0,086	0,083	0,077	0,067	0,058	0,048	0,038	0,029	0,020	0,011
C_9	0,085	0,082	0,076	0,066	0,057	0,047	0,038	0,028	0,019	0,011
C_{11}	0,085	0,081	0,075	0,066	0,056	0,047	0,038	0,028	0,019	0,011
C_6	0,085	0,081	0,075	0,066	0,056	0,047	0,038	0,028	0,019	0,011
C_8	0,078	0,074	0,069	0,060	0,052	0,043	0,034	0,026	0,018	0,010
C_3	0,077	0,074	0,069	0,060	0,051	0,043	0,034	0,026	0,017	0,010
C_4	0,073	0,070	0,065	0,057	0,048	0,040	0,032	0,024	0,016	0,009
C_5	0,072	0,069	0,064	0,056	0,048	0,040	0,032	0,024	0,016	0,009
C_2	0,069	0,066	0,061	0,053	0,046	0,038	0,031	0,023	0,016	0,009

Figure 7. Same results achieved

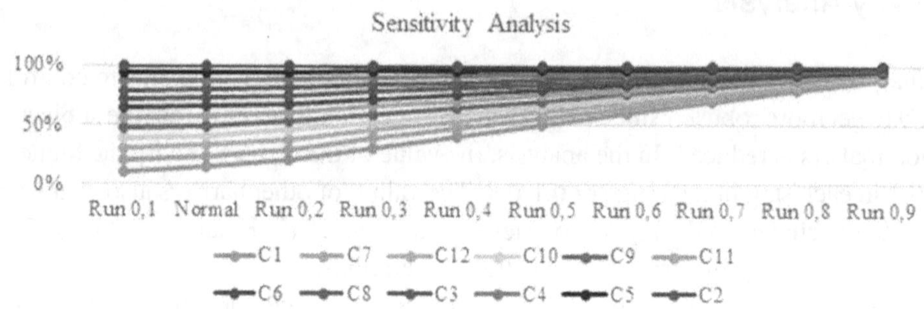

4.3 Discussion

According to the analysis results, "infrastructure setup and adaptation challenges (C_1)" stands out as the most important barrier for food waste management and blockchain integration. When the results of similar studies in the literature are examined, similarities with the findings of this study are striking. Vu et al. (2021) points out blockchain implementation costs as one of the important barriers in their research on blockchain and food supply chains. Khan et al. (2022) identifies the blockchain implication framework as one of the most important barriers in his study in a similar field. In our study, experts state that the cost of hardware required by blockchain technology and insufficient training activities on the adoption of the system by the stakeholders are the most important barriers. Infrastructure costs and implementation challenges are also prominent barriers in similar studies (Müßigmann et al., 2020; Kumar et al., 2022; Choi, 2020). Uncertainties in this area may also be at the root of the difficulties associated with infrastructure setup. The "lack of expertise (C_7)" determined as the second barrier in the analysis may be one of the reasons for these uncertainties. Blockchain is an innovative technology and it still needs time to be widely adopted. The lack of qualified personnel to use the system may reduce the efficiency of the system. Kumar et al. (2022) reach similar findings and conclusions about employees. "Sectoral indifference (C_{12})" takes the third place in the barrier order. Two basic cost items, such as C_1 and C_7, may naturally raise questions about investment. This reduces the attractiveness of food waste management and blockchain integration for the time being.

The barrier named "inadequate government regulations (C_2)" is evaluated as the worst barrier. The government may be waiting for blockchain as a new technology to mature and be used in a wider area. Actually, there is not enough data on the use and possible consequences of digital technologies. The remarkable points in the analysis are that "high energy requirement (C_4)" and "the ever-increasing need for data storage (C_5)" barriers are at the bottom in the order of importance. Today, the search for more efficient systems in energy use and information storage has gained momentum. Decision makers see the potential of these barriers to create problems in integration as low. Results of barrier importance assessments show that economic issues outweigh environmental concerns.

5. CONCLUSION

In this study, a model for food waste management and blockchain integration is presented and possible barriers to this association are examined. Barriers are one of the important factors affecting integration during the application of a new system to a different organization. Preliminary research on barriers contributes to the increase of system efficiency and performance. Indeed, our analysis of the barriers pointed out by decision makers agrees with the results of previous studies. For the analysis, 12 barriers are put forward with the opinions of experts in the field and the support of the relevant literature. Infrastructure setup and adaptation challenges were determined as the most important barrier compared to other barriers. Our analysis allows for the order of importance between the barriers, taking into account the relationship between the barriers. In this way, decision-making processes regarding barriers are facilitated. BWM is a very useful tool for comparing and ranking barriers. The consistency and reliability of the results were tested by sensitivity analysis.

5.1 Theoretical and Managerial Implications

This study is one of the pioneering studies in its field in terms of emphasizing innovative technologies in food waste management. With the bibliometric analysis presented at the beginning of the study, studies in the field of blockchain and food have been extensively researched. This study, which directly deals with the integration of blockchain and food waste management, opens a new field for future studies on this subject. Blockchain-based food waste management system is introduced to the literature as a new conceptual model. BWM, which is one of the multi-criteria decision-making methods, does not need the large number of participants that should be in structural models. Based on the opinions of eight experts, consistent results similar to previous analyzes were reached. BWM provides time and computational advantages over structural models. Practitioners can start their efforts to remove barriers from the point indicated in this study. In this way, limited resources can be used more efficiently with the right planning. The systematic and hierarchical approach presented by our study paves the way for a more effective fight against food waste by helping to remove barriers that threaten system efficiency.

Food waste management is one of the trend topics with economic, environmental and social dimensions. Studies to reduce food waste are supported by many initiatives and government institutions. In this direction, blockchain as an innovative technology emerges as a useful tool for the effective fight against food waste. A blockchain system to be established within the scope of food waste management can monitor food waste and detect the generation points of the waste. This capability gives food waste management stakeholders the ability to reduce or even eliminate waste. In a system where spoilage and other food-related risks are under control, it becomes possible to direct the food and consume it immediately. The ability to observe processes by all stakeholders supports all actors to have certain standards on issues such as distribution, transaction, service, packaging, and information sharing. The system can prevent the manipulation of stakeholders on issues such as cost, transportation, storage, handling, quality assurance. The uninterrupted data flow of the blockchain system reveals big data. Big data, on the other hand, will enable the analysis of processes that generate food waste and the organization of corrective actions.

Efforts on blockchain and various system integrations are currently experimental. For the success of the system, proper management of blockchain-specific features is becoming increasingly important. In this respect, this study offers a starting point for practitioners and government officials who want to make regulations in this field.

5.2 Limitations and Future Research

Since blockchain is an innovative technology, there is currently not enough data on the results of blockchain applications. Moreover, the small number of blockchain experts limits the wide use of these systems. The integration model presented in this study has been discussed conceptually due to the lack of sufficient financial interest and the absence of an expert to perform the system installation. When the widespread use of blockchain and its adoption by the wider community is realized, it will be easier to implement structures such as the model proposed in the study. Blockchain hardware is quite expensive in line with today's technological possibilities. In fact, the experts who expressed their opinions in the study also prioritize the economic dimension of integration. As social sensitivity towards food safety and waste awareness develops, new initiatives on effective food waste management will develop and studies on this subject will increase.

REFERENCES

Ahmad, R. W., Salah, K., Jayaraman, R., Yaqoob, I., Omar, M., & Ellahham, S. (2021). Blockchain-based forward supply chain and waste management for COVID-19 medical equipment and supplies. *IEEE Access : Practical Innovations, Open Solutions, 9*, 44905–44927. doi:10.1109/ACCESS.2021.3066503 PMID:34812386

Aka, S., & Buyukdag, N. (2021). How to prevent food waste behaviour? A deep empirical research. *Journal of Retailing and Consumer Services, 61*, 102560. doi:10.1016/j.jretconser.2021.102560

AlShamsi, M., Al-Emran, M., & Shaalan, K. (2022). A systematic review on blockchain adoption. *Applied Sciences (Basel, Switzerland), 12*(9), 4245. doi:10.3390/app12094245

Ananno, A. A., Masud, M. H., Chowdhury, S. A., Dabnichki, P., Ahmed, N., & Arefin, A. M. E. (2021). Sustainable food waste management model for Bangladesh. *Sustainable Production and Consumption, 27*, 35–51. doi:10.1016/j.spc.2020.10.022

Andoni, M., Robu, V., Flynn, D., Abram, S., Geach, D., Jenkins, D., McCallum, P., & Peacock, A. (2019). Blockchain technology in the energy sector: A systematic review of challenges and opportunities. *Renewable & Sustainable Energy Reviews, 100*, 143–174. doi:10.1016/j.rser.2018.10.014

Angelis, J., & Da Silva, E. R. (2019). Blockchain adoption: A value driver perspective. *Business Horizons, 62*(3), 307–314. doi:10.1016/j.bushor.2018.12.001

Antonucci, F., Figorilli, S., Costa, C., Pallottino, F., Raso, L., & Menesatti, P. (2019). A review on blockchain applications in the agri-food sector. *Journal of the Science of Food and Agriculture, 99*(14), 6129–6138. doi:10.1002/jsfa.9912 PMID:31273793

Aschemann-Witzel, J., De Hooge, I., Amani, P., Bech-Larsen, T., & Oostindjer, M. (2015). Consumer-related food waste: Causes and potential for action. *Sustainability, 7*(6), 6457–6477. doi:10.3390u7066457

Aschemann-Witzel, J., De Hooge, I. E., Rohm, H., Normann, A., Bossle, M. B., Grønhøj, A., & Oostindjer, M. (2017). Key characteristics and success factors of supply chain initiatives tackling consumer-related food waste–A multiple case study. *Journal of Cleaner Production, 155*, 33–45. doi:10.1016/j.jclepro.2016.11.173

Astill, J., Dara, R. A., Campbell, M., Farber, J. M., Fraser, E. D., Sharif, S., & Yada, R. Y. (2019). Transparency in food supply chains: A review of enabling technology solutions. *Trends in Food Science & Technology, 91*, 240–247. doi:10.1016/j.tifs.2019.07.024

Balaji, M., & Arshinder, K. (2016). Modeling the causes of food wastage in Indian perishable food supply chain. *Resources, Conservation and Recycling, 114*, 153–167. doi:10.1016/j.resconrec.2016.07.016

Bamakan, S. M. H., Malekinejad, P., & Ziaeian, M. (2022). Towards blockchain-based hospital waste management systems; applications and future trends. *Journal of Cleaner Production, 349*, 131440. doi:10.1016/j.jclepro.2022.131440

Baralla, G., Ibba, S., Marchesi, M., Tonelli, R., & Missineo, S. (2018, August). A blockchain based system to ensure transparency and reliability in food supply chain. In *European conference on parallel processing* (pp. 379-391). Springer, Cham.

Beck, R., & Müller-Bloch, C. (2017). *Blockchain as radical innovation: a framework for engaging with distributed ledgers as incumbent organization.*

Biswas, B., & Gupta, R. (2019). Analysis of barriers to implement blockchain in industry and service sectors. *Computers & Industrial Engineering*, *136*, 225–241. doi:10.1016/j.cie.2019.07.005

Bumblauskas, D., Mann, A., Dugan, B., & Rittmer, J. (2020). A blockchain use case in food distribution: Do you know where your food has been? *International Journal of Information Management*, *52*, 102008. doi:10.1016/j.ijinfomgt.2019.09.004

Chen, M. F. (2023). Integrating the extended theory of planned behavior model and the food-related routines to explain food waste behavior. *British Food Journal*, *125*(2), 645–661. doi:10.1108/BFJ-07-2021-0788

Chen, S., Liu, X., Yan, J., Hu, G., & Shi, Y. (2021). Processes, benefits, and challenges for adoption of blockchain technologies in food supply chains: A thematic analysis. *Information Systems and e-Business Management*, *19*(3), 909–935. doi:10.100710257-020-00467-3

Choi, T. M. (2020). Supply chain financing using blockchain: Impacts on supply chains selling fashionable products. *Annals of Operations Research*, 1–23. doi:10.100710479-020-03615-7

Coşkun, A., & Özbük, R. M. Y. (2020). What influences consumer food waste behavior in restaurants? An application of the extended theory of planned behavior. *Waste Management (New York, N.Y.)*, *117*, 170–178. doi:10.1016/j.wasman.2020.08.011 PMID:32861079

Creydt, M., & Fischer, M. (2019). Blockchain and more-Algorithm driven food traceability. *Food Control*, *105*, 45–51. doi:10.1016/j.foodcont.2019.05.019

Crosby, M., Pattanayak, P., Verma, S., & Kalyanaraman, V. (2016). Blockchain technology: Beyond bitcoin. *Applied Innovation*, *2*(6-10), 71.

FAO. (2011). Global food losses and food waste – Extent, causes and prevention. Roma.

FAO. (2015). Save Food: Global Initiative on Food Loss and Waste Reduction. Roma.

FAO. (2020). *Grow, Nourish, sustain. Together.* FAO. https://www.fao.org/world-food-day/home/en/

Feng, H., Wang, X., Duan, Y., Zhang, J., & Zhang, X. (2020). Applying blockchain technology to improve agri-food traceability: A review of development methods, benefits and challenges. *Journal of Cleaner Production*, *260*, 121031. doi:10.1016/j.jclepro.2020.121031

França, A. S. L., Neto, J. A., Gonçalves, R. F., & Almeida, C. M. V. B. (2020). Proposing the use of blockchain to improve the solid waste management in small municipalities. *Journal of Cleaner Production*, *244*, 118529. doi:10.1016/j.jclepro.2019.118529

Friedman, N., & Ormiston, J. (2022). Blockchain as a sustainability-oriented innovation?: Opportunities for and resistance to Blockchain technology as a driver of sustainability in global food supply chains. *Technological Forecasting and Social Change*, *175*, 121403. doi:10.1016/j.techfore.2021.121403

Garnett, T. (2011). Where are the best opportunities for reducing greenhouse gas emissions in the food system (including the food chain)? *Food Policy*, *36*, S23–S32. doi:10.1016/j.foodpol.2010.10.010

Ghinoi, S., Silvestri, F., & Steiner, B. (2020). Toward the creation of novel food waste management systems: A network approach. *Journal of Cleaner Production*, *246*, 118987. doi:10.1016/j.jclepro.2019.118987

Gopalakrishnan, P. K., Hall, J., & Behdad, S. (2021). Cost analysis and optimization of Blockchain-based solid waste management traceability system. *Waste Management (New York, N.Y.)*, *120*, 594–607. doi:10.1016/j.wasman.2020.10.027 PMID:33288397

Govindan, K. (2018). Sustainable consumption and production in the food supply chain: A conceptual framework. *International Journal of Production Economics*, *195*, 419–431. doi:10.1016/j.ijpe.2017.03.003

Govindan, K., Nasr, A. K., Karimi, F., & Mina, H. (2022). Circular economy adoption barriers: An extended fuzzy best–worst method using fuzzy DEMATEL and Supermatrix structure. *Business Strategy and the Environment*, *31*(4), 1566–1586. doi:10.1002/bse.2970

Gupta, H., & Barua, M. K. (2016). Identifying enablers of technological innovation for Indian MSMEs using best–worst multi criteria decision making method. *Technological Forecasting and Social Change*, *107*, 69–79. doi:10.1016/j.techfore.2016.03.028

Gupta, P., Anand, S., & Gupta, H. (2017). Developing a roadmap to overcome barriers to energy efficiency in buildings using best worst method. *Sustainable Cities and Society*, *31*, 244–259. doi:10.1016/j.scs.2017.02.005

Heck, S., Campos, H., Barker, I., Okello, J. J., Baral, A., Boy, E., Brown, L., & Birol, E. (2020). Resilient agri-food systems for nutrition amidst COVID-19: Evidence and lessons from food-based approaches to overcome micronutrient deficiency and rebuild livelihoods after crises. *Food Security*, *12*(4), 823–830. doi:10.100712571-020-01067-2 PMID:32839664

Howson, P. (2020). Building trust and equity in marine conservation and fisheries supply chain management with blockchain. *Marine Policy*, *115*, 103873. doi:10.1016/j.marpol.2020.103873

Hughes, L., Dwivedi, Y. K., Misra, S. K., Rana, N. P., Raghavan, V., & Akella, V. (2019). Blockchain research, practice and policy: Applications, benefits, limitations, emerging research themes and research agenda. *International Journal of Information Management*, *49*, 114–129. doi:10.1016/j.ijinfomgt.2019.02.005

Jin, C., Sun, S., Yang, D., Sheng, W., Ma, Y., He, W., & Li, G. (2021). Anaerobic digestion: An alternative resource treatment option for food waste in China. *The Science of the Total Environment*, *779*, 146397. doi:10.1016/j.scitotenv.2021.146397 PMID:33743457

Kamilaris, A., Fonts, A., & Prenafeta-Boldὺ, F. X. (2019). The rise of blockchain technology in agriculture and food supply chains. *Trends in Food Science & Technology*, *91*, 640–652. doi:10.1016/j.tifs.2019.07.034

Kayikci, Y., Subramanian, N., Dora, M., & Bhatia, M. S. (2022). Food supply chain in the era of Industry 4.0: Blockchain technology implementation opportunities and impediments from the perspective of people, process, performance, and technology. *Production Planning and Control*, *33*(2-3), 301–321. doi:10.1080/09537287.2020.1810757

Kayıkcı, Y., Usar, D. D., & Aylak, B. L. (2021). Using blockchain technology to drive operational excellence in perishable food supply chains during outbreaks. *International Journal of Logistics Management*.

Khan, S., Kaushik, M. K., Kumar, R., & Khan, W. (2022). Investigating the barriers of blockchain technology integrated food supply chain: A BWM approach. *Benchmarking*.

Köhler, S., & Pizzol, M. (2020). Technology assessment of blockchain-based technologies in the food supply chain. *Journal of Cleaner Production*, *269*, 122193. doi:10.1016/j.jclepro.2020.122193

Kosseva, M. R. (2013). Introduction: causes and challenges of food wastage. *Food Industry Wastes*, xv-xxiv.

Kouhizadeh, M., Saberi, S., & Sarkis, J. (2021). Blockchain technology and the sustainable supply chain: Theoretically exploring adoption barriers. *International Journal of Production Economics*, *231*, 107831. doi:10.1016/j.ijpe.2020.107831

Kumar, A., Liu, R., & Shan, Z. (2020). Is blockchain a silver bullet for supply chain management? Technical challenges and research opportunities. *Decision Sciences*, *51*(1), 8–37. doi:10.1111/deci.12396

Kumar, S., Raut, R. D., Agrawal, N., Cheikhrouhou, N., Sharma, M., & Daim, T. (2022). Integrated blockchain and internet of things in the food supply chain: Adoption barriers. *Technovation*, *118*, 102589. doi:10.1016/j.technovation.2022.102589

Li, D., Wang, X., Chan, H. K., & Manzini, R. (2014). Sustainable food supply chain management. *International Journal of Production Economics*, *152*, 1–8. doi:10.1016/j.ijpe.2014.04.003

Li, K., Lee, J. Y., & Gharehgozli, A. (2021). Blockchain in food supply chains: A literature review and synthesis analysis of platforms, benefits and challenges. *International Journal of Production Research*, 1–20.

Lo, I. M., & Woon, K. S. (2016). Food waste collection and recycling for value-added products: Potential applications and challenges in Hong Kong. *Environmental Science and Pollution Research International*, *23*(8), 7081–7091. doi:10.100711356-015-4235-y PMID:25772864

Malek, J., & Desai, T. N. (2019). Prioritization of sustainable manufacturing barriers using Best Worst Method. *Journal of Cleaner Production*, *226*, 589–600. doi:10.1016/j.jclepro.2019.04.056

Mangla, S. K., Kazançoğlu, Y., Yıldızbaşı, A., Öztürk, C., & Çalık, A. (2022). A conceptual framework for blockchain-based sustainable supply chain and evaluating implementation barriers: A case of the tea supply chain. *Business Strategy and the Environment*, *31*(8), 3693–3716. doi:10.1002/bse.3027

Masud, M. H., Karim, A., Ananno, A. A., & Ahmed, A. (2020). *Sustainable food drying techniques in developing countries: prospects and challenges*. Springer International Publishing. doi:10.1007/978-3-030-42476-3

McAdams, B., von Massow, M., Gallant, M., & Hayhoe, M. A. (2019). A cross industry evaluation of food waste in restaurants. *Journal of Foodservice Business Research*, *22*(5), 449–466. doi:10.1080/15378020.2019.1637220

Memon, R. A., Li, J. P., & Ahmed, J. (2019). Simulation model for blockchain systems using queuing theory. *Electronics (Basel)*, *8*(2), 234. doi:10.3390/electronics8020234

Mostafaeipour, A., Alvandimanesh, M., Najafi, F., & Issakhov, A. (2021). Identifying challenges and barriers for development of solar energy by using fuzzy best-worst method: A case study. *Energy*, *226*, 120355. doi:10.1016/j.energy.2021.120355

Munir, K. (2022). Sustainable food waste management strategies by applying practice theory in hospitality and food services-a systematic literature review. *Journal of Cleaner Production*, *331*, 129991. doi:10.1016/j.jclepro.2021.129991

Müßigmann, B., von der Gracht, H., & Hartmann, E. (2020). Blockchain technology in logistics and supply chain management—A bibliometric literature review from 2016 to January 2020. *IEEE Transactions on Engineering Management*, *67*(4), 988–1007. doi:10.1109/TEM.2020.2980733

Okorie, O., Russell, J., Jin, Y., Turner, C., Wang, Y., & Charnley, F. (2022). Removing Barriers to Blockchain use in Circular Food Supply Chains: Practitioner Views on Achieving Operational Effectiveness. *Cleaner Logistics and Supply Chain,* 100087.

Rana, R. L., Tricase, C., & De Cesare, L. (2021). Blockchain technology for a sustainable agri-food supply chain. *British Food Journal*, *123*(11), 3471–3485. doi:10.1108/BFJ-09-2020-0832

Raut, R. D., Gardas, B. B., Narwane, V. S., & Narkhede, B. E. (2019). Improvement in the food losses in fruits and vegetable supply chain-a perspective of cold third-party logistics approach. *Operations Research Perspectives*, *6*, 100117. doi:10.1016/j.orp.2019.100117

Rejeb, A., Rejeb, K., Keogh, J. G., & Zailani, S. (2022). Barriers to blockchain adoption in the circular economy: A fuzzy Delphi and best-worst approach. *Sustainability*, *14*(6), 3611. doi:10.3390u14063611

Reyna, A., Martín, C., Chen, J., Soler, E., & Díaz, M. (2018). On blockchain and its integration with IoT. Challenges and opportunities. *Future Generation Computer Systems*, *88*, 173–190. doi:10.1016/j.future.2018.05.046

Rezaei, J. (2015). Best-worst multi-criteria decision-making method. *Omega*, *53*, 49–57. doi:10.1016/j.omega.2014.11.009

Rezaei, J. (2016). Best-worst multi-criteria decision-making method: Some properties and a linear model. *Omega*, *64*, 126–130. doi:10.1016/j.omega.2015.12.001

Rezaei, J., Wang, J., & Tavasszy, L. (2015). Linking supplier development to supplier segmentation using Best Worst Method. *Expert Systems with Applications*, *42*(23), 9152–9164. doi:10.1016/j.eswa.2015.07.073

Stranieri, S., Riccardi, F., Meuwissen, M. P., & Soregaroli, C. (2021). Exploring the impact of blockchain on the performance of agri-food supply chains. *Food Control*, *119*, 107495. doi:10.1016/j.foodcont.2020.107495

Swan, M. (2015). *Blockchain: Blueprint for a new economy*. O'Reilly Media, Inc.

Taylor, P., Steenmans, K., & Steenmans, I. (2020). Blockchain technology for sustainable waste management. *Frontiers in Political Science, 15*.

Teoh, C. W., Koay, K. Y., & Chai, P. S. (2022). The role of social media in food waste prevention behaviour. *British Food Journal, 124*(5), 1680–1696. doi:10.1108/BFJ-04-2021-0368

Tillemann, T., Price, A., Tillemann-Dick, G., & Knight, A. (2019). *The blueprint for blockchain and social innovation*. New America.

Vu, N., Ghadge, A., & Bourlakis, M. (2021). Blockchain adoption in food supply chains: A review and implementation framework. *Production Planning and Control*, 1–18.

Wang, H., Ma, B., Cudjoe, D., Farrukh, M., & Bai, R. (2023). What influences students' food waste behaviour in campus canteens? *British Food Journal, 125*(2), 381–395. doi:10.1108/BFJ-10-2021-1103

Yadav, V. S., Singh, A. R., Raut, R. D., & Govindarajan, U. H. (2020). Blockchain technology adoption barriers in the Indian agricultural supply chain: An integrated approach. *Resources, Conservation and Recycling, 161*, 104877. doi:10.1016/j.resconrec.2020.104877

Zhang, J., Zhang, X., Liu, W., Ji, M., & Mishra, A. R. (2022). Critical success factors of blockchain technology to implement the sustainable supply chain using an extended decision-making approach. *Technological Forecasting and Social Change, 182*, 121881. doi:10.1016/j.techfore.2022.121881

Chapter 15
Sustainable Approach to Food Packaging Materials Using Nanotechnology

Rajni Gautam
School of Basic and Applied Sciences, K.R. Mangalam University, Gurugram, India

ABSTRACT

Ever rising population has resulted in the growth of food packaging industry. Time has come to look for new innovative materials for food packaging which are environment friendly along with improved barrier properties. The advent of Nanotechnology has revolutionized the food packaging industry and has led to extensive research on sustainable food packaging solutions using nanotechnology. The main objective is to reduce the plastic consumption and to look for a better replacement which can be recycled or supports low carbon circular economy apart from having improved antimicrobial/antioxidant, thermal, and mechanical properties. This chapter talks about the recent advancements and contributions of nanotechnology in providing sustainable food packaging solutions. Safety issues linked with the consumption of nanomaterials used in food packaging are also discussed at the end.

1. INTRODUCTION

Food packaging plays a vital role in food industry as it protects the food from contamination occurring due to various physical, chemical and biological sources and increase their shelf life by prevention of diseases and spoilage (Han 2018), (Agarwal 2022). Another important job of food packaging is to keep food safe during transportation. Chemical sources are: gain or loss of moisture, exposure to gases such as oxygen, and light. Biological sources are: microorganisms i.e. pathogens and spoiling agents, insects, rodents, and other animals. Physical sources are: mechanical damage (abrasions, and crushing damage) for example shock and vibration encountered during transportation and distribution. Appropriate packaging becomes even more important for delicate food items such as eggs and fresh fruits. To maintain the standard and quality of food products during transportation and storage, a good quality packaging material is of utmost importance. A good quality packaging material is generally characterized by me-

DOI: 10.4018/978-1-6684-9094-5.ch015

chanical strength, biodegradability, and gas and moisture permeability (Han 2018), (Agarwal 2022). Conventional food packaging materials are: glass, metals (aluminium, foils and laminates, tinplate, and tin-free steel), paper and paperboards, and plastics. Nanotechnology has come up with many new and advanced nanomaterials as alternative food packaging materials (Mohammad 2022). Nanoscience and Nanotechnology has revolutionized many sectors of society such as healthcare, medicine, transportation, energy, food safety, and environmental science, among many others (Jampilek 2019) (Mohammad2022). Nanotechnology deals with the materials having atleast one dimension in nanometer scale (1–100 nanometers). These nanomaterials have many unique properties such as high surface to volume ratio, enhanced catalytic activity, improved antimicrobial properties etc.. The smart food packaging materials based on nanotechnology are gaining a lot of interest owing to their enhanced antimicrobial properties along with biodegradability. Till today, most food packaging materials are based on petro-based plastics which are not biodegradable in nature. The reason for using plastics in packaging is its good performance along with low cost and easy handling during storage and transportation but these are not environment friendly because of the release of greenhouse gases like CO_2 during the production of petroleum-based plastics. Again their disposal is also a problem due to the lack of collection and non-biodegradability, plastics land up in landfills or in water streams contaminating the oceans (Salwa et al., 2019). Although a lot of research is going on to find a more environment friendly method of production of plastic. Overexploitation of resources, environmental degradation even after being an age old problem still continues to be a major global challenge and moving towards sustainable innovations is the only solution. Sustainable innovations include development of novel products and technologies leading to a positive impact on humanity and environment. Different innovations spanning food industry, packaging, transportation and green energy to conserve resources have proven to be a boon in this time of crisis. However, be it in any field, these new innovations need to be ecologically and economically beneficial as well as socially balanced. In order to survive on this planet it is extremely important to look for products and services that could pave the way for more sustainable way of living. Sustainable food packaging is the future of food packaging industry as it uses environment friendly materials that can be used to wrap food items for storage and even for transportation. Sustainable packaging offers so many benefits such as: low carbon footprint, recyclable and biodegradable materials, health benefits, etc. Generally, food packaging materials which are biodegradable suffers from poor barrier and mechanical properties. Thus, there is a trade- off between sustainability and improved barrier properties. This trade-off will be addressed in this chapter. The chapter explains the advantages and drawbacks of using conventional food packaging materials along with recent nanotechnological advancements in food packaging industry which are more environment friendly and sustainable in nature such as edible coatings, bio-polymers and bio- nanocomposites etc. The chapter illustrates the impact of nanotechnology in the field of food packaging along with the discussion on the safety concerns associated with the use of nanotechnology based smart food packaging materials.

Plastics which are made from polymer materials that are petroleum-based like polypropylene (PP), polyester (PET), polyethylene (PE) and polystyrene (PS) (Salwa et al., 2019) is the common choice of food industry for packaging purposes. Owing to its superior properties like good barrier properties, light weight, low cost etc. plastic consumption is far greater than other packaging materials, and it is expected to continue growing in coming years also. But nowadays, researchers have started looking for alternative material for food packaging because of the hazardous effects on plastic on the environment. The development of novel, safe and sustainable packaging is a major concern for the food industry, which aims at conserving food and minimizing waste throughout the distribution chain, granting high-quality products with a longer shelf life. In this context, food industries have been working towards an alternative material such as the bio-based plastics which are abundant, low cost, renewable, and biodegradable as compared to traditional nonrenewable petroleum-based resources (Kawashima et al., 2019). This will not only reduce the use of petroleum resources but will also lead to subsequent reduction of the CO_2 emissions (Salwa et al., 2019). In general, sustainable materials can be classified in three categories:

- Biodegradable material is the material which gets break down or decomposed by bacteria or other microorganism. Thus it decays naturally and does not require any scientific treatment. So disposal of biodegradable waste material is not a problem as they fade down into the soil.
- Compostable materials are those materials which break down completely into non-toxic components such as water, humus, biomass and carbon dioxide. So they are not harmful for the environment. Also, some of the compostable materials can only be composted through industrial processes which are not easily available every time. All compostable materials are biodegradable but all biodegradable materials are not necessarily be compostable.
- Recyclable materials can be processed and reused again. They can be converted into something new. Some of the common examples of recyclable materials are: paper, glass, metals and some types of plastic.

2. CONVENTIONAL SUSTAINABLE FOOD PACKAGING MATERIALS

2.1 Glass

Glass has been in use as a food packaging material since ages. Chemically it is synthesized by heating a mixture of silicates, lime and soda to the fusion point and then cooling (Yarış 2017). The main advantages of glass being used as a food packaging material are: impermeable to gas, moisture and food spoilage causing microorganisms, it is inert and do not react or migrate into food products, suitable for heat processing, transparent to microwaves, reusable and recyclable, transparent and display the contents inside the container, and can be moulded into different types of shapes and colours. The common disadvantages are: susceptibility to fractures, scratches and breakability, high transportation cost due to higher weight and permeability to UV light. Glass containers come in different shapes and sizes such as bottles having narrow necks, Jars and pots having wide openings, tumblers and jugs, vials, and ampoules. Glass bottles are generally used for soft drinks, mineral waters, beers, wines, spirits, and liqueurs etc.. Jar and pots are widely used for instant coffee, dry mixes, spices, processed baby foods, dairy products, sugar preserves (jams and marmalades). Vials and ampoules are small, thin-walled glass containers. They are primarily used in the pharmaceutical industry for drugs and for storing small quantities of very expensive ingredients, such as flavors.

2.2 Metals

Metals like aluminium, tin, tin free steel and stainless steel are the common metals that find applications in food industry for packaging purpose. They can be used in various forms (rigid and semi-rigid) such as cans, foil wraps and retort pouches (Debeaufort 2021). Packaging material based on metals demonstrates excellent barrier properties (impermeable to gas, moisture and light), recyclability, easy conversion into various shapes, high thermal stability, rigid structure, and easy transportation. Although metals are highly recyclable but emission of carbon dioxide by the factories manufacturing metals leading to global warming is a major issue. Another concern is migration of harmful toxic chemicals such as lead, mercury, cadmium, iron, nickel, aluminium, and bisphenol A into the food items from the metal containers. Bulging from cans, dissolution of tin, corrosion and blackening are some other problems related with the use of metal containers.

Thus, a balance between advantages and disadvantages of food packaging based on metals is required using better approach and research.

2.2.1 Aluminium

Aluminium is the most abundantly available metal. Chemically, it is processed by using alumina and sodium hydroxide as precursors at 175 °C. Alumina is converted to aluminium hydroxide ($Al(OH)_3$). The insoluble part is filtered out and soluble $Al(OH)_3$ gets precipitated. Aluminium is a costly metal as compared to steel and also difficult to weld thus mainly preferred for light weight packaging applications. Aluminium when combined with paper and plastics can also be used for foil, cans, laminated sheets. To increase the strength, manganese is generally added to aluminium (Nam 2000). The aluminium foils are thin rolled flat sheets of 99% purity aluminium whose thickness may vary from from 4 to 150 microns. It offers good impermeability to moisture and gas and easy to wrap around food product of any shape. According to some recent studies (Dordevic 2019) use of aluminium foil may be harmful for human health as there is a possibility of leaching out of aluminium from the foil into the food. It is found to be significant in case of acidic and aqueous solutions in comparison to alcoholic and saline solutions.

2.2.2 Tin

Tin is a preferred metal for canning of food items such as beverages and pharmaceutical products as it doesn't react with acidic products and thus provides corrosion resistance (Debeaufort 2021). Corroded substances can contaminate the packaged food and damage the can itself. Thus tin cans are safe for packaging food products having acidic content such as medicines, cosmetics etc for longer time. Sometimes tin is applied as a thin layer on steel as it acts as an oxygen scavenger and corrosion resistant material. The major drawback of tin plated cans is that they are much heavier as compared to aluminum cans. This makes tin-plated cans more expensive and less convenient to transport.

2.2.3 Steel

Stainless steel is an alloy of iron containing chromium and has attractive properties like excillent resistance against corrosion and chemical inertness (Debeaufort 2021). Although chromium itself is not inert in nature but it forms an inert layer of chromium oxide (Cr_2O_3) on steel surface after coming in the contact of oxygen, which provides passivation against corrosion. High corrosion resistance and inertness is the main reason for its wide use as a packaging material specially for storing food items for a longer period of time.

Since it is costly as compared to aluminium and tin therefore it is mostly used for returnable containers such as kegs for beer, wine and soft drinks. However, stainless steel is the choice material for large size containers during transport and storage.

2.2.4 Metal Caps and Closures

Metals are also used for sealing caps and closures for example sealing of beverage glass bottles. Tinplate or tin free steel are more common for making these closures which come in various shapes and types such as press-twist, twist-off, pry off and deep-press. Aluminium foils with polyacrylic or polyvinyl chloride coatings are used as bottle caps and lids on glass jars. Vacuum closures required for sealing of baby foods, milk bottles, lids on jam bottles, pickels, yoghurt and dry powders are also made from aluminium foil.

2.3 Paper

Fig. 1 shows the general classification of paper. Broadly it is divided into two categories. The raw material for making virgin grade paper is the wood pulp whereas *recycled paper* is made from reprocessing of waste paper or used virgin paper. Fine paper has good smoothness and is obtained from bleached pulp whereas coarse paper is made from unbleached pulp. Fine paper is generally used for printing, writing, labelling, books etc. Coarse paper finds applications in food industry for packaging purpose. Since food comes in direct contact of the paper packaging, only virgin grade paper made from unbleached pulp is recommended for the use of food packaging material according to Food Safety and Standard Authority of India (FSSR 2011). The Plain paper is rarely used for packaging purpose as it suffers from poor barrier properties, low heat sealability and strength. So, it is infused with some additives to enhance its properties. It is also laminated with aluminium or plastic to make it useful for food packaging application.

Figure 1. Classification of paper

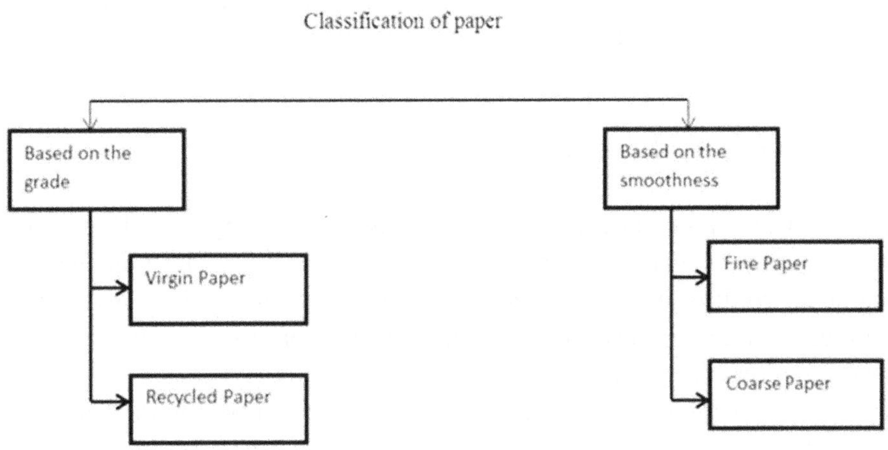

2.3.1 Kraft Paper

Kraft process is used to make Kraft paper which is also based on chemical conversion of wood into wood pulp. Lignin and cellulose are the basic building blocks of wood. Good quality paper is made by separating lignin from the cellulose. Wood chips are mixed with white liquor made of sodium sulphide and sodium hydroxide. Lignin and cellulose are broken apart, leaving behind a solid wood pulp and a black liquid. Larger pieces are now removed from the pulp followed by washing in order to remove any residual liquor. Now the wood pulp is ready for making paper products out of it. Kraft paper is different from normal paper because of the reduced lignin and high sulphur content. Also Kraft paper is of coarse texture and is stronger as compared to normal paper as it doesn't involve extensive bleaching, which reduces the strength of the paper. Bleaching also add to the manufacturing cost. Kraft paper is used for food items which require high strength of the packaging material such as flour, sugar, and dried fruits etc.

2.3.2 Bleached Paper

Bleaching is a chemical process to make the paper more brighter and whiter. The main chemical agents that are being used in this process are: chlorine, chlorine dioxide, hydrogen peroxide, ozone etc. The amount of bleaching is generally indicated by Kappa number. Lower kappa number means use of bleaching chemicals in lesser amounts (Deshwal 2019). Bleaching process makes the paper soft and whiter but it decreases the strength of paper. Bleached paper is usually used where appearance is more important than the strength. Since, bleaching has adverse effects on the environment as it leads to release of hazardous chemicals into water bodies, new environmental friendly ways are being researched upon. For example, an eco-friendly approach for bleaching called bio-bleaching of paper pulp has been developed which uses ligninolytic enzymes like xylanase, laccase and manganese peroxidase (Saleem et al. 2018). Clay coatings can also be used to augment the appearance of bleached paper for food.

2.3.3 Wrapping Tissue Paper or Tissues

Tissue paper is a very lightweight paper made from recycled pulp. There are various categories of tissue paper for different purposes like hygienic tissue paper, facial tissues, paper towels, as packing material, among other. In the field of food packaging, tissue paper is majorly used as wrapping material for bakery products, sandwich, box liner, tea and coffee bags. (Macarthur 2017).

2.3.4 Greaseproof Paper and Waxed Paper

As the name suggests, Greaseproof paper is generally used for packaging of food items that contain oil or fat such as butter. Greaseproof paper has a very good resistance against oil and grease. They are translucent and hydrated. It is made by prolonged beating of the pulp which causes breaking down of the cellulose fibres resulting in enhancement of their water absorption capacity and finally leading to their superficial gelatinization and stickiness (Deshwal 2019).

Waxed paper is the paper coated with wax to improve the barrier property against liquid and gases. Movement of moisture and gases into paper is inhibited by applying wax onto the paper which also provides heat sealability. Waxed paper finds applications in packaging of bread, biscuits, dairy products, sandwich, cakes, and sunflower oil.

3. ROLE OF NANOTECHNOLOGY IN FOOD PACKAGING: SUSTAINABLE FOOD PACKAGING MATERIALS

Food products are subjected to constant biological, chemical and physical change during storage and distribution (Sean, 2015). There are some changes which are highly desirable such as ripening of fruits but other changes may degrade the quality of the food such as the loss or absorption of moisture. The major role of food packaging material is to provide a barrier to these changes thus packaging provides protection against moisture, air, light, disease causing pathogens like bacteria, and food spoiling decomposition and harmful reactions.

To improve food packaging properties i.e. mechanical strength, moisture and gas permeability, and heat resistance, researchers and scientists are working towards the development of new and advanced nanomaterials such as Polymer nanocomposites, edible coatings etc. By employing active antimicrobial packaging, the growth of spoilage causing unwanted microorganisms on the surface of food items can be suppressed. Some of the examples are: Silver, copper, and other metal oxide nanoparticles *Titanium oxide* and *Zinc oxide*.

3.1 Edible Coatings

Edible films or coatings have revolutionized the food packaging industry. A layer of edible material which covers the food items act as a barrier to food damaging mechanisms i.e. gas and moisture. One of the important methods to prepare edible films is Nanolamination. It consists of multilayers with each layer having nano-dimensions (i.e. 1–100 nm). Such type of edible films are generally present on food items like biscuits, chocolates, fruits, vegetables, meats, and other baked items, etc. Along with improvised textural properties, they add to the flavors and nutrients along with better antimicrobial properties.

Edible nano caotings (Jung 2014) result in reduction in loss of water, volatile flavors and aromas, restriction of movement of oxygen, moisture, fats, oils and gas diffusion leading to improved structural properties(appearance of the food products) and less mould and fungal growth along with easy incorporation of food additives, flavoring agents and pigments.

Edible coatings have to meet a number of functional requirements. The most important functions are: satisfactory barrier properties to water, other substances and gases, solubility in water and fat, adequate color and appearance, mechanical and rheological properties, non-toxicity. Those properties are determined by the type of applied material, production method and manner of application on the food product.

Edible and biodegradable coatings are generally obtained from naturally occurring polymers, such as polysaccharides, proteins and lipids (Basumatary 2022). Natural polymers can be classified into three categories according to the method of production:

1. Polymers obtained directly from natural materials (i.e. plants). Examples include polysaccharides, such as starch and cellulose, and proteins such as casein and wheat gluten.
2. Polymers produced in the process of classical chemical synthesis from monomers of biological materials. Examples include polyactide, biopolyester from lactic acid monomers. These monomers are obtained in the process of fermentation of hydrocarbon chemical components.
3. Polymers produced by microorganisms or genetically modified bacteria. The best-known example of this polymer is polyhydroxyisobutyrate, hydroxyisobutyrate /HB/ andhydroxyvalerate /HV/ copolymers.

Fig. 2 shows the classification of materials used for making edible coatings. Polysaccharide-based: The most common polysaccharides used for edible coating of fruits and vegetables are chitosan, starch, alginate, cellulose, pullulan, carrageenan, gellan gum etc., (Basumatary 2022). Polysaccharides based edible coatings having poor moisture barrier properties, it is water soluble. But it contains moderately low O_2 permeability. Mostly, polysaccharides based edible coatings are applied on fresh and minimally processed fruits and vegetables, by creation of modified atmosphere condition to reduce their respiration rate. This leads to improved mechanical handling property and additives carrying capacity. Polysaccharides give crispiness, hardness, compactness, thickening quality, adhesiveness and viscosity to a variety of edible coatings. Polysaccharides are made up of polymer chain, having excellent gas barrier properties, resulting in desirable modified atmosphere that extend the shelf life of the fruits and vegetables without forming anaerobic conditions (Basumatary 2022).

Figure 2. Classification of materials used for making edible coatings

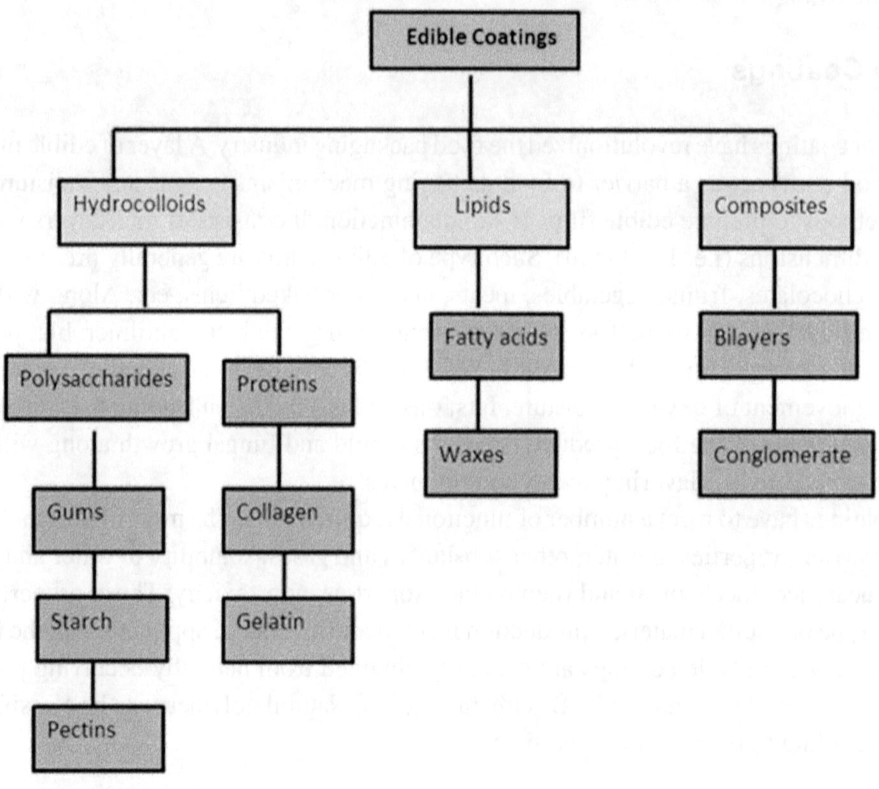

Edible coatings based on polysaccharides (Kocira 2021) (cellulose and its derivatives, starch and its derivatives, resins) or proteins (gelatin, zein, milk proteins, gluten) are generally have good mechanical properties but suffer from relatively low air and misture barrier properties.

On the other hand, coatings comprising lipids (wax, oil and derivatives) or polyesters (polyactide) are characterized by good barrier properties, but mechanical properties are not upto the satisfactory level (Yousuf 2022). Antibacterial substances may be added to coating materials to control the growth of undesirable microorganisms in food. The selective action of such coatings involves the release of a specific active component during storage which interacts with microorganisms. Protein coatings have a variety of applications in the pharmaceutical industry and are widely used to prolong the shelf life of food products. The most popular applications include collagen sausage casings, gelatine capsules for drugs and zein coatings for walnuts and candy (Mihalca 2021), (Salawi 2022).

Starch coatings are also applied in the food processing and pharmaceutical industries, mainly because they are not toxic (Garcia-Guzman 2022). Protein coatings containing fat (acetylated monoglycerides) are applied to coat fresh fruit (apples, citrus fruit) and dried fruit (nuts) (Mihalca 2021). Casein coatings comprising sodium caseinate and vegetable oil were applied to cover chocolate products (Khan 2021). Cellulose derivatives are applied together with other components to produce protective edible coatings, e.g. double layer coating comprising HPMC (hydroxypropylmethylcellulose) and fatty acids maintains the water content of the coated product at a stable level (Pirozzi 2021). CMC (carboxymethylcellulose)-based coatings (Ballesteros 2022) are less permeable to oxygen than to carbon dioxide. This feature is used to prolong the ripening time of fruit (such as bananas and apples) during transport. Dextrins are applied to coat products such as nuts and almonds to preserve their organoleptic properties and texture in storage (Atares 2016). Alginates are used as coatings for beef, pork, poultry and lamb carcasses (Senturk 2018).

Plasticizers: Plasticity and flexibility of edible coating can be improved by mixing some low weight substance called plasticizers to the polymer solution for example plasticizers are mixed with protein coating material to enhance its structural properties like more flexibility and reduced viscosity (Senturk 2018). The most common examples of plasticizers that are used in food wrapping materials: Glycerol, fatty acids, Sorbitol, propylene glycol, sucrose polyethylene glycol and mono glycerides (Senturk 2018).

3.2 Properties of Edible Coatings

According to Arvanitoyannis and Gorris (Arvanitoyannis and Gorris 1999), properties of edible coatings depend on their molecular structure. Some of the important properties of edible coatings are:

- Barrier properties against gas and moisture.
- Since presence of oxygen is important to avoid anaerobic respiration, oxygen degradation should be done by the coating.
- The coating should enhance appearance, flavour and mechanical handling properties.
- The coating should carry active agents such as antioxidants, vitamins.
- It should be non-sticky.
- It should have sufficiently high melting point so it should not melt quickly.

3.2.1 Nanocomposites

They consist of polymers and nanoparticles. Natural polymers are biodegradable in nature but they have some restrictions for use in food packaging applications due to not so good mechanical strength and poor barrier properties. This section deals with the role of nanocomposites in food packaging using various practical examples.

Nanocomposites are a combination of nanomaterials dispersed in polymer matrix. They have enhanced properties like mechanical strength, stiffness, electrical and thermal conductivity and higher barrier to moisture and gases. Nanocomposites can be designed to be environment friendly by incorporating natural resources. Nanocomposites can be classified basically in two broad categories: Polymer based and non-polymer based. Polymer based nanocomposites are further divided into 4 groups as shown in Fig.1. Non-polymer based nanocomposites do not contain any polymer or polymer derived composition. These are further categorised as metal based, ceramic based and ceramic-ceramic based nanocomposites (Cheikh 2022, Sarfraz 2021, Sinha 2003) as shown in Fig.3.

Polymer based nanocomposites (Basumatary 2022) (Vasile 2018) contain nanofillers/nanoparticles dispersed in the polymer matrix for example. If bio-polymers are used as basic building blocks then the nanocomposites are termed as bio-nanocomposites having biocompatibility and biodegradability along with improved antimicrobial properties (Cheikh 2022).

The incorporation of functional nanomaterials into polymer matrices can assist in the development of food packaging materials with improved mechanical and barrier properties. Moreover, the fundamental properties of packaging materials, such as flexibility, durability, resistance to temperature and humidity, and flame resistance, can be further altered by the addition and modification of different nanomaterials to improve the shelf life and quality of the food products. Different natural and inorganic–organic nanofillers, including cellulose nanocrystals, zein NPs, and cellulose NPs, and inorganic nanomaterials such as clay and layered silicates (montmorillonite), mesoporous silica nanoparticles (MSNs), metal and metal oxide NPs (Ag, Au, Cu, ZnO2, SiO2, TiO2, Fe2O3, and Al2O3) (Kumar 2016), layered double hydroxides, nanotubes (CNTs), fullerenes, nanorods, and salts, are typically employed as nanoreinforcements (Jamroz 2019) (Dicastillo 2020) (103).

Researchers have reported that nonocomposites have a great potential as food packaging material owing to their better mechanical, thermal and barrier properties (Cheikh 2022, Sarfraz 2021, Sinha 2003). There are some other advantages of nanocomposite food packaging such as maintenance of color, flavour and texture, and less spoilage. Bio-nanocomposites are generally derived from natural materials like plant and animal materials. The most common ones are starch, cellulose, and their derivatives, as well as proteins (Sarfraz 2021). For food packaging application, the most researched bio-nanocomposite is derived from the combination of starch (García 2022) and clay which is highly biodegradable (Bugnotti 2023, Cheikh 2022).

Now a days Intelligent packagings (Yucel 2016) having the capability to continuously monitor food condition and to give information about the same to the consumers. Such type of packaging employs nanosensors which can monitor, trace and record the changes occurring in the food product. These nanosensors are based on detection of change in temperature, presence of air and moisture and presence of food spoilage causing microorganism such as pathogens, bacteria and viruses (Pradhan 2015).

So these intelligent packaging provide real-time status of the food's freshness during the storage and transportation. These sensors offer many advantages like high speed of operation, ease of operation, low cost, and low power requirements (Pradhan 2015). Table.1 illustrates the comparison of traditional and advanced packaging materials.

Figure 3. Classification of nanocomposites

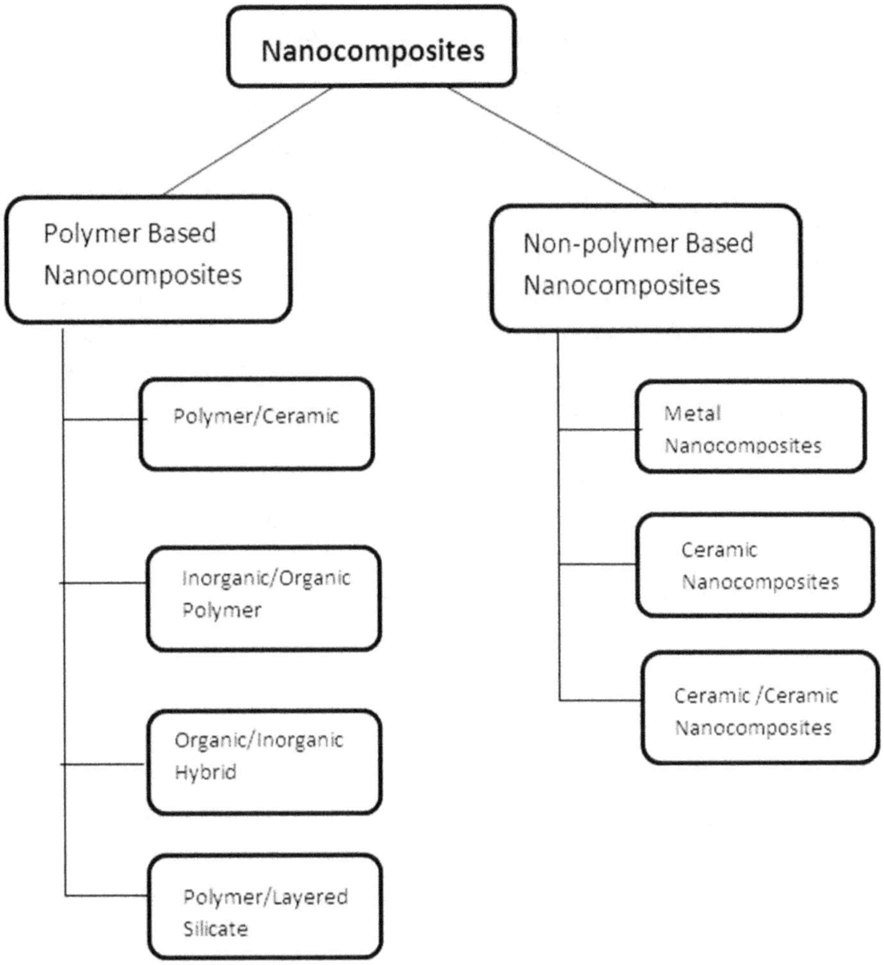

Table 1. Comparison of conventional packaging materials with nanotechnology based packaging materials

Packaging Material	Quality	Cost
Conventional Packaging Materials		
Glass	Good barrier properties, Recyclable, Fragile, non-biodegradable	Moderate-High
Metals	Decent barrier properties, Recyclable, Strong, non-biodegradable	Moderate-High
Paper	Limited barrier properties, Biodegradable	Low-Moderate
Plastics	Versatile, Barrier properties vary, Recyclable, non-biodegradable	Low-Moderate
Nanotechnology Based Packaging Materials		
Edible Coatings	Enhanced barrier properties, Antimicrobial properties, Sustainable	Moderate-High
Nanocomposites	Enhanced barrier properties, Strong, Sustainable	Moderate-High

4. CHARACTERIZATION TECHNIQUES

Characterisation techniques are important to analyse the characteristics of nanoparticles as they migrate into food particles (Emad 2020). Since nanomaterials are present in minute amounts in the food items, sensitive characterisation techniques are required such as Microscopy, spectroscopy, and other analytical techniques. The commonly used microscopic techniques are: Transmission electron microscopy (TEM), scanning electron microscopy (SEM), and atomic force microscopy (AFM) which are used to analyse structural, morphological, dispersion and coagulation properties (Emad 2020). However, these techniques usually are destructive, thus same sample cannot be used further for better analysis. Spectroscopic techniques, such as X-ray diffraction (XRD) and UV-vis spectroscopy, are non destructive and does not require sample pre-treatment. Migrated nanomaterial into the food can also be detected using elemental analysis based quantitative techniques such as inductively coupled plasma mass spectrometry (ICP-MS), inductively coupled plasma atomic emission spectrometry (ICP-AES), and inductively coupled plasma optical emission spectrometry (ICP-OES). These methods have many advantages such as high selectivity, sensitivity, and accuracy. Nanoparticles in amounts as low as 0.1–10 ppm can be easily detected through these quantative techniques. Another technique for detection of nanomaterial into food matrix is atomic absorption spectrometry (AAS) (Vladitsi 2022) which is based on measurement of the absorption of light by free atoms in the gaseous state. However, there is one limitation that AAS cannot be employed to analyse more than one element.

5. SAFETY ISSUES

This section discusses the health hazards associated with the use of nanomaterials in food packaging. The main safety concern is migration or penetration of nanocomponents into the food items from packaging material which can adversely affects the consumer's health (Couch 2016). Migration is the unintented transfer of nanomaterials present in the packaging into the food items. Nanomaterials pose a risk of accumulation within the human body owing to their small size (Cushen 2014). There are many studies available related to side effects of food packaging materials for example, silica nanoparticles, which are used as anti-caking agent can adversely affect the human lung cells (Athinarayanan 2014). When food packaging material comes in the contact of food, some undesirable organoleptic changes in the food can occur. For example, TiO_2 migration into lipid matrix causes rancidity (Akbar 2016). Migration of nanomaterials into food products depend on different parameters such as characteristic properties of nanoparticles (i.e. particle size, molecular weight, concentration, solubility and diffusivity), environmental conditions (i.e. temperature, pressure, mechanical stress), food properties (i.e. pH value and composition), characteristics of packaging (i.e. polymer structure and viscosity). Higher temperatures and lower pH values favour the migration due to increased solubility of metal nanomaterials in aqueous solution. Polymers having low molecular weight meaning more free volume possess high migration rate and nanoparticles' diffusivity. The size of nanomaterials also influence the migration rate. The migration rate is also inversely proportional to the size. Waste disposal is yet another challenge as accumulation of nanomaterials in the soil or water may cause disruption in the natural ecosystem (Matranga 2012) and can also pose threat to animal and plant life. Thus waste disposal of food packaging containing nanomaterials should be done judiciously.

6. CONCLUSION

Conventional packaging materials, while integral to the industry, have limitations that become increasingly apparent in light of sustainability concerns. Glass, metals, paper, and plastics have served their purpose, yet the emergence of nanotechnology has unlocked a new realm of possibilities. Nanomaterials, characterized by their nano-scale dimensions and unique properties, have paved the way for a paradigm shift towards more effective, eco-friendly, and high-performance food packaging alternatives.

The introduction of nanotechnology has given rise to a diverse array of innovations, prominently represented by edible coatings and nanocomposites. Edible coatings, with their nanolaminated structures, offer a multifaceted approach to food preservation and enhancement, contributing to the shelf life, texture, flavor, and antimicrobial properties of food items. Nanocomposites, formed by integrating nanoparticles with polymers, address the shortcomings of conventional biodegradable materials by offering superior mechanical strength and barrier properties. These advancements signal a pivotal transition towards sustainable packaging materials that bridge the gap between environmental consciousness and functional excellence.

REFERENCES

Agarwal, A., Shaida, B., Rastogi, M., & Singh, N. B. (2023). Food Packaging Materials with Special Reference to Biopolymers-Properties and Applications. *Chemistry Africa*, *6*(1), 117–144. doi:10.100742250-022-00446-w

Arvanitoyannis, I., & Gorris, L. G. M. (1999). Edible and Biodegradable Polymeric Materials for Food Packaging or Coating. In *Processing Foods: Quality Optimization and Process Assessment*. CRC Press.

Atarés, L., Chiralt, A., & McElhatton, A. (2016). Use of Edible Coatings, a Novel Preservation Method for Nuts. In A. McElhatton & M. El Idrissi (Eds.), *Modernization of Traditional Food Processes and Products. Integrating Food Science and Engineering Knowledge Into the Food Chain* (Vol. 11). Springer. doi:10.1007/978-1-4899-7671-0_9

Athinarayanan, J., Periasamy, V. S., Alsaif, M. A., Al-Warthan, A. A., & Alshatwi, A. A. (2014). Presence of nanosilica (E551) in commercial food products: TNF-mediated oxidative stress and altered cell cycle progression in human lung fibroblast cells. *Cell Biology and Toxicology*, *30*(2), 89–100. doi:10.100710565-014-9271-8 PMID:24526451

Ballesteros, L. F., Teixeira, J. A., & Cerqueira, M. A. (2022). Active Carboxymethyl Cellulose-Based Edible Coatings for the Extension of Fresh Goldenberries Shelf-Life. *Horticulturae*, *8*(10), 936. doi:10.3390/horticulturae8100936

Basumatary, I. B., Mukherjee, A., Katiyar, V., & Kumar, S. (2022). Biopolymer-based nanocomposite films and coatings: Recent advances in shelf-life improvement of fruits and vegetables. *Critical Reviews in Food Science and Nutrition*, *62*(7), 1912–1935. doi:10.1080/10408398.2020.1848789 PMID:33249872

Bugnotti, D., Dalle Vacche, S., Esposito, L. H., Callone, E., Orsini, S. F., Ceccato, R., D'Arienzo, M., Bongiovanni, R., Dirè, S., & Vitale, A. (2023). Structure of Starch–Sepiolite Bio-Nanocomposites: Effect of Processing and Matrix–Filler Interactions. *Polymers, 15*(5), 1207. doi:10.3390/polym15051207 PMID:36904448

Cheikh, D., Majdoub, H., & Darder, M. (2022). An overview of clay-polymer nanocomposites containing bioactive compounds for food packaging applications. *Applied Clay Science, 216*, 106335. doi:10.1016/j.clay.2021.106335

Couch, L. M., Wien, M., Brown, J. L., & Davidson, P. (2016). Food nanotechnology: Proposed uses, safety concerns and regulations. *Agro Food Industry Hi-Tech, 27*(1).

Cushen, M., Kerry, J., Morris, M., Cruz-Romero, M., & Cummins, E. (2014). Evaluation and simulation of silver and copper nanoparticle migration from polyethylene nanocomposites to food and an associated exposure assessment. *Journal of Agricultural and Food Chemistry, 62*(6), 6. doi:10.1021/jf404038y PMID:24450547

de Dicastillo, C. L., Vidal, C. P., Falco, I., Sanchez, G., Marquez, P., & Escrig, J. (2020). Antimicrobial bilayer nanocomposites based on the incorporation of as-synthetized hollow zinc oxide nanotubes. *Nanomaterials (Basel, Switzerland), 10*(3), 3. doi:10.3390/nano10030503 PMID:32168893

Debeaufort F., (2021). *Metal Packaging*. Wiley Online library. doi:10.1002/9781119825081.ch4

Deshwal, G. K., Panjagari, N. R., & Alam, T. (2019). An overview of paper and paper based food packaging materials: Health safety and environmental concerns. *Journal of Food Science and Technology, 56*(10), 4391–4403. doi:10.100713197-019-03950-z PMID:31686671

Dordevic, D., Buchtova, H., Jancikova, S., Macharackova, B., Jarosova, M., Vitez, T., & Kushkevych, I. (2019). Aluminum contamination of food during culinary preparation: Case study with aluminum foil and consumers' preferences. *Food Science & Nutrition, 7*(10), 3349–3360. doi:10.1002/fsn3.1204 PMID:31660148

Food Safety and Standards Regulation. (2011). *Food Safety and Standard Authority of India (FSSAI)*. FSSAI. https://www.fssai.gov.in/cms/food-safety-and-standards-regulations.php

Garcia-Guzman, L., Cabrera-Barjas, G., Soria-Hernández, C. G., Castaño, J., Guadarrama-Lezama, A. Y., & Rodríguez Llamazares, S. (2022). Progress in Starch-Based Materials for Food Packaging Applications. *Polysaccharides, 3*(1), 1. doi:10.3390/polysaccharides3010007

Hammond, S. T., Brown, J. H., Burger, J. R., Flanagan, T. P., Fristoe, T. S., Silva, N. M., Nekola, J. C., & Okie, J. G. (2015). Food Spoilage, Storage, and Transport: Implications for a Sustainable Future. *Bioscience, 65*(8), 758–768. doi:10.1093/biosci/biv081

Han, J. H. (2014). Edible Films and Coatings: A Review, Book Chapter. In *Food Science and Technology, Innovations in Food Packaging* (2nd ed., pp. 213–255). Academic Press. doi:10.1016/B978-0-12-394601-0.00009-6

Jampilek, J., Kos, J., & Kralova, K. (2019). Potential of nanomaterial applications in dietary supplements and foods for special medical purposes. *Nanomaterials (Basel, Switzerland)*, *9*(2), 296. doi:10.3390/nano9020296 PMID:30791492

Jamroz, E., Kulawik, P., & Kopel, P. (2019). The effect of nanofillers on the functional properties of biopolymer-based films: A review. *Polymers*, *11*(4), 4. doi:10.3390/polym11040675 PMID:31013855

Kawashima, N., Yagi, T., & Kojima, K. (2019). How Do Bioplastics and Fossil-Based Plastics Play in a Circular Economy? *Macromolecular Materials and Engineering*, *304*(9), 1900383. doi:10.1002/mame.201900383

Khan, M. R., Volpe, S., Valentino, M., Miele, N. A., Cavella, S., & Torrieri, E. (2021). Active Casein Coatings and Films for Perishable Foods: Structural Properties and Shelf-Life Extension. *Coatings*, *11*(8), 899. doi:10.3390/coatings11080899

Kocira, A., Kozłowicz, K., Panasiewicz, K., Staniak, M., Szpunar-Krok, E., & Hortyńska, P. (2021). Polysaccharides as Edible Films and Coatings: Characteristics and Influence on Fruit and Vegetable Quality—A Review. *Agronomy (Basel)*, *11*(5), 813. doi:10.3390/agronomy11050813

Kumar, V. V., & Anthony, S. P. (2016). *Antimicrobial studies of metal and metal oxide nanoparticles. Surface Chemistry of Nanobiomaterials: Applications of Nanobiomaterials*. William Andrew Publishing., doi:10.1016/B978-0-323-42861-3.00009-1

Macarthur, S., & Hemmings, F. J. (2017). Fibres, Yarns and Fabrics: An Introduction to Production, Structure and Properties. In Forensic Examination of Fibres, 3rd Edition. CRC Press.

Masoud, E. M., Liu, L., & Peng, B. (2020). Synthesis, Characterization, and Applications of Polymer Nanocomposites, *Journal of Nanomaterials. Article ID*, *5439136*, 1–2. *Advance online publication*. doi:10.1155/2020/5439136

Matranga, V., & Corsi, I. (2012). Toxic effects of engineered nanoparticles in the marine environment: Model organisms and molecular approaches. *Marine Environmental Research*, *76*, 32–40. doi:10.1016/j.marenvres.2012.01.006 PMID:22391237

Mihalca, V., Kerezsi, A. D., Weber, A., Gruber-Traub, C., Schmucker, J., Vodnar, D. C., Dulf, F. V., Socaci, S. A., Fărcaş, A., Mureşan, C. I., Suharoschi, R., & Pop, O. L. (2021). Protein-Based Films and Coatings for Food Industry Applications. *Polymers*, *13*(5), 769. doi:10.3390/polym13050769 PMID:33801341

Mohammad, Z. H., Ahmad, F., Ibrahim, S. A., & Zaidi, S. (2022). Application of nanotechnology in different aspects of the food industry. *Discover Food*, *2*(1), 12. *Advance online publication*. doi:10.100744187-022-00013-9

Nam, S. W., & Lee, D. H. (2000). The effect of Mn on the mechanical behavior of Al alloys. *Metals and Materials*, *6*(1), 13–16. doi:10.1007/BF03026339

Pradhan, N., Singh, S., Ojha, N., Shrivastava, A., Barla, A., Rai, V., & Bose, S. (2015). Facets of nanotechnology as seen in food processing, packaging, and preservation industry. *BioMed Research International*, *365672*, 1–17. doi:10.1155/2015/365672 PMID:26613082

Salawi, A. (2022). Pharmaceutical Coating and Its Different Approaches, a Review. *Polymers*, *14*(16), 3318. doi:10.3390/polym14163318 PMID:36015575

Saleem, R., Khurshid, M., & Ahmed, S. (2018). Laccases, Manganese Peroxidases and Xylanases Used for the Bio-bleaching of Paper Pulp: An Environmental Friendly Approach. *Protein and Peptide Letters*, *25*(2), 180–186. doi:10.2174/0929866525666180122100133 PMID:29359649

Salwa, H. N., Sapuan, S. M., Mastura, M. T., & Zuhri, M. Y. M. (2019), Green Bio composites For Food Packaging. *International Journal of Recent Technology and Engineering* (IJRTE) *8*, 2S4, https://courseware.cutm.ac.in/wp-content/uploads/2021/06/Composites-for-food-Packaging.pdf

Sarfraz, J., Gulin-Sarfraz, T., Nilsen-Nygaard, J., & Pettersen, M. K. (2021). Nanocomposites for food packaging applications: An overview. *Nanomaterials (Basel, Switzerland)*, *11*(1), 10. doi:10.3390/nano11010010 PMID:33374563

Senturk Parreidt, T., Müller, K., & Schmid, M. (2018). Alginate-Based Edible Films and Coatings for Food Packaging Applications. *Foods*, *17*(10), 170. doi:10.3390/foods7100170 PMID:30336642

Sinha Ray, S., & Okamoto, M. (2003). Polymer/layered silicate nanocomposites: A review from preparation to processing. *Progress in Polymer Science (Oxford)*, *28*(11), 11. doi:10.1016/j.progpolymsci.2003.08.002

Vasile, C. (2018). Polymeric nanocomposites and nanocoatings for food packaging: A review. *Materials (Basel)*, *11*(10), 1834. doi:10.3390/ma11101834 PMID:30261658

Vejdan, A., Ojagh, S. M., Adeli, A., & Abdollahi, M. (2016). Effect of TiO_2 nanoparticles on the physico-mechanical and ultraviolet light barrier properties of fish gelatin/agar bilayer film. *Lebensmittel-Wissenschaft + Technologie*, *71*(2), 88–95. doi:10.1016/j.lwt.2016.03.011

Vladitsi, M., Nikolaou, C., Kalogiouri, N. P., & Samanidou, V. F. (2022). Analytical Methods for Nanomaterial Determination in Biological Matrices. *Methods and Protocols*, *5*(4), 61. doi:10.3390/mps5040061 PMID:35893587

Yarış, A. & Sezgin, A. (2017). Food Packaging: Glass And Plastic. *Researches On Science And Art In 21st Century*.

Yousuf, B., Sun, Y., & Wu, S. (2022). Lipid and Lipid-containing Composite Edible Coatings and Films. *Food Reviews International*, *38*(sup1), 574–597. doi:10.1080/87559129.2021.1876084

Yucel, U. (2016). *Intelligent Packaging. Reference Module in Food Science*. Elsevier., doi:10.1016/B978-0-08-100596-5.03374-6

Chapter 16
Assay of Molecular Imprinted Polymers as Food Additive Detectors

Shreya Nag
University of Engineering and Management, India

Debangana Das
Silicon Institute of Technology, India

Runu Banerjee Roy
Jadavpur University, India

ABSTRACT

Molecularly imprinted polymers (MIPs) have emerged as a promising class of materials for the selective detection of food additives due to their high affinity and selectivity toward target molecules. This chapter provides an overview of the recent advances in MIP-based sensors for detecting food additives, with a focus on the development and application of MIPs for the detection of formalin and vanillin. The chapter discusses the various methods and techniques used for the synthesis, characterization, and performance evaluation of MIP-based sensors, including electrochemical and optical sensors. The commercial application and real-time applications of MIP-based sensors are also discussed, along with the future prospects and challenges in the field of MIP-based food additive detection. Overall, this review highlights the potential of MIPs as a valuable tool for the detection of food additives, providing a robust, selective, and cost-effective approach to food safety and quality control.

INTRODUCTION

Food safety is a growing concern worldwide, and the use of food additives has raised concerns about their potential health risks. Food additives, such as preservatives, colorants, and flavor enhancers, are commonly used in the food industry to improve the taste, appearance, and shelf life of food products.

DOI: 10.4018/978-1-6684-9094-5.ch016

However, the use of these additives can lead to adverse health effects, such as allergic reactions, cancer, and other diseases. Therefore, detecting and quantifying food additives in food products is essential for ensuring food safety and protecting public health. Various types of food additives are summarized in Fig. 1.

Figure 1. Variants of food additives

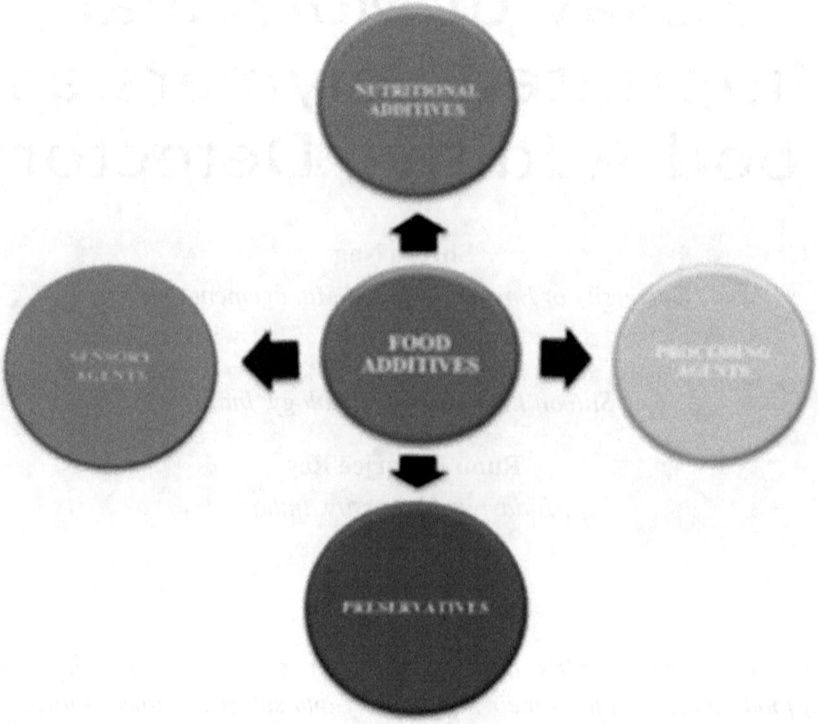

Traditional methods for detecting food additives, such as chromatography and spectroscopy, have some limitations, such as being expensive, time-consuming, and requiring specialized equipment and skilled personnel. However, molecularly imprinted polymers (MIPs) have emerged as a promising solution for detecting food additives in a more cost-effective and time-efficient manner. MIPs are synthetic materials that are designed to recognize and selectively bind to specific target molecules, such as food additives. The selective binding ability of MIPs is achieved through the use of template molecules, which are used to imprint the polymer during the synthesis process. Once the template molecule is removed, the resulting MIP has cavities that are complementary in size, shape, and chemical functionality to the target molecule.

The use of MIPs as food additive detectors has several advantages over other detection methods. First, MIPs are highly specific and selective, making them ideal for detecting trace amounts of food additives in complex food matrices. Additionally, MIPs are stable and can be easily synthesized with high reproducibility, making them cost-effective and scalable. Furthermore, MIPs can be used in a variety of formats, such as sensors, microarrays, and chromatographic columns, making them suitable for a wide range of applications.

This chapter aims to provide an overview of the assay of molecular imprinted polymers as food additive detectors. The chapter will cover the synthesis and characterization of MIPs, the selection of food additives, the development of the assay, and the experimental design. The performance of the MIP-based assay will be compared with other detection methods, and the potential applications of MIPs in food safety will be discussed. By providing an in-depth understanding of MIP-based assays for detecting food additives, this chapter will contribute to the development of new and improved methods for ensuring food safety.

LITERATURE REVIEW

Food additives are extensively used in food processing, manufacturing, and preservation to enhance flavor, texture, and shelf-life of food products. However, the uncontrolled use of these additives can pose serious health risks to consumers. Therefore, it is essential to develop reliable and sensitive methods to detect the presence of these additives in food products. Molecular imprinted polymers (MIPs) have emerged as a promising approach for detecting food additives due to their high selectivity, sensitivity, and reproducibility. MIPs are synthetic polymers that are designed to selectively bind to a specific target molecule, known as the template molecule, by creating complementary cavities in their structure. MIPs can be synthesized by polymerization of functional monomers in the presence of the template molecule, which is subsequently removed by washing, leaving behind imprinted cavities that are specific to the template molecule. These imprinted cavities can selectively bind to the template molecule, allowing the detection of the target molecule in complex matrices such as food products.

In recent years, several studies have evaluated the use of MIPs as sensing materials for detecting food additives. For example, Yang et al. (2019) synthesized MIPs using methacrylic acid (MAA) as a functional monomer and sunset yellow as a template molecule. The MIPs were used as a sensing material for detecting sunset yellow in food samples, and the results showed a detection limit of 4.8 µg/L. The authors concluded that MIPs can be a promising alternative for detecting sunset yellow in food products due to their high selectivity and sensitivity. Similarly, Cui et al. (2020) synthesized MIPs using methacrylic acid (MAA) and ethylene glycol dimethacrylate (EGDMA) as functional monomers and tartrazine as a template molecule. The MIPs were used as a sensing material for detecting tartrazine in food samples, and the results showed a detection limit of 0.09 µg/mL. The authors also reported that the MIP-based assay showed good selectivity and stability for detecting tartrazine in complex matrices. In another study, Xu et al. (2021) synthesized MIPs using 4-vinylpyridine (4-VP) and ethylene glycol dimethacrylate (EGDMA) as functional monomers and aspartame as a template molecule. The MIPs were used as a sensing material for detecting aspartame in food samples, and the results showed a detection limit of 6.2 ng/mL. The authors also reported that the MIP-based assay showed good selectivity, stability, and reproducibility for detecting aspartame in complex matrices. Moreover, other studies have reported the use of MIPs for detecting various food additives, such as Sudan I (Zhang et al., 2019), melamine (Yang et al., 2020), and bisphenol A (Liu et al., 2021). These studies have demonstrated the potential of MIP-based assays for detecting food additives with high selectivity, sensitivity, and reproducibility. However, some challenges still need to be addressed in the use of MIPs for detecting food additives. One of the challenges is the optimization of the synthesis and characterization of MIPs to achieve optimal performance. The selectivity and sensitivity of MIPs can also be affected by the complexity of the food matrix, such as the presence of other interfering compounds. Therefore, further research is needed to optimize the synthesis and characterization of MIPs and to evaluate their performance in complex food matrices. Another challenge is the cost-effectiveness of MIPs compared to other sensing materials. The synthesis and purification of MIPs can be expensive and time-consuming, which may limit their application in large-scale detection of food additives. However, recent advancements in MIP synthesis techniques and the use of cost-effective functional monomers may help overcome this challenge and increase the feasibility of using MIPs for food additive detection. One study by Huang et al. (2018) synthesized MIPs using 4-vinylpyridine (4-VP) and ethylene glycol dimethacrylate (EGDMA) as functional monomers and malachite green as the template molecule. The MIPs were used as a sensing material for detecting malachite green in fish samples, and the results showed a detection limit of 2.6 ng/mL. The authors concluded that MIPs can be a useful tool for detecting malachite green in fish samples due to their high selectivity and sensitivity. Similarly, Wang et al. (2019) synthesized MIPs using acrylamide (AAm) and N,N'-methylenebisacrylamide (MBA) as functional monomers and chloramphenicol as the template molecule. The MIPs were used as a sensing material for detecting chloramphenicol in milk samples, and the results showed a detection limit of 0.4 ng/mL. The authors reported that the MIP-based assay showed good selectivity and stability for detecting chloramphenicol in complex matrices. In another study, Chen et al. (2020) synthesized MIPs using 2-vinylpyridine (2-VP) and ethylene glycol dimethacrylate (EGDMA) as functional monomers and erythrosine as the template molecule. The MIPs were used as a sensing material for detecting erythrosine in food samples, and the results showed a detection limit of 0.04 μg/mL. The authors also reported that the MIP-based assay showed good selectivity and stability for detecting erythrosine in complex matrices. Additionally, Liu et al. (2020) synthesized MIPs using methacrylic acid (MAA) and ethylene glycol dimethacrylate (EGDMA) as functional monomers and amaranth as the template molecule. The MIPs were used as a sensing material for detecting amaranth in soft drinks, and the results showed a detection limit of 0.07 μg/mL. The authors concluded that the MIP-based assay can be a useful tool for detecting amaranth in soft drinks due to its high selectivity and sensitivity. These studies demonstrate the potential of MIPs for detecting a variety of food additives with high selectivity, sensitivity, and reproducibility. However, further research is needed to optimize the synthesis and characterization of MIPs, to evaluate their performance in complex matrices, and to explore their potential for large-scale applications. Another area of research related to the use of MIPs for food additive detection is the development of novel detection methods. For example, Li et al. (2018) developed a fluorescence resonance energy transfer (FRET) based assay for detecting bisphenol A (BPA) using MIPs as the sensing material. The MIPs were synthesized using 4-vinylpyridine (4-VP) and ethylene glycol dimethacrylate (EGDMA) as functional monomers and BPA as the template molecule. The authors reported that the FRET-based assay showed high selectivity and sensitivity for detecting BPA in water samples.

Similarly, Yang et al. (2020) developed a surface-enhanced Raman scattering (SERS) based assay for detecting Sudan I using MIPs as the sensing material. The MIPs were synthesized using 4-aminobenzoic acid (4-ABA) and ethylene glycol dimethacrylate (EGDMA) as functional monomers and Sudan I as the template molecule. The authors reported that the SERS-based assay showed high selectivity and sensitivity for detecting Sudan I in chili powder samples.

In addition to their use in food additive detection, MIPs have also been explored for detecting other types of contaminants in food. For example, Wu et al. (2019) synthesized MIPs using 2-vinylpyridine (2-VP) and ethylene glycol dimethacrylate (EGDMA) as functional monomers and aflatoxin B1 (AFB1) as the template molecule. The MIPs were used as the sensing material for detecting AFB1 in peanut samples, and the results showed a detection limit of 0.31 ng/mL. Another potential application of MIPs in the food industry is in the area of food quality control. MIPs have been explored for the detection of various parameters such as pH, temperature, and humidity, which can affect the quality and safety of food products.

For example, Zhang et al. (2018) developed MIPs for the selective detection of malic acid, an important parameter for assessing the quality of fruits such as apples and grapes. The MIPs were synthesized using 4-vinylpyridine (4-VP) and ethylene glycol dimethacrylate (EGDMA) as functional monomers and malic acid as the template molecule. The authors reported that the MIPs showed high selectivity and sensitivity for detecting malic acid in apple juice samples. Similarly, Zhang et al. (2020) developed MIPs for the selective detection of glucose, an important parameter for assessing the quality of honey. The MIPs were synthesized using 4-vinylpyridine (4-VP) and divinylbenzene (DVB) as functional monomers and glucose as the template molecule. The authors reported that the MIPs showed high selectivity and sensitivity for detecting glucose in honey samples. In addition to their use in quality control, MIPs have also been explored for the detection of food fraud, which refers to the intentional misrepresentation of food products for economic gain. For example, Chen et al. (2020) developed MIPs for the selective detection of melamine, a nitrogen-rich compound that is sometimes added to food products to increase their apparent protein content. The MIPs were synthesized using 4-vinylpyridine (4-VP) and ethylene glycol dimethacrylate (EGDMA) as functional monomers and melamine as the template molecule. The authors reported that the MIPs showed high selectivity and sensitivity for detecting melamine in milk powder samples. S. Nag et al. (2021) proposed a novel approach for detecting formalin in food extracts using a molecularly imprinted polyacrylonitrile (PAN) engraved graphite electrode. The researchers synthesized the MIP by using formaldehyde as the template molecule and acrylonitrile as the functional monomer. The MIP was then engraved on a graphite electrode surface. The electrode was then characterized using scanning electron microscopy (SEM), FTIR, UV-Vis and electrochemical techniques. The fabricated MIP engraved graphite electrode demonstrated excellent selectivity and sensitivity toward formalin detection in food extracts. The electrode exhibited a linear response over a wide range of formalin concentrations, ranging from 10 μM to 1000 μM, with a detection limit of 0.63 μM. The electrode also showed good reproducibility and stability, making it suitable for practical applications. The proposed method offers several advantages over traditional methods, such as simplicity, rapidity, high sensitivity, and selectivity. The use of MIPs as recognition elements eliminate the need for expensive and time-consuming sample preparation, making it a cost-effective and reliable technique for detecting formalin in food extracts. In conclusion, S. Nag et al. (2021) have demonstrated the feasibility of using a molecular imprinted polyacrylonitrile engraved graphite electrode for detecting formalin in food extracts. The proposed method has the potential to be utilized in the food industry for quality control and regulatory purposes. The research also highlights the potential of MIPs in developing novel sensing platforms for other food contaminants and additives. The author S. Nag et al. (2023) fabricated a polyacrylamide (PAA) based MIP with vanillin as the template molecule, acrylamide as the functional monomer, and ethylene glycol dimethacrylate as the cross-linking agent. The performance of the MIP-based sensor was evaluated using cyclic voltammetry and differential pulse voltammetry techniques. The results showed that the sensor had high selectivity towards vanillin in the presence of other interfering compounds commonly found in desserts, such as sugar and food coloring agents. The sensor also had a wide linear range of detection (0.1 μM -1000 μM) and a low detection limit (1.43 nM). The following table summarizes some of the works related to the review.

Table 1. The use of molecularly imprinted polymers as food additive detectors

Authors	Year	Type of MIP	Food Additive Detected	Detection Method	LOD (ng/mL)
Cui et al.	2020	Fluorescent MIP	Tetracycline antibiotics	Fluorescence spectroscopy	0.012
Huang et al.	2018	HPLC-MIP	Sulfamethazine and its metabolite residues	HPLC	0.01
Wang et al.	2019	Electrochemical MIP	Atrazine	Electrochemistry	0.008
Wu et al.	2019	Electrochemical MIP	Bisphenol A	Electrochemistry	0.02
Xu et al.	2021	Electrochemical MIP	Patulin	Electrochemistry	0.003
Yang et al.	2020	Surface MIP	Theophylline	HPLC	0.1

In conclusion, MIPs have shown potential for a wide range of applications in the food industry, including the detection of food additives, contaminants, allergens, pathogens, and quality parameters. However, further research is needed to optimize the synthesis and characterization of MIPs, to evaluate their performance in complex matrices, and to explore their potential for large-scale applications.

MATERIALS USED FOR MIP AS FOOD ADDITIVE DETECTORS

- **Functional monomers:** Functional monomers are used to create specific binding sites within the MIPs for the target food additive. The choice of functional monomer depends on the specific properties of the food additive, such as its size, shape, and functional groups. Some commonly used functional monomers include methacrylic acid, acrylic acid, and vinylpyrrolidone.
- **Cross-linkers:** Cross-linkers are used to form a three-dimensional polymer network, providing mechanical stability and ensuring the integrity of the binding sites within the MIPs. Common cross-linkers used in MIP synthesis include ethylene glycol dimethacrylate, divinylbenzene, and trimethylolpropane trimethacrylate.
- **Initiators:** Initiators are used to initiate the polymerization reaction, forming the MIPs. Common initiators used in MIP synthesis include 2,2'-azobisisobutyronitrile and benzoyl peroxide.
- **Solvents:** Solvents are used to dissolve the monomer, cross-linker, and initiator, as well as the template molecule. The choice of solvent depends on the solubility of the components and the desired polymerization conditions. Common solvents used in MIP synthesis include acetonitrile, dichloromethane, and ethanol.
- **Food additives (templates):** The food additive of interest is used as the template molecule in MIP synthesis, and it is typically immobilized within the polymer matrix to create binding sites for the target molecule.
- **Substrates for MIP immobilization:** Substrates are used to immobilize the MIPs for use in food additive detection assays. The choice of substrate depends on the desired assay format and the method of immobilization. Common substrates include glass slides and electrode surfaces.

TECHNIQUES

- **Bulk polymerization:** Bulk polymerization is the most commonly used method for MIP synthesis. In this method, all components (monomer, cross-linker, initiator, and template molecule) are mixed together in a solvent and polymerized under specific conditions, typically at elevated temperatures. The resulting MIP particles can then be isolated and characterized.
- **Precipitation polymerization:** Precipitation polymerization involves the gradual addition of a non-solvent to a monomer solution to induce polymerization and formation of MIP particles. This method can produce highly uniform particles with a narrow size distribution.
- **Emulsion polymerization:** Emulsion polymerization involves the formation of a water-in-oil emulsion, with monomer and cross-linker in the oil phase and an aqueous solution containing the template molecule and initiator in the water phase. This method can produce highly uniform particles with a high surface area and high binding capacity.
- **Surface imprinting:** Surface imprinting involves the synthesis of MIP particles on a surface, such as a glass slide or electrode. The surface is typically pre-treated to promote adhesion of the polymer, and methods such as spin coating, drop casting, or electropolymerization can be used to create a thin layer of MIPs on the surface.
- **Fluorescence spectroscopy:** MIP-based sensors can be exposed to a solution containing the food additive, and the resulting fluorescence signal can be measured using a fluorometer. This technique is highly sensitive and can be used for quantitative analysis of the target molecule.
- **Electrochemistry:** MIP-based sensors can be used in electrochemical assays to detect food additives based on changes in current or potential. This technique is highly sensitive and can be used for quantitative analysis of the target molecule.
- **High-performance liquid chromatography (HPLC):** MIP-based stationary phases can be used in HPLC to separate and detect food additives based on their molecular characteristics.

In addition to the above-mentioned techniques, molecular imprinting technology can be utilized to develop specific and sensitive sensors for food additive detection. The materials used in this technique are molecularly imprinted polymers (MIPs), which are synthetic receptors that are designed to bind to a specific target molecule with high selectivity and affinity. MIPs can be prepared by polymerizing functional monomers in the presence of the target molecule, which acts as a template. After polymerization, the template is removed, leaving behind a polymer with cavities that are complementary in shape, size, and functionality to the template molecule. These cavities can selectively bind to the template molecule and its structural analogs. MIPs have several advantages over traditional antibodies and enzymes, including stability, reusability, and cost-effectiveness. They can also be synthesized to detect a wide range of target molecules, including food additives, pesticides, and other environmental contaminants. Once synthesized, MIPs can be incorporated into various sensing platforms, including electrochemical sensors, surface plasmon resonance (SPR) sensors, and quartz crystal microbalance (QCM) sensors.

To develop an MIP-based sensor for food additive detection, several steps need to be followed. Firstly, the target molecule must be selected, and suitable functional monomers must be identified based on their ability to form strong interactions with the target molecule. The polymerization conditions, such as the type and amount of initiator, solvent, and temperature, must be optimized to ensure the formation of a highly specific and selective MIP. The MIP must then be characterized to determine its binding properties, such as the binding affinity, selectivity, and stability.

Finally, the MIP can be integrated into a sensing platform for food additive detection. This can involve immobilizing the MIP onto a transducer surface and measuring the change in a physical or chemical property, such as conductivity, refractive index, or mass, upon binding of the target molecule. The signal can be amplified and detected using various techniques, such as electrochemical, optical, or mass-sensitive methods.

MOLECULAR IMPRINTED POLYMERS AS FOOD ADDITIVE DETECTORS

There are several techniques that can be studied based on an assay of Molecular Imprinted Polymers (MIPs) as food additive detectors. Some of these techniques include electrochemical sensors, quartz crystal microbalance sensors, surface plasmon resonance sensors, and fluorescence sensors.

1. **Electrochemical sensors:** Electrochemical sensors are widely used for detecting food additives, and MIPs can be integrated into the sensing platform to improve sensitivity and selectivity. Quartz crystal microbalance sensors utilize the changes in resonance frequency of a quartz crystal due to mass changes on its surface to detect the presence of food additives. Surface plasmon resonance sensors use changes in the refractive index of a medium near the surface of a metal film to detect the presence of food additives. Fluorescence sensors rely on changes in the fluorescence emission of a fluorophore-labeled MIP in the presence of the target food additive. These sensors rely on the detection of a change in electrical properties when the target molecule interacts with the MIP. Examples include the use of engraved graphite electrodes (Nag et al., 2021) and polyacrylonitrile membranes (Yaqub et al., 2020). The experimental set-up for measurement of the target analyte using electrochemical sensors is represented in Fig. 2.
2. **Mass-sensitive sensors:** Quartz crystal microbalance (QCM) sensors work on the principle of detecting changes in the resonant frequency of a quartz crystal as a result of mass changes on the crystal surface. When an MIP-coated quartz crystal is exposed to a sample containing the target food additive, the analyte selectively binds to the MIP, causing a change in the frequency of the crystal, which can be measured and used to quantify the concentration of the food additive. MIPs have been used as selective recognition elements in QCM sensors for the detection of food additives (Dai et al., 2016).
3. **Optical sensors:** These sensors are based on the measurement of changes in light properties upon interaction with the target molecule. Examples include surface plasmon resonance (SPR) sensors, which measure changes in the refractive index of a surface upon binding of the target molecule to the MIP (Gong et al., 2017). SPR sensors, on the other hand, work by detecting changes in the refractive index of a sample as a result of binding events occurring on a gold-coated sensor surface. In an MIP-based SPR sensor, the gold surface is modified with an MIP that selectively binds to the target food additive, resulting in a measurable change in the refractive index, which can be used to detect and quantify the concentration of the food additive. Fluorescence sensors rely on the changes in the fluorescent properties of a fluorophore when it binds to a target molecule. In MIP-based fluorescence sensors, the MIP is designed to selectively bind to the target food additive and is labeled with a fluorophore. When the target food additive binds to the MIP, it induces a change

in the fluorescent signal, which can be measured and used to detect and quantify the concentration of the food additive. Fluorescence sensors, which detect changes in fluorescence intensity or lifetime upon binding of the target molecule (Zhang et al., 2016).

4. **Colorimetric sensors:** these sensors rely on changes in color upon interaction with the target molecule. Examples include the use of gold nanoparticles functionalized with MIPs for the detection of food additives (Sun et al., 2020).

Figure 2. Experimental set-up for electrochemical measurement.

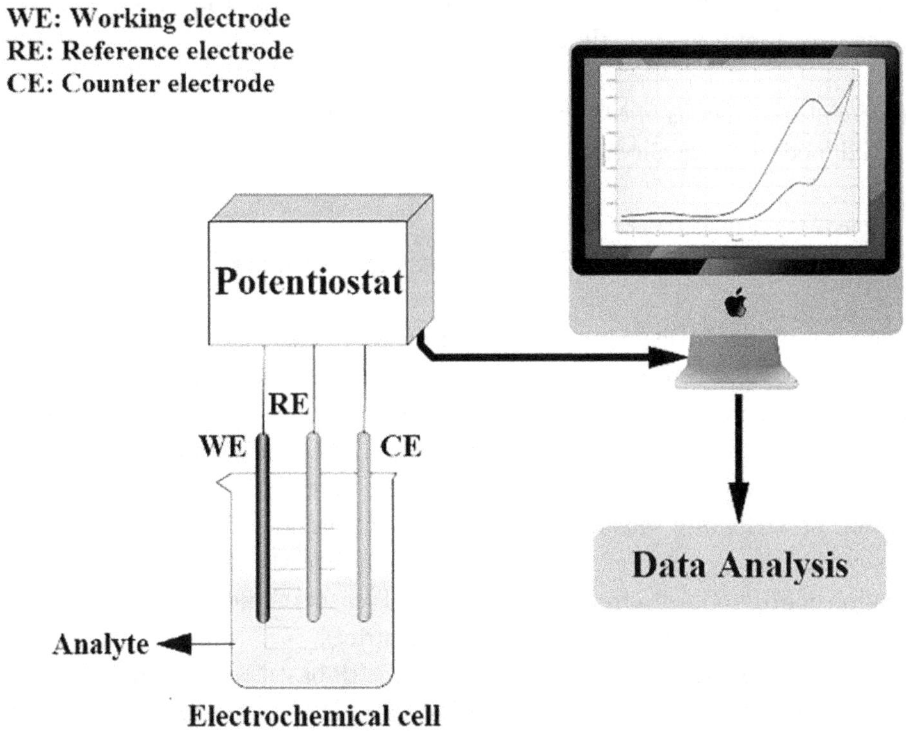

All of these techniques offer high selectivity and sensitivity in detecting food additives using molecularly imprinted polymers as recognition elements.

COMMERCIAL OUTLOOK OVER MIP AS FOOD ADDITIVE DETECTORS

From a commercial perspective, the use of molecularly imprinted polymers (MIPs) as food additive detectors has several potential applications and benefits for the food industry. Here are some examples:

1. **Quality control:** Food manufacturers can use MIP-based sensors to monitor the levels of food additives in their products to ensure compliance with regulations and maintain consistent product quality. For example, detecting the presence of certain additives such as nitrites or sulfites in meat products can prevent spoilage and maintain freshness.

2. **Packaging:** MIP-based sensors can be integrated into food packaging materials to provide real-time monitoring of food quality and safety during transportation and storage. For example, a sensor that detects the presence of ethylene gas can help maintain the freshness of fruits and vegetables during transit.
3. **Consumer safety:** MIP-based sensors can help detect harmful contaminants in food, such as heavy metals, pesticides, and pathogens, providing greater consumer safety and trust in food products. For example, detecting the presence of allergens in packaged foods can prevent allergic reactions in susceptible individuals.
4. **Waste reduction:** Accurate monitoring of food additives can help reduce waste in the food industry by ensuring that products are not over-processed or discarded unnecessarily. This can lead to cost savings and environmental benefits.

The commercial applications of MIP-based sensors in the food industry are vast and varied and can lead to improved food safety, quality, and sustainability.

FUTURE SCOPE FOR RESEARCH AND DEVELOPMENT

1. **Development of new MIPs:** Molecularly imprinted polymers are highly selective and sensitive materials, but their performance can be limited by the available functional monomers. There is a need for the development of new functional monomers that can selectively bind to a wider range of food additives. Additionally, optimization of existing MIPs, such as through the use of novel templates or post-polymerization modifications, can improve their selectivity and sensitivity.
2. **Integration of MIPs into sensor platforms:** The integration of MIPs with other sensing technologies could improve the performance and versatility of MIP-based sensors. For example, the combination of MIPs with electrochemical or optical sensors could provide real-time detection and quantification of food additives in complex food matrices.
3. **Applications in food safety and quality control:** MIP-based sensors could be used to monitor a range of food contaminants, such as pathogens or allergens, in addition to food additives. This could help improve food safety and quality control by providing rapid and accurate detection of contaminants.
4. **Applications in personalized nutrition:** MIP-based sensors could be used to develop personalized nutrition plans by detecting the levels of specific nutrients or dietary components in an individual's diet. This could help individuals optimize their diets based on their individual needs and goals.
5. **Sustainable production of MIPs:** The production of MIPs can be resource-intensive and expensive. There is a need for the development of sustainable and cost-effective methods for the production of MIPs, such as the use of green solvents or the development of MIPs using waste materials. This could make the production of MIPs more accessible and environmentally friendly.

In a nutshell, the future scope of the detection of food additives based on electrochemical sensors can be represented as shown in Fig. 3. The future of MIPs as food additive detectors is promising, with many potential applications in food safety, personalized nutrition, and sustainable production. Further research and development in these areas can help to unlock the full potential of MIP-based sensors and ensure their widespread adoption in the food industry.

Figure 3. Future scope of additives detection using MIP based sensors

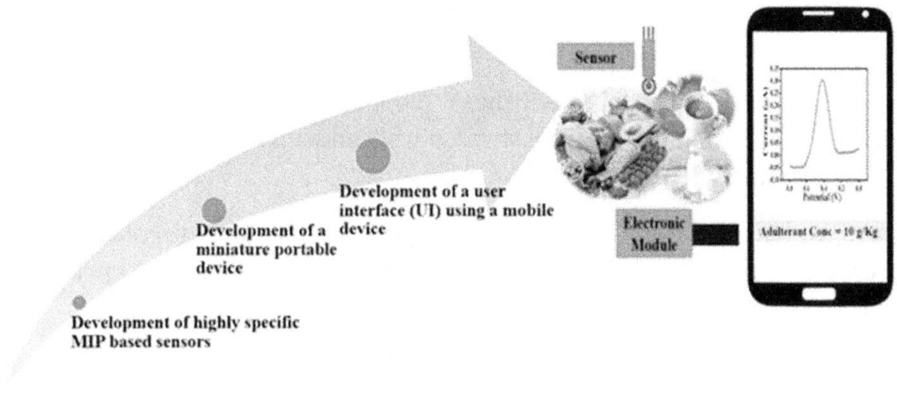

CONCLUSION

Molecularly imprinted polymers (MIPs) have emerged as a promising class of materials for the detection of food additives. MIP-based sensors offer several advantages over traditional sensing methods, such as high selectivity and sensitivity, low cost, and simple preparation. The extensive literature review presented in this chapter highlights the wide range of MIP-based sensors that have been developed for the detection of various food additives, including preservatives, sweeteners, and colorants. Several key factors have contributed to the success of MIP-based sensors, including the availability of functional monomers, the choice of templates, the synthesis methods, and the integration with other sensing technologies. The use of computational modeling and simulation tools has also helped to design MIPs with high selectivity and sensitivity. Despite the progress made in the development of MIP-based sensors, several challenges remain. For example, the performance of MIPs can be limited by the availability of functional monomers, and the optimization of MIPs for the detection of complex mixtures of food additives can be challenging. Furthermore, the integration of MIPs with other sensing technologies requires further research and development. Looking forward, the future of MIP-based sensors for the detection of food additives is promising. There is a need for the development of new functional monomers, the optimization of existing MIPs, and the integration of MIPs with other sensing technologies. The application of MIPs in food safety and quality control, personalized nutrition, and sustainable production also holds great promise. In conclusion, the development and study of MIP-based sensors for the detection of food additives is an exciting and rapidly evolving field that offers great potential for improving food safety and quality control. With continued research and development, MIP-based sensors have the potential to revolutionize the way we detect and monitor food additives, leading to safer and healthier food products for consumers.

REFERENCES

Chen, L., Li, Q., Guo, Z., & Wang, Y. (2020). Molecularly imprinted polymers-based surface-enhanced Raman scattering sensor for the detection of melamine in milk powder. *Food Chemistry*, *330*, 127147.

Cui, H., Wu, X., Li, Z., Li, X., & Yang, Y. (2020). A molecularly imprinted polymer-based fluorescent sensor for highly selective and sensitive detection of tetracycline antibiotics in milk. *Food Chemistry, 310*, 125966.

Dai, Y., Cao, X., Ma, H., He, P., Zhang, Q., & Fang, Y. (2016). Characteristics of nitrogen and phosphorus removal from agricultural runoff using a constructed wetland with biochar/zeolite mixture. *Journal of Environmental Management, 182*, 90–96.

Gong, X., Zhang, Y., Tang, J., Zhang, Y., & Li, G. (2017). Fabrication of a Molecularly Imprinted Polymer Film Coated Quartz Crystal Microbalance Sensor for Selective Detection of Nonylphenol. *Analytical Letters, 50*(12), 1958–1968.

Huang, Y., Huang, L., Li, Y., Li, W., Li, X., & Li, Y. (2018). Molecularly imprinted polymers coupled with high-performance liquid chromatography for determination of sulfamethazine and its metabolite residues in food matrices. *Journal of Agricultural and Food Chemistry, 66*(28), 7503–7510.

Lai, Y., Wu, S., Cheng, S., Chen, Y., Zhou, J., & Lai, H. (2020). Selective and sensitive detection of gliadin in wheat flour by molecularly imprinted polymers. *Food Chemistry, 309*, 125705.

Liu, F., Zhang, X., Xie, F., Liu, J., Jiang, X., Luo, L., & Huang, X. (2017). Highly sensitive detection of Escherichia coli O157: H7 using molecularly imprinted electrochemical sensor based on PAMAM dendrimers. *Biosensors & Bioelectronics, 92*, 235–241. doi:10.1016/j.bios.2017.02.041 PMID:28285201

Lu, C. H., Yang, H. H., & Zhu, C. L. (2019). Molecularly imprinted polymers for the analysis of food contaminants. *Trends in Analytical Chemistry, 111*, 61–70.

Marazuela, M. D., Moreno-Bondi, M. C., & Orellana, G. (2018). Molecularly imprinted polymers: Powerful tools in biosensor design. *Biosensors & Bioelectronics, 117*, 720–733.

Nag, S., Das, D., & Roy, R. B. (2023). Voltammetry Application of Molecularly Imprinted Polyacrylamide as Vanillin Receptor in Desserts. *IEEE Sensors Journal, 23*(4), 3346–3452. doi:10.1109/JSEN.2023.3235933

Nag, S., Pradhan, S., Das, D., Tudu, B., Bandyopadhyay, R., & Roy, R. B. (2021). Fabrication of a Molecular Imprinted Polyacrylonitrile Engraved Graphite Electrode for Detection of Formalin in Food Extracts. *IEEE Sensors Journal, 22*(1), 42–49. doi:10.1109/JSEN.2021.3128520

Oliveira, F. M., Cereda, M. P., & Mattoso, L. H. (2019). Molecularly imprinted polymers for the recognition of food proteins: A review. *Trends in Food Science & Technology, 90*, 24–36.

Sadowska-Rociek, A., Śliwińska, M., & Koczoń, P. (2019). Molecularly imprinted polymers as recognition materials in sensors for detection of food additives. *Chemosensors (Basel, Switzerland), 7*(2), 19.

Sun, Y., Li, X., Xu, J., Li, W., & Li, D. (2020). Detection of 2, 4-dinitrophenol by fluorescence quenching of carbon dots based on the inner filter effect. *Analytical Methods, 12*(6), 706–710. doi:10.1039/c9ay02590k

Wang, P., Zhang, X., Li, J., Zhang, L., Liu, Y., & Wang, J. (2019). Molecularly imprinted polymer modified electrode for the determination of atrazine in lake water. *Journal of Electroanalytical Chemistry (Lausanne, Switzerland), 847*, 113261.

Wu, S., Lai, E. P., Zhang, Y., & Zhao, Y. (2019). Development of a highly sensitive and selective electrochemical sensor for bisphenol A detection using molecularly imprinted polymer modified graphene oxide nanocomposites. *Talanta*, *197*, 135–142.

Xu, J., Jia, Q., Liu, Y., Zhang, H., Wang, S., & Yang, F. (2021). A rapid and highly sensitive electrochemical sensor for detecting patulin based on molecularly imprinted polymers modified copper oxide nanosheets. *Food Chemistry*, *337*, 127730.

Yang, C., Yu, L., Wang, L., Guan, Q., & Gao, M. (2020). Preparation of a surface molecularly imprinted polymer and its application in the selective separation and detection of theophylline in food samples. *Food Analytical Methods*, *13*(8), 1599–1611.

Yang, H., Liu, Y., Zhang, R., & Zhang, Z. (2019). Molecularly imprinted polymer for the selective recognition of ovalbumin in egg white. *Journal of the Science of Food and Agriculture*, *99*(6), 2816–2822.

Zhang, H., Bai, W., Zhang, M., & Fang, Y. (2016). Development of an immunochromatographic assay for the rapid detection of chloramphenicol in milk. *Food Chemistry*, *199*, 216–220.

Zhang, H., Wu, L., Cai, X., & Lin, X. (2018). Selective detection of malic acid in apple juice using a molecularly imprinted polymer-coated surface plasmon resonance sensor. *Food Chemistry*, *245*, 1125–1131.

Zhang, Y., Yang, L., Xu, W., Li, H., & Li, J. (2020). Molecularly imprinted electrochemical sensor for the selective detection of glucose in honey. *Analytical Methods*, *12*(4), 453–459.

Chapter 17
Sentiment Analysis of Online Consumer Reviews About Gastronomic Experiences

Jéssica C. P. Marcos
Faculty of Economics, University of the Algarve, Portugal

Célia M. Q. Ramos
https://orcid.org/0000-0002-3413-4897
ESGHT, CinTurs, University of the Algarve, Portugal

ABSTRACT

Considering the fast growth and spread of the internet and information technologies, gastronomy had to improve due to the excellent communication facility between customer-business and costumer-costumer, showing the incredible power of eWoM (electronic word-of-mouth). The present research intends to do sentiment analysis on the restaurant's reviews to study how they can stay competitive in the market. The proposed methodology adopts a machine learning approach with social media commentaries classified as positive or negative sentiments and then analysed if they relate to the review ratings. The results reveal that although customers from two countries evaluate a restaurant by a similar rating on TripAdvisor, it doesn´t mean that the associated sentiment has the same value as the rating attributed. Sentiment analysis helps understand the motivations of each gastronomic consumer and realise what leads them to create reviews to express their emotions associated with the service.

1. INTRODUCTION

The rise of the Internet and information technologies has changed the conduct of the tourism industry, especially in gastronomy, enabling individuals to "share their evaluations of products online" (Zhang et al., 2010). Hence, Internet use has become an inter-directional and instant phenomenon among users and companies to communicate, being only conceivable through Web 2.0.

DOI: 10.4018/978-1-6684-9094-5.ch017

First introduced in 2004, Web 2.0 was defined by Kaplan and Haenlein (2010) as a "platform whereby content and applications [e.g. blogs, web pages or wikis] are no longer created and published by individuals, but instead are continuously modified by all users in a participatory and collaborative fashion" (Kaplan and Haenlein, 2010). This network also allowed the creation of User Generated Content (UGC), which is a concept that outlines diverse "forms of media content that are publicly available and created by end-users" (Kaplan and Haenlein, 2010). As a result, there has been an increasing introduction of online consumer rating and review platforms for many categories of products (e.g. hotels, restaurants, activities, and electronic goods), where consumers come together to share and seek opinions and feelings, information, photos and videos (Wang, 2011).

Online restaurant review websites play an essential role in users' electronic word-of-mouth (e-WoM) and decision-making process due to the great connection and information provided between and for them. For instance, Trip Advisor is an online consumer rating and review platform that provides contact and assistance to users since it offers a variety of filtering services to narrow the search selection, information about the products or services (e.g. location, menus and photos) and comments posted by the consumers about their experience and the "service, food quality, price, (…) on the restaurant" (Yan et al., 2015; Yu and Zhang, 2020).

In light of the above, the present paper aims to provide a better understanding of electronic word-of-mouth (e-WOM) for restaurants and its motivations and how it can influence the behaviour of other consumers, which will be accomplished through the sentiment analysis of reviews in one of the most extensive review platforms regarding gastronomy, the TripAdvisor. Nonetheless, investigating a possible relationship between sentiment and review rating is also one of the goals of this present paper.

Additionally, the present research is organized as follows. Section 2 reviews eWOM and related literature about it, and then it is followed by the methodology description (data collection and analysis) in section 3. After that, section 4 presents the results of the data analyses. Finally, section 5 concludes the paper and suggests future research.

2. LITERATURE REVIEW

2.1 Word-of-Mouth (WOM) and Electronic Word-of-Mouth (e-WOM)

Word-of-mouth (WOM) is defined as a personal communication channel among consumers regarding their own personal experiences (Arndt, 1967; Ritchins, 1983). Such communication must be informal, interpersonal, selfless, and spontaneous to influence the consumers' purchase behaviour in the presence of products or services (Zhang et al, 2010).

With regard to the purchase of travel-related products and services, the WOM is recognized as an important external source of information (Zhang et al, 2010), plays an important role by sharing customers' experiences, information and advice about "products and services without engaging in [the] first hand" (Lei and Law, 2015; Litvin et al, 2008). For this reason, customers trust and rely on WOM to reduce the level of perceived risk and choose the right product or service to fulfil their personal needs and wants.

With the advent of the Internet, there are available several online consumer reviews and rating platforms (e.g. TripAdvisor, Couch Surfing, Google for Business, Yelp and many others) which empower both, consumers and producers, in terms of sharing information, opinions and recommendations "both from Business to Consumer, and from Consumer to Consumer" (Litvin et al, 2008). Hence, e-WOM is referred to as an informal communication, which can be positive or negative, intended and made by potential, actual or former customers through Internet-based technology, being easily accessible by everyone that has a technological device with an Internet connection and, in fact, 50% of travellers visited a forum or online community before taking the decision and buying travel products online (Zhang et al, 2010).

Even though WOM and e-WOM have the same purpose, there are some differences between both concepts. Firstly, WOM requires face-to-face communication in a specific moment while e-WOM is online through a "written form in cyberspace which can be retrieved, linked, and searched such that its effects last longer" (Wang, 2011). Secondly, e-WOM reaches a wider audience in real-time and may establish personal connections with its users. Thirdly, e-WOM gives its users the opportunity to read comments and recommendations anywhere and anytime. Finally, the misuse of e-WOM may lead to the manipulation of online information on web pages to influence the customer's purchaser behaviour of products or services, since "consumers rely more and more on the Internet to search information" (Wang, 2011).

In recent years, the importance of e-WOM has increased (Li et al, 2021), both on the demand side in terms of the decision-making process for the purchase of tourism products, and on the supply side in terms of service providers, contributing to the definition of marketing strategies to improve automatic tasks, which boosts e-wom among tourists in order to increase sales.

Furthermore, the e-WoM is also considered an indicator that contributes to influencing tourist behaviour, enabling the improvement of the perception of quality associated with services in order to encourage consumption of the product (Lee, et al, 2022). On the other hand, e-WOM has enabled the response of service providers to contribute to the promotion, highlight the professionalism of the staff, and the quality of services, and control the online reputation caused by negative comments (Kim and Velthuis, 2021).

2.2 e-WOM Motivations

Considering a study carried out by Hennig-Thurau et al (2004, pp. 42-43) based on the traditional WOM motivation studies and Balasubramanian and Mahajan (2001) framework, it was identified five main motives of e-WOM communication:

- **Focus-Related Utility:** It refers to the addition of value through the post of "reviews and commentary on products and services of interest to other community members" (Hennig-Thurau et al, 2004). Thus, this main motivation may be subdivided into four other motivations: concern for other consumers, company assistance, social benefit, and extension of (collective) power over companies.
- **Consumption Utility:** It is the gain of the value added on those platforms over the comments written. Furthermore, this utility enables the consumers to obtain "more specific and useful feedback than simply anonymously reading comments" (Hennig-Thurau et al, 2004) and motivated them to interact to better understand or use the product.
- **Approval Utility:** Besides the satisfaction received by reading the reviews that were approved, this motivation has another element as motivations, the self-enhancement and economic rewards.
- **Moderator-Related Utility:** It refers to the "existence of a moderator in the consumer-to-consumer interaction process" (Hennig-Thurau et al, 2004) with the main purpose of solving the consumer's problems, being also more convenient to them.

- **Homeostase Utility:** Taking into account the people's desire of having balance in their lives, this utility is based on two main motives, positive emotions and negative feelings. That means that the feelings felt will influence the consumption experiences of consumers.

In the context of virtual communication, they even suggest eleven motivations of consumer engagement in e-WOM communication, such as awareness of the impact on other consumers, contribution to helping the company, allowing social benefits, enabling the exercise of power over companies, allowing sharing the shopping experience that contributes to councils for other consumers, self-improvement, rewards with monetary values, convenience in seeking redress, hope that some moderation will occur, expression of positive emotions and release of negative feelings (Hennig-Thurau et al, 2004).

Emotions play a leading role in experiences, as they can help to make cognitive assessments of tourist experiences that provoke behavioural responses, as mentioned by Volo (2017) that arise at different stages in the decision-making process, during the holidays and after the experience, consequently, they have an effect on the choice of products and services, create memorable experiences and contribute to redefining future intentions of re-experiencing experiences, which influences positive feelings in the establishment of emotional connections. All of these aspects connected facilitate the involvement between the tourist and the experience.

As considered in the work of Lai et al (2021) which studies the relationship between customer sentiment and online hotel ratings from the perspective of customers' motives in the context of eWOM, and in the Research of Gonçalves et al (2018), which investigate what motivates consumers to produce reviews and how can be more and better eWOM for hotels and, consequently, more reservations or in the investigation of Kang (2018) that intend to analyze the role of overall satisfaction in the relationship between active participation and consumer behaviours that benefit a tourism company.

The motivation for using e-WOM associated with travel arises from the accommodation, quality of services, the image of destinations, transport, gastronomy and restaurants have been investigated by several authors (Bu et al 2021; Dossena and Mochi, 2020; Filieri and McLeay, 2014; Gonçalves et al 2018; Kang, 2018; Lai et al, 2021a; Lai et al, 2021b; Prayogo and Kusumawardhani, 2017; Raguseo and Vitari, 2017; Seo et al, 2020). However, according to the investigation of Lai et al (2021b), the tourist experience is related to gastronomy influencing emotions, which is the major source affecting tourists' satisfaction with the memory of the destination experience.

2.3 Gastronomy e-WOM and Restaurant Experiences

Since 87% of consumers read online reviews, 94% of U.S. diners are influenced by online reviews and "41% of restaurant review readers subsequently visited a restaurant" (Murphy, 2020). The industry associated with restaurants is the activity where the reviews are most viewed (Murphy, 2020), followed by accommodation and then by medical/healthcare. The gastronomy e-WOM has a huge influence on users because it allows them to interact and get reviews on restaurants. Thus, the restaurant ratings and reviews platforms with more affluence are Google My Business, Yelp and Facebook (Erin, 2021; Zhang et al, 2010).

Regarding the restaurant experience, its quality is the most perceived measurement by the customers. In a general way, experience "occurs when a company intentionally uses services as the stage, and goods as props, to engage individual customers in a way that creates a memorable event" (Pine and Gilmore, 1998, p. 98). Correspondingly, the experience in a restaurant is obtained through the customers' wise observation of the restaurant attributes in the course of the dining experience, which may result in positive or negative outcomes reflected on online reviews (Jeong and Jang, 2011; Lai et al, 2021b). Nevertheless, the main review factors considered crucial when taking a decision are: star rating, legitimacy, recency, sentiment and quantity (Murphy, 2020).

Figure 1. The review factors consumers care most about
Source: Adapted from Murphy (2020).

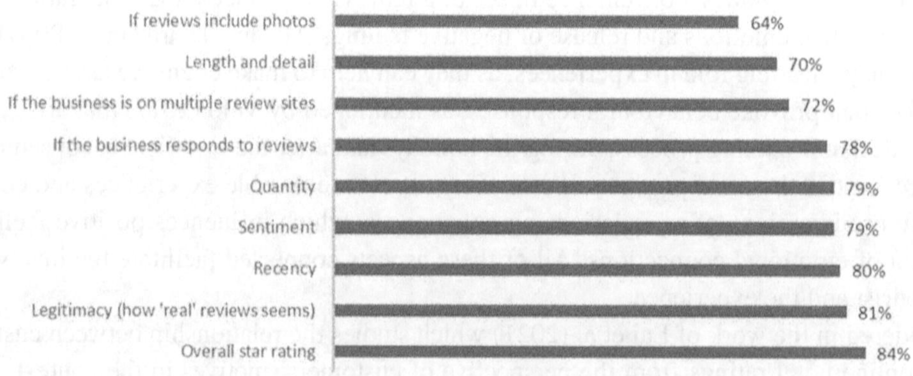

In terms of sentiments, the way customers feel about online reviews and, according to it, negative reviews has a notable impact on consumers behaviour (92% were less likely to buy or use the product or company, respectfully), is also a factor that should be analyzed to understand how the behaviour of the customer can be influenced by the polarity of the review, as presented in Figure 2. For this reason, the consumers of nowadays are more easily influenced by friends, relatives or unfamiliar individuals' online recommendations than what is published in formal media (e.g. newspapers or guidebooks). In fact, the study carried out by Murphy (2020) confirms the above because it takes an average of 10 reviews for a customer to be able to trust a company.

Additionally, the recency of online reviews is another element that impacts severely the customers decision-making process. The more recent it is, the more willing they are to trust the company (86% of customers consider reviews from the past three months and 73% just take into account comments from the last month) (Murphy, 2020). Even so, the great majority focus on the reviews from the last two weeks.

In light of the above, online reviews are a tool of e-WOM that can influence opinions, the customer's purchase behaviour, or even avoid a tourism company. Thus, the companies must have a rich and positive review profile in order to engage customers and keep the business successful, contributing to a valuable online reputation/performance that highlights the quality of products and/or services and contributes to its notoriety. For instance, a higher score on the Yelp platform results in the selling of 19% more tables during the peak seasons. (Anderson & Magruder, 2012; Murphy, 2020).

Figure 2. How do customers feel about online reviews?
Source: Adapted from Murphy (2020)

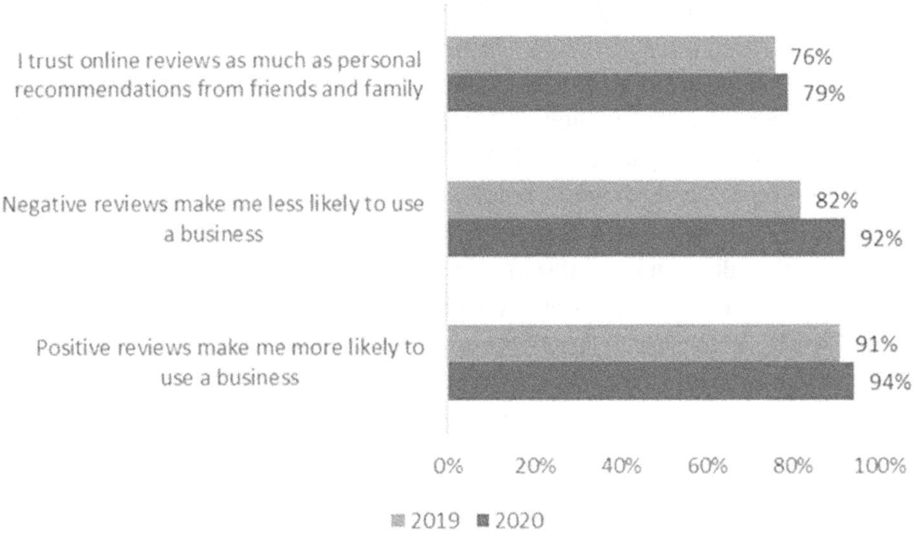

2.4 Restaurant Online Reputation

The performance of an industry is defined by reputation (Hearn, 2010), a concept that arose throw the appearance of web 2.0 and then, in social media. The participation of tourist consumers in online evaluations and reviews allows tourists (and residents) to express their emotions in this medium (Leung and Wen, (2021).

The "reputation" is a noun used to underline the companies or brands' quality levels and the term means "the opinion that people, in general, have about someone or something, or how much respect or admiration someone or something receives, based on past behaviour, character [or experiences]" (Cambridge dictionary, 2021), which may be good or bad.

Online reputation (e-reputation) can be defined as a co-construction process involving both companies and consumers [that relies on] information, brand presence on social media and transparency (Stenger, 2014). With the capacity to boost or destroy a company, the online reputation must be managed in order to: (i) raise positive online reviews and maintain good results; (ii) prevent the negative online reputation; (iii) increase the online interaction, community behaviour and e-WOM; and (iv) extend the customer loyalty, trust and satisfaction. Online reputation can be considered a measurement of the performance of a company, brand or tourist destination (Hearn, 2010). This concept emerged with the appearance of social media and contribute to an indiscriminate change in social relations and in the nature of capitalism. Moreover, consumers exhibit personal emotions, create their digital memories, and generate testimonies about the authenticity of tourist sites and the felt affection. Gastronomy, which is the main tourism experiences component, leads to the creation of emotional memories associated with the places (Lai et al, 2021).

Noteworthy, is that the "restaurant online reputation" management is relatively recent but it must be taken into account due to the great reliability of costumer's on online reviews, which impacts and influences their decisions (Mandić et al, 2020). When considering the "restaurant online reputation"

(Mandić et al, 2020), social media plays an important role in it because it may positively or negatively influence the consumption in a specific restaurant due to the great trust given in the online reviews. As a result, it creates an online image (reputation) through all the existing online information (comments, images, videos, articles, blog posts and many other elements).

The online opinions expressed and the associated reputation system contribute to defining the sociotechnical capital of a restaurant's reputation, which can be interpreted in the digital environment as an economy of digital reputation (Gandini 2016). The online reputation can be very sensitive to tourist experiences since they represent the tourist's emotions, having the power to influence others in the environment. For this reason, an increase or decrease in the economic value of the companies may happen. This effect on economic value can be analyzed using sentiment analysis techniques (Thelwall 2019).

The combination of gastronomic experiences reviews and sentiment analysis can contribute to knowledge creation and service innovation related to gastronomy (Shamim, et al, 2021) but "few review studies were found on ML applications for sentiment analysis" (Jain et al, 2021), and the number of studies which investigate Machine Learning (ML) applications in customer sentiment analysis in the domain of gastronomy is very limited.

3. METHODOLOGY

To create sentiment analysis is necessary to consider a machine learning methodology in five steps (El-Masri, et al, 2017): (i) data collection, (ii) preprocessing the data, (iii) extracting the main features, (iv) applying the analysis (machine learning based), and (v) evaluating the results.

3.1 Data Collection

The present research undertakes a content analysis approach of consumer reviews from TripAdvisor.com on six restaurants presented in Table 1, in which all of which are located in the Algarve region and have 1 to 2 Michelin stars (https://guide.michelin.com/en/restaurants). Moreover, these restaurants have different types of cuisine and prices.

The consumer reviews were collected from the TripAdvisor website and stored in an Excel file, only considering the reviews in English. The structure of the data extracted considered the following variables presented in Table 2.

Table 1. Algarve Michelin restaurants

RestaurantID	City	Cuisine
1	Porches	european, portuguese, contemporary
2	Albufeira	mediterranean, european
3	Almancil	european
4	Portimão	european, portuguese, mediterranean
5	Vila Nova de Cacela	seafood, meditarrean, portuguese
6	Carvoeiro	seafood, portuguese

Source: Own elaboration adapted from TripAdvisor (2022)

Table 2. Restaurants variables considered from reviews website

Variable	Description
RestaurantID	An identifier that represents in an unique way the analyzed restaurant
RestaurantName	The name of the analyzed restaurant
City	The city where is located the restaurant
Cuisine	The type of cuisine prepared in the restaurant
Price	The price level (domain: three levels: cheap, medium and expensive)
Star Numbers	The number of stars defined on the Michelin website.

Source: Own elaboration.

With regard to the structure of data collected, associated with each comment and for each restaurant, the attributes were considered: *RestaurantID* (restaurant identifier), *ReviewRating* (the rating associated to the review), *PublishedDate* (the published date), *Title* (the review title), *FullText* (the complete review text), *AuthorCity* (the city of the author of the review), *AuthorCountry* (the country of the author of the review), *AuthorName* (the name of the author of the review), *AuthorAge* (optional) and *AuthorGender* (optional), whenever it was available, the age and gender of the author of the comment were also collected, but since it was not even available, it was considered optional in nature.

3.2 Preprocessing

In order to analyze the collected data, this study used one platform of business intelligence (Power BI), which is an interactive reporting platform for data and business analytics, and it is necessary to clean the data from useless components, the ETL (extract, transform, load) process, where initially the data was obtained from the Excel and then, it was transformed through the editor Power Query, incorporated in the Power BI Desktop application. The transformation refers to the processing and cleaning of data, which consisted of the normalization of data, replacing strange characters, correcting typos, and treating the different definitions of information. Finally, the load step allowed the ordination, aggregation and consolidation of data to be a part of the Data Warehouse (Inmon, 1996).

3.3 Extract the Main Features

The collection of integrated, non-volatile, subject-oriented data, recorded over time, used to support the decision-making process that consists of the datamart created for the study identified the selected features that will be considered to study the sentiment analysis.

3.4 Applying the Sentiment Analysis (Machine Learning Based)

According to El-Masri et al (2017), there are three major approaches to applying sentiment analysis methods: machine learning, lexicon-based and combined methods also called hybrid methods. So, in the present study considered the machine learning method once it is the most common method used in sentiment analysis (El-Masri et al, 2017).

The open-source machine learning visual language Orange was used since it is a visual tool that uses data mining algorithms for statistical analysis, visual programming and data visualization. Through the Orange platform, this tool was used to identify relationships, patterns or models that are implicit in the data stored (Santos and Ramos, 2009).

3.5 Evaluating the Results and Analytical Reports

The last step connected the data from both platforms and presented information to facilitate the analyses of the restaurant reviews. This information was shown through dashboards, tables and graphics (Inmon, 2016). An aspect that should be considered, is that some data was transformed along the analyses since some errors were found and the normalization and transformation of it were needed with the purpose of providing truthful and correct data analysis.

4. RESULTS OF SENTIMENT ANALYSIS APPLIED TO THE RESTAURANT'S REVIEWS

Considering the collected data, belonging to the six Algarve restaurants with at least one Michelin star, as presented in table 1, was obtained 564 online reviews that were collected during the June and July months of 2021, which were revealed by the TripAdvisor platform, the present research examines the consumer's online reviews. Before applying the sentiment analysis method was considered three types of approach: (i) word cloud, (ii) Topic modelling, and (iii) sentiment analysis.

4.1 Word Cloud

The Word Cloud was generated from the reviews corpus, presented in Figure 3, where the size represents the frequency of the word in the reviews, helping to understand the terms that will be considered in the sentiment analysis. Hence, the word cloud (figure 3) exposed the words that appeared most frequently in the restaurants' reviews, such as food (408), restaurant (393), menu (347), service (338), course (220), Michelin (204), staff (201), good (182), great (179), star (171), chef (164), meal (159) and among many others. This shows that the great majority of the reviews are positive, highlighting the high quality of the restaurant and its meals, services and professionals (staff and chefs).

In the word cloud, it is possible to verify that the positive emotions appear more frequently and with a higher frequency when compared with the negative feelings, also highlighting that "staff", "beautiful", and "friendly" are words used to talk about the human resources of the restaurants, also appealing to an aesthetic component of the service, while contributes to engaging the customer when they feel a friendly environment in this restaurant, which contributes to increase the customer satisfaction and potentiate the loyalty.

Figure 3. Consumers' reviews wordcloud
Source: Own elaboration.

4.2 Topic Modelling

The Topic Modelling algorithm permit to the discovery of abstract themes in a corpus based on clusters of words found in each review and their respective frequency. Figure 4 presents the ten-topic modelling where the positive related to a topic appear in green and the negative words appear in red. It gives more emphasis to the positive reviews due to the large number of positive words related to the tasteful dishes and wonderful service. On the other hand, the negative words underline the sea and sushi dishes and even, the restaurant service.

4.3 Sentiment Analysis

The sentiment analysis algorithm can be used to predict sentiment from a collection of reviews, where the VADER method will be considered once use a lexicon and rule-based sentiment analysis (Hutto and Gilbert, 2014). The VADER (Valence Aware Dictionary for Sentiment Reasoning (Hutto and Gilbert, 2014) consider a combination of quantitative and qualitative methods to identify and validate the sentiment lexicon adequate for microblogs context. Then, the linguistic features are used to identify the sentiment intensity. As the author refers (Hutto and Gilbert, 2014), the lexicon considered in this method is exceptionally good for the social media domain, which is adequate for our study.

To omit the identification of the analyzed data and ensure their protection, a code was randomly assigned to each of the restaurants. The value of sentiments obtained from -1 to 1 was normalized to the range from 1 to 5 so that it was possible to compare with the ratings. Table 3 shows the number of comments for each restaurant, the average rating for ratings and the average sentiment analysis. Regarding the characterization of the sample, 449 comments were considered globally, an average of review rating 4.72, with a standard deviation of 0.73; regarding the average sentiment associated with the reviews, a value of 4.73 was obtained, with a standard deviation of 0.54. We can conclude that the variability of opinion in review ratings is higher than that associated with the feelings expressed in the reviews. In terms of restaurants, the variability in ratings is higher in those with a lower average (#B and #E); with regard to feelings, the variability of opinions does not identify a relationship with the feelings expressed in the comments.

Figure 4. Positive and negative words of the consumers' reviews
Source: Own elaboration.

restaurant food menu service course michelin good dish star dishes

2 good, dish, food, little, service, flavour, small, restaurant, fish, strawberry

3 restaurant, course, menu, tasting, michelin, wonderful, service, food, bar, beach

4 food, menu, course, service, michelin, star, bar, sea, great, tasting

5 well, menu, restaurant, star, food, truly, dinner, cheese, dining, great

6 star, dinner, dish, nice, great, orange, dishes, gusto, jelly, meal

7 food, bar, beach, dish, service, dishes, sushi, small, sea, top

8 truly, service, cooked, fish, restaurant, little, desert, nice, well, restaurants

9 food, fish, best, ocean, dining, taste, great, dinner, nice, fine

10 great, dinner, course, dish, dishes, surprise, good, plate, ocean, list

Table 3. Restaurants variables considered from reviews website

Restaurant Code	Review Numbers	Average Review Rating	Standard Desviation Review Rating	Average Review Sentiment	Standard Desviation Review Sentiment
A	85	4.85	0.45	4.75	0.49
B	86	4.47	0.98	4.50	0.49
C	79	4.80	0.49	4.79	0.29
D	81	4.77	0.77	4.78	0.43
E	74	4.65	0.83	4.80	0.39
F	44	4.86	.50	4.83	0.22

Source: Own elaboration

Since most of the consumer's reviews are of British (Figure 5) and Portuguese (Figure 6) diners in the six chosen Michelin-star restaurants, it is crucial to analyze and understand their rating and sentiment.

Figure 5. Sentiment analysis reviews per country and restaurant for United Kingdom consumers
Source: Own elaboration.

Code	Average Rating Review	Average Sentiment Review
F	4,71	4,93
D	4,96	4,89
E	4,87	4,83
A	4,97	4,82
C	4,90	4,77
B	4,44	4,44
Total	**4,77**	**4,71**

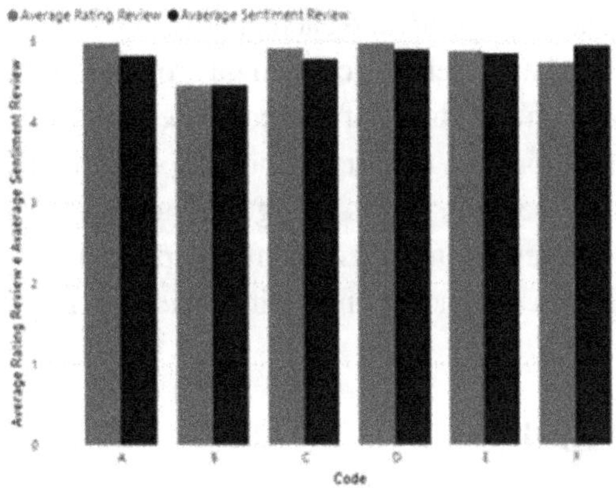

Figure 6. Sentiment analysis reviews per country and restaurant for Portugal consumers
Source: Own elaboration.

Code	Average Rating Review	Average Sentiment Review
C	4,80	4,85
F	4,96	4,81
E	4,57	4,79
B	4,50	4,66
D	4,27	4,54
A	4,62	4,42
Total	**4,67**	**4,71**

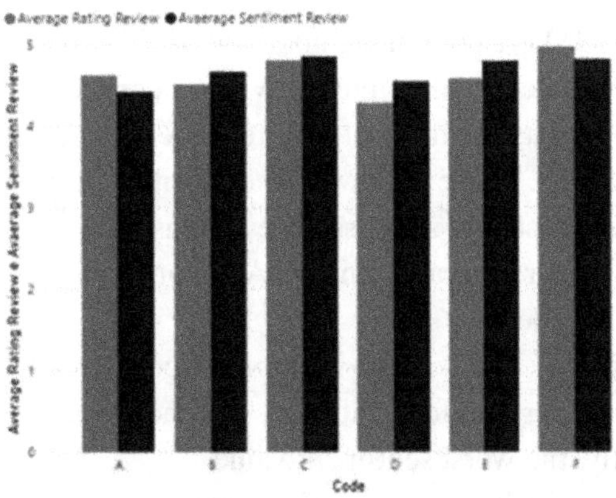

Firstly, the British reviews reveal similar values with rates of 4.44 and 0.78 in rating and sentiment averages, respectively. Conversely, there is a significant discrepancy in the minimum sentiment values. This states that even if the restaurants have good ratings and positive reviews, the negative ones are detrimental. Furthermore, restaurant code #A is the best rated (4.97); however, restaurant Code #F has a better sentiment value (4.93). So, although restaurant #A is considered the best restaurant in several aspects (quality, service, price and many others), restaurant #F has more positive comments. In contrast, restaurant #B has the worst rating (4.44) and sentiment values (4.44).

Secondly, the Portuguese reviews present lower ratings and minimum sentiment values than the British ones, but they are alike when considering the sentiment average. It means that the Portuguese comments are more negative and pessimistic. Regarding each restaurant, the restaurant #F is considered the best due to the attributed rating (4.96) however, it is the restaurant #C that has a higher sentiment value, 4.85 (in contrast with the British reviews, since they considered the restaurant #A just the best in the rating variable). Moreover, restaurant #B has the worst review rating (4.50) and #A and review sentiment (4.42), respectively.

In light of the above, how consumers of each country rate restaurants in terms of review rating are different, in this case between the UK and Portugal. It means that it should be rated for each country, as well as for each age group and each gender, with a view to defining strategies that meet an environment of personalized experiences for each gastronomic consumer profile. On the other hand, the feeling expressed in the comments is not related to the rating values, as otherwise they would be identified for each restaurant and for each country, which implies the need to analyze in detail the most negative/positive comments to assess the points strengths, and thus to promote the service and the weaknesses in order to detect its weaknesses and consequently improve it to give more value to the consumer, paying attention to their opinions and modifying their services to improve the quality offered.

5. CONCLUSION

The Internet and information systems development led to new transformations in several sectors, especially in the gastronomy field. For this reason, restaurants need to constantly improve their services, be aware of what the competitors are doing and understand the customer's behaviour related to their needs and wants.

The e-WOM is a communicative tool, which arises from the Internet and IS growth, that gives power for customers to constantly interact with other potential customers and businesses. For instance, TripAdvisor is a big platform for everyone to search for services, and activities, and even post reviews about a lived experience. Those reviews, either positive or negative, may influence the decision-making process of anyone interested in some service and, as a result, the present research demonstrates the consumer's reviews of four restaurants on TripAdvisor, giving insights into the positive and negative reviews of those establishments.

Regarding the British and Portuguese costumer's reviews, it is possible to conclude that the rating variable may not interfere with the sentiment variables since some restaurants got the best rate and, subsequently, the worst sentiment value.

Regarding the limitations of the present research, they also may provide some orientation for future research. First, the current work's lack of information and analysis about the gastronomy E-WOM, restaurant experience and online reputation was a challenge. However, examining it is exciting and may be a good topic for future research. Second, the study doesn't consider the variables that determine the ratings (food quality, service, environment and price), so future research could include these variables. Third, it would be interesting to research the possible connection between the ratings and sentiments and even the variables that determine a positive or negative review and, consequently, good or bad sentiment values with the primary purpose of analyzing the relationship between gastronomic experience and Sentiment Analysis assessment, for a different generation, genders and countries.

ACKNOWLEDGMENT

This paper is financed by National Funds provided by FCT- Foundation for Science and Technology through project UIDB/04020/2020.

REFERENCES

Anderson, M., & Magruder, J. (2012). Learning from the crowd: Regression discontinuity estimates of the effects of an online review database. *Economic Journal (London), 122*(563), 957–989. doi:10.1111/j.1468-0297.2012.02512.x

Arndt, J. (1967). Role of product-related conversations in the diffusion of a new product. *JMR, Journal of Marketing Research, 4*(3), 291–295. doi:10.1177/002224376700400308

Balasubramanian, V. M., & Mahajan, V. (2001). The economic leverage of the virtual community. *International Journal of Electronic Commerce, 5*(3), 103–138. doi:10.1080/10864415.2001.11044212

Bu, Y., Parkinson, J., & Thaichon, P. (2021). Digital content marketing as a catalyst for e-WOM in food tourism. *Australasian Marketing Journal, 29*(2), 142–154. doi:10.1016/j.ausmj.2020.01.001

Cambridge Dictionary. (2021). *Reputation*. Available at: https://dictionary.cambridge.org/pt/dicionario/ingles/reputation

Dossena, C., & Mochi, F. (2020). 'Managing Online Communities and E-WOM: Prosumers' Characteristics and Behaviors in the Food Service Sector. In *Digital Business Transformation* (pp. 205–217). Springer. doi:10.1007/978-3-030-47355-6_14

El-Masri, M., Altrabsheh, N., & Mansour, H. (2017). Successes and challenges of Arabic senti-ment analysis research: A literature review. *Social Network Analysis and Mining, 7*(1), 1–22. doi:10.100713278-017-0474-x

Erin, B. (2021). *11 Best Restaurants Review Sites*. Available at: https://restaurantclicks.com/restaurant-review-sites/

Filieri, R., & McLeay, F. (2014). E-WOM and accommodation: An analysis of the factors that influence travelers' adoption of information from online reviews. *Journal of Travel Research, 53*(1), 44–57. doi:10.1177/0047287513481274

Gandini, A. (2016). *The reputation economy: Understanding knowledge work in digital society*. Palgrave Macmillan. doi:10.1057/978-1-137-56107-7

Gonçalves, H. M., Silva, G. M., & Martins, T. G. (2018). Motivations for posting online reviews in the hotel industry. *Psychology and Marketing, 35*(11), 807–817. doi:10.1002/mar.21136

Hearn, A. (2010). Structuring feeling: Web 2.0, online ranking and rating, and the digital reputation economy. *Ephemera, 10*, 421–438.

Hennig-Thurau, T., Gwinner, K. P., Walsh, G., & Gremler, D. D. (2004). Electronic word-of-mouth via consumer-opinion platforms: What motivates consumers to articulate themselves on the internet? *Journal of Interactive Marketing*, *18*(1), 38–52. doi:10.1002/dir.10073

Hutto, C., & Gilbert, E. (2014). VADER: A Parsimonious Rule-Based Model for Sentiment Analysis of Social Media Text. *Proceedings of the International AAAI Conference on Web and Social Media*, *8*(1), 216-225. Available at: https://ojs.aaai.org/index.php/ICWSM/article/view/14550

Inmon, W. H. (1996). The data warehouse and data mining. *Communications of the ACM*, *39*(11), 49–51. doi:10.1145/240455.240470

Jain, P. K., Pamula, R., & Srivastava, G. (2021). *A systematic literature review on machine learning applications for consumer sentiment analysis using online reviews*. Elsevier BV. doi:10.1016/j.cosrev.2021.100413

Jeong, E., & Jang, S. S. (2011). Restaurant experiences triggering positive electronic word-of-mouth (eWOM) motivations. *International Journal of Hospitality Management*, *30*(2), 356–366. doi:10.1016/j.ijhm.2010.08.005

Kang, J. (2018). Effective marketing outcomes of hotel Facebook pages: The role of active participation and satisfaction. *Journal of Hospitality and Tourism Insights*, *1*(2), 106–120. doi:10.1108/JHTI-10-2017-0003

Kaplan, A. M., & Haenlein, M. (2010). Users of the world, unite! The challenges and opportunities of Social Media. *Business Horizons*, *53*(1), 59–68. doi:10.1016/j.bushor.2009.09.003

Kim, B., & Velthuis, O. (2021). From reactivity to reputation management: Online consumer review systems in the restaurant industry. *Journal of Cultural Economics*, *14*(6), 675–693. doi:10.1080/17530350.2021.1895280

Lai, I. K. W., Liu, Y., & Lu, D. (2021b). The effects of tourists' destination culinary experience on electronic word-of-mouth generation intention: The experience economy theory. *Asia Pacific Journal of Tourism Research*, *26*(3), 231–244. doi:10.1080/10941665.2020.1851273

Lai, X., Wang, F., & Wang, X. (2021a). Asymmetric relationship between customer sentiment and online hotel ratings: The moderating effects of review characteristics. *International Journal of Contemporary Hospitality Management*, *33*(6), 2137–2156. doi:10.1108/IJCHM-07-2020-0708

Lee, W. L., Liu, C. H., & Tseng, T. W. (2022). The multiple effects of service innovation and quality on transitional and electronic word-of-mouth in predicting customer behavior. *Journal of Retailing and Consumer Services*, *64*, 102791. doi:10.1016/j.jretconser.2021.102791

Lei, S., & Law, R. (2015). Content analysis of TripAdvisor reviews on restaurants: A case study of Macau. *Journal of Tourism*, *16*(1), 17-28.

Leung, X. Y., & Wen, H. (2021). How emotions affect restaurant digital ordering experiences: A comparison of three ordering methods. *Journal of Hospitality and Tourism Technology*, *12*(3), 439–453. doi:10.1108/JHTT-05-2020-0117

Li, Y.-H., Zheng, J., Yue, S.-T., & Fan, Z.-P. (2021). Capturing and analyzing e-WOM for travel products: A method based on sentiment analysis and stochastic dominance. *Kybernetes*. Advance online publication. doi:10.1108/K-02-2021-0146

Litvin, S. W., Goldsmith, R. E., & Pan, B. (2008). Electronic word-of-mouth in hospitality and tourism management. *Tourism Management*, *29*(3), 458–468. doi:10.1016/j.tourman.2007.05.011

Mandić, A., Pivčević, S., & Petrić, L. (2020). Restaurant online reputation and destination competitiveness: insight into TripAdvisor data. In *Gastronomy for Tourism Development* (pp. 155–184). Emerald Publishing Limited. doi:10.1108/978-1-78973-755-420201011

Murphy, R. (2020). *Local Consumer Review Survey*. Available in: https://www.brightlocal.com/research/local-consumer-review-survey/

Pine, B. J., & Gilmore, J. H. (1998). *Welcome to the experience economy*. Harvard Business Review.

Prayogo, R. R., & Kusumawardhani, A. (2017). Examining relationships of destination image, service quality, e-WOM, and revisit intention to Sabang Island, Indonesia. *APMBA*, *5*(2), 89–102.

Raguseo, E., & Vitari, C. (2017). The effect of brand on the impact of e-WOM on hotels' financial performance. *International Journal of Electronic Commerce*, *21*(2), 249–269. doi:10.1080/10864415.2016.1234287

Richins, M. L. (1983). Negative word-of-mouth by dissatisfied consumers: A pilot study. *Journal of Marketing*, *47*(1), 68–78. doi:10.1177/002224298304700107

Seo, E. J., Park, J. W., & Choi, Y. J. (2020). The effect of social media usage characteristics on e-WOM, trust, and brand equity: Focusing on users of airline social media. *Sustainability*, *12*(4), 1691. doi:10.3390u12041691

Shamim, S., Yang, Y., Zia, N. U., & Shah, M. H. (2021). Big data management capabilities in the hospitality sector: Service innovation and customer generated online quality ratings. *Computers in Human Behavior*, *121*, 121. doi:10.1016/j.chb.2021.106777

Stenger, T. (2014). Social media and online reputation management as practice: First steps towards social CRM? *International Journal of Technology and Human Interaction*, *10*(4), 49–64. doi:10.4018/ijthi.2014100104

Thelwall, M. (2019). *Sentiment analysis for tourism. Big Data and Innovation in Tourism, Travel, and Hospitality*. Springer.

Volo, S. (2017). Emotions in Tourism: From Exploration to Design. In D. Fesenmaier & Z. Xiang (Eds.), *Design Science in Tourism. Tourism on the Verge* (pp. 31–40). Springer. doi:10.1007/978-3-319-42773-7_3

Wang, H. Y. (2011). Exploring the factors of gastronomy blogs influencing readers' intention to taste. *International Journal of Hospitality Management*, *30*(3), 503–514. doi:10.1016/j.ijhm.2010.07.009

Yan, X., Wang, J., & Chau, M. (2015). Customer revisit intention to restaurants: Evidence from online reviews. *Information Systems Frontiers*, *17*(3), 645–657. doi:10.100710796-013-9446-5

Yu, C. E., & Zhang, X. (2020). The embedded feelings in local gastronomy: A sentiment analysis of online reviews. *Journal of Hospitality and Tourism Technology*, *11*(3), 461–478. doi:10.1108/JHTT-02-2019-0028

Zhang, Z., Ye, Q., Law, R., & Li, Y. (2010). The impact of e-word-of-mouth on the online popularity of restaurants: A comparison of consumer reviews and editor reviews. *International Journal of Hospitality Management*, *29*(4), 694–700. doi:10.1016/j.ijhm.2010.02.002

KEY TERMS AND DEFINITIONS

Consumer Behavior: This is the focus of studies of consumers and the processes they employ to choose use (i.e., consume), and dispose of products and services.

Electronic Word-of-Mouth: Shared content on social media that conveys an opinion about a product, which becomes a recommendation for future users looking for testimonials about the same product to support them in the decision-making process.

Online Review: Comments made by visitors about a service or tourist product, shared on social networks or travel websites, that convey a testimony of the experience lived during the visit.

Sentiment Analysis: Through the application of text mining algorithms, it allows analyzing textual data, for example, online reviews, to understand whether the sentiment expressed in the text is positive, negative or neutral.

Social Media: These aggregates of online communications channels can be considered tools that can be used to define new business models strategically, taking into consideration analyses of community user-generated content and information shared with other members of these online communities.

Tourism Experience: This is a set of activities in which individuals engage on their personal terms, such as pleasant and memorable places, allowing each tourist to build his or her own travel experiences so that these satisfy a wide range of personal needs, from pleasure to a search for meaning.

Tourism Industry: One of the largest industries in the world. It emerged to satisfy the human need to travel to and see different places as part of the service sector, including hospitality (e.g., accommodation, restaurants), transportation (e.g., airlines, car rental), travel facilitation and information (e.g., tour operators, tourist information centers), and attractions and entertainment (e.g., heritage sites and traditional and cultural events).

Compilation of References

8th Wall. (2021). *Pizza Hut turns special edition pizza boxes into an augmented reality arcade.* 8TH Wall. https://www.8thwall.com

Abbasi, S., & Azari, S. (2009). Novel microwave-freeze drying of onion slice. *International Journal of Food Science & Technology, 44*(5), 947–979. doi:10.1111/j.1365-2621.2008.01774.x

Abdul- Kareem, M. M. (2021). Gana'da Tarım İşletmelerinde Gıda Kaybının Ekonomik Ölçümü ve Etkileri. Ankara: Ankara Üniversitesi Fen Bilimleri Enstitüsü.

Abdul Zahra, M. M. (2022). In Y. Zhang (Ed.), *'Internet of Things-Based Smart and Connected Supply Chain: A Review', International Journal of Antennas and Propagation* (p. 8182813)., doi:10.1155/2022/8182813

Abideen, A.Z. (2021). Food Supply Chain Transformation through Technology and Future Research Directions - A Systematic Review. *Logistics*. . doi:10.3390/logistics5040083

ACF. (2014). *Postharvest Losses and Strategies to Reduce Them.* ACF International.

Aday, S., & Aday, M.S. (2020). *Üç Boyutlu Yazıcıların Gıda Endüstrisinde Kullanımı.* Türkiye 13. Gıda Kongresi, 21-23 Ekim 2020, Çanakkale, Türkiye, 444 s.

Aday, M. S., & Aday, S. (2021). Gıda Kayıp ve İsrafının Azaltılmasında Gıda Bankacılığı. *IBAD Sosyal Bilimler Dergisi, 9*(9), 291–310. doi:10.21733/ibad.844971

Ademoğlu, A. (2021). Atık Gıda Yönetimi ve Sürdürülebilirlik. In Y. Oğan (Ed.), *Gastronomi Araştırmaları* (pp. 64–73). Çizgi Kitabevi Yayınları.

Adenuga, O. T., Mpofu, K., & Modise, R. K. (2020). An Approach for Enhancing Optimal Resource Recovery From Different Classes of Waste in South Africa: Selection of Appropriate Waste to Energy Technology. *Sustainable Futures, 2*(2), 100033.

Agarwal, A., Shaida, B., Rastogi, M., & Singh, N. B. (2023). Food Packaging Materials with Special Reference to Biopolymers-Properties and Applications. *Chemistry Africa, 6*(1), 117–144. doi:10.100742250-022-00446-w

Agbai, C. M. (2020). Application of artificial intelligence (AI) in the food industry. *GSC Biol. Pharm. Sci, 13*(1), 171–178. doi:10.30574/gscbps.2020.13.1.0320

Agrawal, T. K., Kumar, V., Pal, R., Wang, L., & Chen, Y. (2021). Blockchain-based framework for supply chain traceability: A case example of textile and clothing industry. *Computers & Industrial Engineering, 154*, 107130. doi:10.1016/j.cie.2021.107130

Aguilar, G. (2017). *Pérdida y Desperdicio de Alimentos en México.* México, Cd.MX: Caso de estudio para el Banco Mundial.

Ahmad, R. W., Salah, K., Jayaraman, R., Yaqoob, I., Omar, M., & Ellahham, S. (2021). Blockchain-based forward supply chain and waste management for COVID-19 medical equipment and supplies. *IEEE Access : Practical Innovations, Open Solutions, 9*, 44905–44927. doi:10.1109/ACCESS.2021.3066503 PMID:34812386

Ahmed, W. A. H., & MacCarthy, B. L. (2021). Blockchain-Enabled Supply Chain Traceability in the Textile and Apparel Supply Chain: A Case Study of the Fiber Producer, Lenzing. *Sustainability, 13*(19), 10496. doi:10.3390u131910496

Aiplusinfo. (2023). *AI and Recipes*. https://www.aiplusinfo.com/blog/ai-and-recipes/

Akan, S. (2022). *Gıda Kaybı ve İsraf Sorunu*. Gaziantep Üniversitesi Sosyal Bilimler Enstitüsü.

Akar, E. (2022). *Yalın Üretim Anlayışının Otel İşletmelerinde Uygulanabilirliği: Gıda İsrafı ve Kaybı Üzerine Bir Çalışma*. Mersin Üniversitesi Sosyal Bilimler Enstitüsü.

Aka, S., & Buyukdag, N. (2021). How to prevent food waste behaviour? A deep empirical research. *Journal of Retailing and Consumer Services, 61*, 102560. doi:10.1016/j.jretconser.2021.102560

Akbulut, B. A., & Aksoy, M. (2016). Restoranlardaki Teknolojik Yeniliklerin Deneyim Pazarlaması Açısından Değerlendirilmesi (Assessment of the Technological Innovations in Restaurants in terms of Experience Marketing. *International Congress on Cultural Heritage and Tourism*, 1–10.

Akın, G. (2014). İnsan Sağlığı ve Çevre Etkileşimi. *Ankara Üniversitesi Dil ve Tarih-Coğrafya Fakültesi Dergisi, 54*(1), 105–116.

Akkerman, R., Farahani, P., & Grunow, M. (2010). Quality, Safety and Sustainability in Food Distribution: A Review of Quantitative Operations Management Approaches and Challenges. *OR-Spektrum, 32*(4), 863–904. doi:10.100700291-010-0223-2

Al Amin, M., Arefin, M. S., Sultana, N., Islam, M. R., Jahan, I., & Akhtar, A. (2021). Evaluating the customers dining attitudes, e-satisfaction and continuance intention toward mobile food ordering apps (MFOAs): Evidence from Bangladesh. *European Journal of Management and Business Economics, 30*(2), 211–229. doi:10.1108/EJMBE-04-2020-0066

Albayrak, M. S., Oner, A., Atakli, I. M., & Ekenel, H. K. (2019). Personalized training in fast-food restaurants using augmented reality glasses. *Proceedings - 2019 International Symposium on Educational Technology, ISET 2019*, (pp. 129–133). IEEE. 10.1109/ISET.2019.00035

Alfaro, L., Siramard, S., Chouljenko, A., & Sathivel, S. (2018). Effects of liquid nitrogen pretreatment on the osmotic dehydration and quality of cryogenically frozen blueberries (*Vaccinium angustifolium Ait.*). *Food Bioscience, 22*, 165–169. doi:10.1016/j.fbio.2018.02.006

Ali, M. M., Hashim, N., Abd Aziz, S., &Lasekan, O. (2021). Quality inspection of food and agricultural products using artificial intelligence. *Advances in Agricultural and Food Research Journal, 2*(2).

Ali, A., Xia, C., Ouattara, N., Mahmood, İ., & Faisal, M. (2021). Economic and Environmental Consequences' of Post-harvest Loss Across Food Supply Chain in the Developing Countries. *Journal of Cleaner Production, 323*, 129–146. doi:10.1016/j.jclepro.2021.129146

Alıntaş, O. (2021). *Gastronomi ve Mutfak Sanatları Öğrencilerinin Gıda İsrafı ve Gıda Güvenliği Hakkında Görüşleri: İstanbul İli Örneği*. İstanbul Gelişim Üniversitesi Lisansüstü Eğitim Enstitüsü.

AlShamsi, M., Al-Emran, M., & Shaalan, K. (2022). A systematic review on blockchain adoption. *Applied Sciences (Basel, Switzerland), 12*(9), 4245. doi:10.3390/app12094245

Altenburg, T. (2016) *Making retail modernisation in developing countries inclusive: a development policy perspective.*

Altınel, H. (2014). *Menü Yönetimi ve Menü Planlama*. Detay Yayıncılık.

America, C. G. T. N. (2019). Nature is the key to a happy kitchen for an eco-friendly chef in Ecuador. Retrieved from https://www.youtube.com/watch?v=O0OT4UYyxkE Access Date: 26.06.2023.

Ammendolia, M. G., Iosi, F., De Berardis, B., Guccione, G., Superti, F., Conte, M. P., & Longhi, C. (2014). Listeria monocytogenes behaviour in presence of non-UV-irradiated titanium dioxide nanoparticles. *PLoS One*, *9*(1), e84986. doi:10.1371/journal.pone.0084986 PMID:24416327

Ananno, A. A., Masud, M. H., Chowdhury, S. A., Dabnichki, P., Ahmed, N., & Arefin, A. M. E. (2021). Sustainable food waste management model for Bangladesh. *Sustainable Production and Consumption*, *27*, 35–51. doi:10.1016/j.spc.2020.10.022

Anderson, M., & Magruder, J. (2012). Learning from the crowd: Regression discontinuity estimates of the effects of an online review database. *Economic Journal (London)*, *122*(563), 957–989. doi:10.1111/j.1468-0297.2012.02512.x

Andoni, M., Robu, V., Flynn, D., Abram, S., Geach, D., Jenkins, D., McCallum, P., & Peacock, A. (2019). Blockchain technology in the energy sector: A systematic review of challenges and opportunities. *Renewable & Sustainable Energy Reviews*, *100*, 143–174. doi:10.1016/j.rser.2018.10.014

Angelis, J., & Da Silva, E. R. (2019). Blockchain adoption: A value driver perspective. *Business Horizons*, *62*(3), 307–314. doi:10.1016/j.bushor.2018.12.001

Anton, N., & Vandamme, T. F. (2011). Nano-emulsions and micro-emulsions: Clarifications of the critical differences. *Pharmaceutical Research*, *28*(5), 978–985. doi:10.100711095-010-0309-1 PMID:21057856

Antonucci, F., Figorilli, S., Costa, C., Pallottino, F., Raso, L., & Menesatti, P. (2019). A review on blockchain applications in the agri-food sector. *Journal of the Science of Food and Agriculture*, *99*(14), 6129–6138. doi:10.1002/jsfa.9912 PMID:31273793

Arah, I. K., Ahorbo, G. K., Anku, E. K., Kumah, E. K., & Amaglo, H. (2016). Postharvest Handling Practices and Treatment Methods for Tomato Handlers in Developing Countries: A Mini Review. *Advances in Agriculture*, *2016*, 1–9. doi:10.1155/2016/6436945

Arndt, J. (1967). Role of product-related conversations in the diffusion of a new product. *JMR, Journal of Marketing Research*, *4*(3), 291–295. doi:10.1177/002224376700400308

Artés, F., Vallejo, F., & Martinez, J. (2001). Quality of broccoli as influenced by film wrapping during shipment. *European Food Research and Technology*, *213*(6), 480–483. doi:10.1007002170100390

Arvanitoyannis, I., & Gorris, L. G. M. (1999). Edible and Biodegradable Polymeric Materials for Food Packaging or Coating. In *Processing Foods: Quality Optimization and Process Assessment*. CRC Press.

Aschemann-Witzel, J., De Hooge, I. E., Rohm, H., Normann, A., Bossle, M. B., Grønhøj, A., & Oostindjer, M. (2017). Key characteristics and success factors of supply chain initiatives tackling consumer-related food waste–A multiple case study. *Journal of Cleaner Production*, *155*, 33–45. doi:10.1016/j.jclepro.2016.11.173

Aschemann-Witzel, J., De Hooge, I., Amani, P., Bech-Larsen, T., & Oostindjer, M. (2015). Consumer-related food waste: Causes and potential for action. *Sustainability*, *7*(6), 6457–6477. doi:10.3390u7066457

Aschemann-Witzel, J., Giménez, A., & Ares, G. (2018). Consumer in-store choice of suboptimal food to avoid food waste: The role of food category, communication and perception of quality dimensions. *Food Quality and Preference*, *68*, 29–39. doi:10.1016/j.foodqual.2018.01.020

Ashokkumar, M., Haskaracharya, R., Entish, S., Lee, J., Almer, M., & Bogdan, Z. (2010). The ultrasonic processing of dairy products: An overview. *EDP Sciences*, *90*(2-3), 147–168. doi:10.1051/dst/2009044

Ashok, Y., & Satya, S. (2014). Osmotic dehydration of fruits and vegetables: A review. *Journal of Food Science and Technology*, *51*(9), 1654–1673. doi:10.100713197-012-0659-2 PMID:25190823

Astill, J., Dara, R. A., Campbell, M., Farber, J. M., Fraser, E. D. G., Sharif, S., & Yada, R. Y. (2019). Transparency in food supply chains: A review of enabling technology solutions. *Trends in Food Science & Technology*, *91*, 240–247. doi:10.1016/j.tifs.2019.07.024

Atarés, L., Chiralt, A., & McElhatton, A. (2016). Use of Edible Coatings, a Novel Preservation Method for Nuts. In A. McElhatton & M. El Idrissi (Eds.), *Modernization of Traditional Food Processes and Products. Integrating Food Science and Engineering Knowledge Into the Food Chain* (Vol. 11). Springer. doi:10.1007/978-1-4899-7671-0_9

Athinarayanan, J., Periasamy, V. S., Alsaif, M. A., Al-Warthan, A. A., & Alshatwi, A. A. (2014). Presence of nanosilica (E551) in commercial food products: TNF-mediated oxidative stress and altered cell cycle progression in human lung fibroblast cells. *Cell Biology and Toxicology*, *30*(2), 89–100. doi:10.100710565-014-9271-8 PMID:24526451

Attarin, S., & Attaran, M. (2020). Food Printing: Evolving technologies, challenges, opportunities, and best adoption strategies. *Journal of International Technology and Information Management*, *29*(1), 25–55. https://scholarworks.lib.csusb.edu/jitim/vol. doi:10.58729/1941-6679.1442

Auty, S. (1992, July 1). Consumer choice and segmentation in the restaurant industry. *Service Industries Journal*, *12*(3), 324–339. doi:10.1080/02642069200000042

Awan, S. (2021). In S. Khan (Ed.). *IoT with BlockChain: A Futuristic Approach in Agriculture and Food Supply Chain. Wireless Communications and Mobile Computing* (p. 5580179)., doi:10.1155/2021/5580179

Ayan, B., Güner, E. & Son-Turan, S. (2022). Blockchain Technology and Sustainability in Supply Chains and a Closer Look at Different Industries: A Mixed Method Approach. *Logistics*. . doi:10.3390/logistics6040085

Aydın, Ş., & Uçkan Çakır, M. (2022). Gastronomi ve Dijitalleşme. *Nevşehir Hacı Bektaş Veli Üniversitesi SBE Dergisi*, *12*(4), 2143–2159. doi:10.30783/nevsosbilen.1123324

Bablon, G., Bellamy, W. D., Bourbigot, M. M., Daniel, F. B., Dore, M., Erb, F., Gordon, G., Langlais, B., Laplanche, A., Legube, B., Martin, G., Masschelein, W. J., Pacey, G., Reckhow, D. A., & Ventresque, C. (1991). Fundamental aspects. In G. Langlais, D. A. Reckhow, & D. R. Brink (Eds.), *Ozone in water treatment: Application and engineering* (pp. 11–132). Lewis Publishers.

Babu Loganathan, G. (2021). Recent Scope for AI in the Food Production Industry Leading to the Fourth Industrial Revolution. *Webology*, *18*(2), 1066–1080. doi:10.14704/WEB/V18I2/WEB18375

Bai, Y. X., Yang, G. J., Hu, Y. C., & Qu, M. (2012). Physical and sensory properties of electrohydrodynamic (EHD) dried scallop muscle. *Journal of Aquatic Food Product Technology*, *21*(3), 238–247. doi:10.1080/10498850.2011.590271

Balaji, M., & Arshinder, K. (2016). Modeling the causes of food wastage in Indian perishable food supply chain. *Resources, Conservation and Recycling*, *114*, 153–167. doi:10.1016/j.resconrec.2016.07.016

Balasubramaniam., Ramaswamy, R., Rodriguez, R. I., Mustafa, V. and Yousef, A. (2014). Ozone Technology Fact Sheet for Food Processors. pp 1-3.

Balasubramanian, V. M., & Mahajan, V. (2001). The economic leverage of the virtual community. *International Journal of Electronic Commerce*, *5*(3), 103–138. doi:10.1080/10864415.2001.11044212

Ballesteros, L. F., Teixeira, J. A., & Cerqueira, M. A. (2022). Active Carboxymethyl Cellulose-Based Edible Coatings for the Extension of Fresh Goldenberries Shelf-Life. *Horticulturae, 8*(10), 936. doi:10.3390/horticulturae8100936

Bamakan, S. M. H., Malekinejad, P., & Ziaeian, M. (2022). Towards blockchain-based hospital waste management systems; applications and future trends. *Journal of Cleaner Production, 349*, 131440. doi:10.1016/j.jclepro.2022.131440

Banerjee, R., & Verma, A. (2015). Minimally Processed Meat and Fish Products. In M. W. Siddiqui & M. S. Rahman (Eds.), *Minimally Processed Foods* (pp. 193–250). Springer International Publishing. doi:10.1007/978-3-319-10677-9_10

Bansal, V., Siddiqui, M., & Rahman, M. (2015). Minimally Processed Foods: Overview. In M. W. Siddiqui & M. S. Rahman (Eds.), *Minimally Processed Foods* (pp. 1–15). Springer International Publishing. doi:10.1007/978-3-319-10677-9_1

Ban, Y., Kearney, K., Sundahl, B., Medina, L., Bonnecaze, R. T., & Chopra, M. J. (2021, February). Fast etch recipe creation with automated model-based process optimization. In *Advanced Etch Technology and Process Integration for Nanopatterning X* (Vol. 11615, pp. 66–76). SPIE. doi:10.1117/12.2583868

Baptista, P., Campos, I., Pires, I., & Sofia, V. (2012). *Do Campo ao Garfo.* Desperdício Alimentar em Portugal.

Baralla, G., Ibba, S., Marchesi, M., Tonelli, R., & Missineo, S. (2018, August). A blockchain based system to ensure transparency and reliability in food supply chain. In *European conference on parallel processing* (pp. 379-391). Springer, Cham.

Baratharaj, V. (2009). *Use of Ozone in Food Processing and Cold Storage.* OTSIL. www.otsil.net

Barcelona Visio. (2009). "Slow Food". A boost to ecogastronomy. Retrieved from https://www.youtube.com/watch?v=QWVgcd_rqVk Access Date: 10.06.2023.

Barkworth, H. (1938). Taints and off-flavors of milk. *Dairy Inds, 3*, 367.

Basumatary, I. B., Mukherjee, A., Katiyar, V., & Kumar, S. (2022). Biopolymer-based nanocomposite films and coatings: Recent advances in shelf-life improvement of fruits and vegetables. *Critical Reviews in Food Science and Nutrition, 62*(7), 1912–1935. doi:10.1080/10408398.2020.1848789 PMID:33249872

Batra, D., Diwan, N., Upadhyay, U., Kalra, J. S., Sharma, T., Sharma, A. K., & Bagler, G. (2020). Recipedb: A resource for exploring recipes. *Database (Oxford)*, 2020. PMID:33238002

Beck, D. E., Crandall, P. G., O'Bryan, C. A., & Shabatura, J. C. (2016). *Taking food safety to the next level—An augmented reality solution.* Taylor and Francis.. doi:10.1080/15378020.2016.1185872

Beck, R., & Müller-Bloch, C. (2017). *Blockchain as radical innovation: a framework for engaging with distributed ledgers as incumbent organization.*

Bedane, T. F., Chen, L., Marra, F., & Wang, S. (2017). Experimental study of radio frequency (RF) thawing of foods with movement on conveyor belt. *Journal of Food Engineering, 201*, 17–25. doi:10.1016/j.jfoodeng.2017.01.010

Bendegul, O., Ahmet, B. O., & Anil, B. (2021). Generation Y's dining out behavior. *International Hospitality Review, 35*(1), 41–56. doi:10.1108/IHR-07-2020-0023

Benko, A., & Lányi, C. S. (2009). History of artificial intelligence. In *Encyclopedia of Information Science and Technology* (2nd ed., pp. 1759–1762). IGI Global. doi:10.4018/978-1-60566-026-4.ch276

Berezina, K., Ciftci, O., & Cobanoglu, C. (2019). Robots, artificial intelligence, and service automation in restaurants. In Robots, artificial intelligence, and service automation in travel, tourism and hospitality, 185-219. doi:10.1108/978-1-78756-687-320191010

Berryman, D. R. (2012). Augmented Reality. *RE:view*, *31*(2), 212–218. Http://Dx.Doi.Org/10.1080/02763869.2012.670604. doi:10.1080/02763869.2012.670604 PMID:22559183

Bestdestination (2022). Macao Gastronomy City and Sustainable Gastronomy Day. Retrieved from https://www.youtube.com/watch?v=YxZ55B4XtdM Access Date: 10.06.2023.

Betz, A., Buchli, J., Göbel, C., & Müller, C. (2015). Food Waste İn The Swiss Food Service İndustry – Magnitude And Potential For Reduction. *Waste Management (New York, N.Y.)*, *35*, 218–226. doi:10.1016/j.wasman.2014.09.015 PMID:25305683

Bhandari, B., Datta, N., & Howes, T. (1997). Problems Associated with Spray Drying Of Sugar-Rich Foods. *Drying Technology*, *15*(2), 671–684. doi:10.1080/07373939708917253

Bhatia, S., Stevanovic, J. T., & Ratti, C. (2020). Freeze-Drying of Plant-Based Foods. *Foods*, *9*(1), 87. doi:10.3390/foods9010087 PMID:31941082

Bhatnagar, S. K. (2002). *Front Office Management* (2nd ed.). Frank Brothers and Company.

Bień, M., Gilski, M., Maciejewska, M., Taisner, W., Wisniewski, D., & Lawrynowicz, A. (2020, December). RecipeNLG: A cooking recipes dataset for semi-structured text generation. In *Proceedings of the 13th International Conference on Natural Language Generation* (pp. 22-28). IEEE.

Bintsis, T., Litopoulou-Tzanetaki, E., & Robinson, R. (2000). Existing and potential applications of the ultraviolet light in the food industry- A critical review. *Journal of the Science of Food and Agriculture*, *80*(6), 637–645. doi:10.1002/(SICI)1097-0010(20000501)80:6<637::AID-JSFA603>3.0.CO;2-1 PMID:29345786

Biswas, B., & Gupta, R. (2019). Analysis of barriers to implement blockchain in industry and service sectors. *Computers & Industrial Engineering*, *136*, 225–241. doi:10.1016/j.cie.2019.07.005

Bitividi (2020). Sustainable Gastronomy Day. Retrieved from https://www.youtube.com/watch?v=6cGxObtNaBU Access Date: 22.06.2023.

Blaxter, M. (2003). *Health and lifestyles*. Routledge. doi:10.4324/9780203393000

Bodó, B., Gervais, D., & Quintais, J. P. (2018). Blockchain and smart contracts: The missing link in copyright licensing? *International Journal of Law and Information Technology*, *26*(4), 311–336. doi:10.1093/ijlit/eay014

Bonanni, L., Lee, C.-H., & Selker, T. (2005). CounterIntelligence: Augmented Reality Kitchen. MIT Media Laboratory, 2239(45).

Bonanni, L. (2010). *Beyond transparency : collective engagement in sustainable design.*

Bontoux, L., & Leone, F. (1997). *The Legal Definition of Waste and its Impact on Waste Management in Europe*. European Commission - Joint Research Centre Institute for Prospective Technological Studies WTC.

Bos, M. (2019). *3D printing meat: The potential of the food industry's new frontier*. WUR. https://www.wur.nl/en/newsarticle/3D-printing-meat-The-potential-of-the-food-industrys-new-frontier.htm

Bothra, K. (2021, December). Consumer's Perception Regarding Online Food Delivery Applications in India. *International Journal of Research Publication and Reviews*, *2*(12), 948–953.

Bourke, P., & O'Beirne, D. (2004). Effects of packaging type, gas atmosphere and storage temperature on survival and growth of Listeria spp. in shredded dry coleslaw and its components. *International Journal of Food Science & Technology*, *39*(5), 509–523. doi:10.1111/j.1365-2621.2004.00811.x

Bowen, J., & Morosan, C. (2018). Beware hospitality industry: The robots are coming. *Worldwide Hospitality and Tourism Themes*, *10*(6), 726–733. doi:10.1108/WHATT-07-2018-0045

Bowie, D. (2018). Innovation and 19th century hotel industry evolution. *Tourism Management*, *64*, 314–323. doi:10.1016/j.tourman.2017.09.005

Bozkurt, R., & Yüksel, A. Y. (2019). Günümüzde Balık Atıklarının Helal Sektörde Kullanım Alanları. *Türk Doğa ve Fen Dergisi*, *8*(2), 10–16.

Bratovčić, A., Odobašić, A., Ćatić, S., & Šestan, I. (2015). Application of polymer nanocomposite materials in food packaging. *Croatian Journal of Food Science and Technology*, *7*(2), 86–94. doi:10.17508/CJFST.2015.7.2.06

Bräutigam, K.-R., Jörissen, J., & Priefer, C. (2014). The Extent of Food Waste Generation Across EU-27: Different Calculation Methods and the Reliability of Their Results. *Waste Management & Research*, *32*(8), 683–694. doi:10.1177/0734242X14545374 PMID:25161274

Breakthrough Institute. (2017). Sustainable Gastronomy. Retrieved from https://www.youtube.com/watch?v=s6dkVmsCyUw Access Date: 15.06.2023.

Breffi, N. M. (2012). Applying Modern Robotics Technologies to Demand Prediction and Production Management in the Quice-Service Restaurant Sector. *Cornell Hospitality Quarterly*, *53*(2), 122–133. doi:10.1177/1938965511434112

Bugnotti, D., Dalle Vacche, S., Esposito, L. H., Callone, E., Orsini, S. F., Ceccato, R., D'Arienzo, M., Bongiovanni, R., Dirè, S., & Vitale, A. (2023). Structure of Starch–Sepiolite Bio-Nanocomposites: Effect of Processing and Matrix–Filler Interactions. *Polymers*, *15*(5), 1207. doi:10.3390/polym15051207 PMID:36904448

Bumblauskas, D., Mann, A., Dugan, B., & Rittmer, J. (2020). A blockchain use case in food distribution: Do you know where your food has been? *International Journal of Information Management*, *52*, 102008. doi:10.1016/j.ijinfomgt.2019.09.004

Burton-Hughes, L. (2020, February 7). www.highspeedtraining.co.uk. Retrieved from High Speed Training: https://www.highspeedtraining.co.uk/hub/improve-sustainability-in-restaurants/

Bu, Y., Parkinson, J., & Thaichon, P. (2021). Digital content marketing as a catalyst for e-WOM in food tourism. *Australasian Marketing Journal*, *29*(2), 142–154. doi:10.1016/j.ausmj.2020.01.001

Buzrul, S. (2015). Multi-pulsed high hydrostatic pressure treatment of foods. *Food*, *4*(4), 173–183. doi:10.3390/foods4020173 PMID:28231197

Caleb, O. J., Opara, U. L., & Witthuhn, C. R. (2012). Modified atmosphere packaging of pomegranate fruit and arils: A review. *Food and Bioprocess Technology*, *5*(1), 15–30. doi:10.100711947-011-0525-7

Caleb, O., Mahajan, P., Al-Said, F. A.-J., & Opara, U. L. (2013). Modified Atmosphere Packaging Technology of Fresh and Fresh-cut Produce and the Microbial Consequences—A Review. *Food and Bioprocess Technology*, *6*(2), 303–329. doi:10.100711947-012-0932-4 PMID:32215166

Callaghan, V., Miller, J., Yampolskiy, R., & Armstrong, S. (Eds.). (2017). *The technological singularity: managing the journey*. Springer. doi:10.1007/978-3-662-54033-6

Cambridge Dictionary. (2021). *Reputation*. Available at: https://dictionary.cambridge.org/pt/dicionario/ingles/reputation

Campbell, B., Khachatryan, H., Behe, B., Dennis, J., & Hall, C. (2015). Consumer perceptions of eco-friendly and sustainable terms. *Agricultural and Resource Economics Review*, *44*(1), 21–34. doi:10.1017/S1068280500004603

Canküll, D., Doğan, A., & Sönmez, B. (2018). Innovation And Augmented Reality Applications in Food And Beverage Companıes. *İşletme Araştırmaları Dergisi Journal of Business Research--Turk*, *10*, 576–591.

Caria, M., Sara, G., Todde, G., Polese, M., & Pazzona, A. (2019). Exploring Smart Glasses for Augmented Reality: A Valuable and Integrative Tool in Precision Livestock Farming. *Animals 2019, 9*(11), 903. doi:10.3390/ani9110903

Casari, F. A., Navab, N., Hruby, L. A., Kriechling, P., Nakamura, R., Tori, R., de Lourdes dos Santos Nunes, F., Queiroz, M. C., Fürnstahl, P., & Farshad, M. (2021). Augmented Reality in Orthopedic Surgery Is Emerging from Proof of Concept Towards Clinical Studies: A Literature Review Explaining the Technology and Current State of the Art. *Current Reviews in Musculoskeletal Medicine*, *14*(2), 192–203. doi:10.100712178-021-09699-3 PMID:33544367

Castellanos, D. A., Cerisuelo, J. P., Hernández-Muñoz, P., Herrera, A. O., & Gavara, R. (2016). Modelling the evolution of O2 and CO2 concentrations in MAP of a fresh product: Application to tomato. *Journal of Food Engineering*, *168*, 84–95. doi:10.1016/j.jfoodeng.2015.07.019

Castillo, O., & Meliif, P. (1970). Automated quality control in the food industry combining artificial intelligence techniques with fractal theory. *WIT Transactions on Information and Communication Technologies, 10.*

Castillo, G., Spinella, K., Poturnayová, A., Šnejdárková, M., Mosiello, L., & Hianik, T. (2015). Detection of aflatoxin b1 by aptamer-based biosensor using pamam dendrimers as immobilization platform. *Food Control*, *2015*, 529–18. doi:10.1016/j.foodcont.2014.12.008

Causer, C. (2009). They've got a golden ticket. *Potentials IEEE*, *28*(4), 42–44. doi:10.1109/MPOT.2009.933608

CEC. (2017). *Characterization and Management of Food Loss and Waste in North America.* Commission for Environmental Cooperation. https://rescuefood.ca/wp-content/uploads/2021/04/11772-characterization-and-management-food-loss-and-waste-in-north-america-en.pdf

Ceryes, C. A., Antonacci, C. C., Harvey, S. A., Spiker, M. L., Bickers, A., & Neff, R. A. (2021). "Maybe It's Still Good?" A Qualitative Study of Factors Influencing Food Waste and Application of the E.P.A. Food Recovery Hierarchy in U.S. Supermarkets. *Appetite*, *161*, 105–111. doi:10.1016/j.appet.2021.105111 PMID:33482300

Çetin, H. C. (2022). *Zincir Restoran İşletmelerinde Gıda Atığı Yönetimi: İstanbul Örneği.* Akdeniz Üniversitesi Sosyal Bilimler Enstitüsü.

Ceylan, Z., Bulkan, S., & Tozan, H. (2017). Tek ve Çok Dönemli Envanter Kontrol Modelleri. *Mühendislik Bilimleri ve Tasarım Dergisi*, *5*(2), 441–455.

CGTN. (2019). What is sustainable gastronomy? Retrieved from https://www.youtube.com/watch?v=isGKvrauay4 Access Date: 11.06.2023.

Chai, J. J. K., O'Sullivan, C., Gowen, A. A., Rooney, B., & Xu, J. L. (2022). Augmented/mixed reality technologies for food: A review. *Trends in Food Science & Technology*, *124*, 182–194. doi:10.1016/j.tifs.2022.04.021

Chalak, A., Abou-Daher, C., Chaaban, J., & Abiad, M. G. (2016). The Global Economic and Regulatory Determinants of Household Food Waste Generation: A Cross-Country Analysis. *Waste Management (New York, N.Y.)*, *48*, 418–422. doi:10.1016/j.wasman.2015.11.040 PMID:26680687

Chang, S. N., & Chen, W. L. (2017). Does visualize industries matter? A technology foresight of global Virtual Reality and Augmented Reality Industry. *Proceedings of the 2017 IEEE International Conference on Applied System Innovation: Applied System Innovation for Modern Technology, ICASI 2017*, (pp. 382–385). IEEE. 10.1109/ICASI.2017.7988432

Chaplin, M. (2004). *What are biosensors?* London south bank university. www.lsbu.ac.uk/biology/enztech/biosensors.html

Charcosset, C. (2009). Preparation of emulsions and particles by membrane emulsification for the food processing industry. *Journal of Food Engineering*, *92*(3), 241–249. doi:10.1016/j.jfoodeng.2008.11.017

Charles, F., Sanchez, J., & Gontard, N. (2003). Active modified atmosphere packaging of fresh fruits and vegetables: Modeling with tomatoes and oxygen absorber. *Journal of Food Science*, *68*(5), 1736–1742. doi:10.1111/j.1365-2621.2003.tb12321.x

Cha, S. S., & Seo, B. K. (2020). The effect of food delivery application on customer loyalty in restaurant. *Journal of Distribution Science*, *18*(4), 5–12.

Cheikh, D., Majdoub, H., & Darder, M. (2022). An overview of clay-polymer nanocomposites containing bioactive compounds for food packaging applications. *Applied Clay Science*, *216*, 106335. doi:10.1016/j.clay.2021.106335

Chemat, F., Zill-e-Huma, & Khan, M. K. (2011). Applications of ultrasound in food technology: Processing, preservation, and extraction. *Ultrasonics Sonochemistry*, *18*(4), 813–835. doi:10.1016/j.ultsonch.2010.11.023 PMID:21216174

Cheng, K., Li, Z., He, Y., Guo, Q., Lu, Y., Gu, S., & Wu, H. (2023). Potential use of artificial intelligence in infectious disease: Take ChatGPT as an example. *Annals of Biomedical Engineering*, *51*(6), 1–6. doi:10.100710439-023-03203-3 PMID:37074486

Cheng, Y., Liu, L., Zhao, G., Shen, C., Yan, H., Guan, J., & Yang, K. (2015). The effects of modified atmosphere packaging on core browning and the expression patterns of PPO and PAL genes in 'Yali' pears during cold storage. *Lebensmittel-Wissenschaft + Technologie*, *60*(2), 1243–1248. doi:10.1016/j.lwt.2014.09.005

Chen, J., Chen, D., Xie, Y., Yuan, T., & Chen, X. (2013). Progress of Microfluidics for Biology and Medicine. *Nano-Micro Letters*, *5*(1), 66–80. doi:10.1007/BF03354852

Chen, K. H., Hsieh, K. J., Chang, F. H., & Chen, N. C. (2015). The customer citizenship behaviors of food blog users. *Sustainability*, *7*(9), 12502–12520. doi:10.3390u70912502

Chen, L. D. (2008). A model of consumer acceptance of mobile payment. *International Journal of Mobile Communications*, *6*(1), 32–52. doi:10.1504/IJMC.2008.015997

Chen, L., Li, Q., Guo, Z., & Wang, Y. (2020). Molecularly imprinted polymers-based surface-enhanced Raman scattering sensor for the detection of melamine in milk powder. *Food Chemistry*, *330*, 127147.

Chen, L., Remondetto, G. E., & Subirade, M. (2006). Food protein-based materials as nutraceutical delivery systems. *Trends in Food Science & Technology*, *17*(5), 272–283. doi:10.1016/j.tifs.2005.12.011

Chen, L., Wei, X., Irmak, S., Chaves, B. D., & Subbiah, J. (2019). Inactivation of Salmonella enterica and Enterococcus faecium NRRL B-2354 in cumin seeds by radiofrequency heating. *Food Control*, *103*, 59–69. doi:10.1016/j.foodcont.2019.04.004

Chen, M. F. (2023). Integrating the extended theory of planned behavior model and the food-related routines to explain food waste behavior. *British Food Journal*, *125*(2), 645–661. doi:10.1108/BFJ-07-2021-0788

Chen, S. L. (1804). R, Z. M., & Si, Y. (2021). Mobile communications for tourism and hospitality: A review for hospitality evolution, prsent status, and future trends. *Electronics (Basel)*, *10*(5).

Chen, S., Liu, X., Yan, J., Hu, G., & Shi, Y. (2021). Processes, benefits, and challenges for adoption of blockchain technologies in food supply chains: A thematic analysis. *Information Systems and e-Business Management*, *19*(3), 909–935. doi:10.100710257-020-00467-3

Choi, T. M. (2020). Supply chain financing using blockchain: Impacts on supply chains selling fashionable products. *Annals of Operations Research*, 1–23. doi:10.100710479-020-03615-7

Cho, M., Bonn, M. A., & Li, J. J. (2019). Differences in perceptions about food delivery apps between single-person and multi-person households. *International Journal of Hospitality Management, 77*, 108–116. doi:10.1016/j.ijhm.2018.06.019

Christidis, K., & Devetsikiotis, M. (2016). Blockchains and Smart Contracts for the Internet of Things. *IEEE Access : Practical Innovations, Open Solutions, 4*, 2292–2303. doi:10.1109/ACCESS.2016.2566339

Chui, M. (2017). Artificial intelligence is the next digital frontier. *McKinsey and Company Global Institute, 47*(3.6).

Cil, A. Y., Abdurahman, D., & Cil, I. (2022). Internet of Things enabled real time cold chain monitoring in a container port. *Journal of Shipping and Trade, 7*(1), 9. doi:10.118641072-022-00110-z

Cirişoğlu, E., & Akoğlu, A. (2021). Restoranlarda Oluşan Gıda Atıkları ve Yönetimi: İstanbul İli Örneği. *Akademik Gıda, 19*(1), 38–48.

Clarifai. (n.d.) The *world's AI: Clarifai Computer Vision AI & Machine Learning Platform*. Clarifai. https://www.clarifai.com/

Clark, J., Crandall, P., & Shabatura, J. (2018). Wearable technology effects on training outcomes of restaurant food handlers. *Journal of Food Protection, 81*(8), 1220–1226. doi:10.4315/0362-028X.JFP-18-033 PMID:29963922

Cohen, D. L., Lipton, J. I., Cutler, M., Coulter, D., Vesco, A., & Lipson, H. (2009, September). Hydrocolloid Printing: A Novel Platform for Customized Food Production. In *2009 International Solid Freeform Fabrication Symposium*. University of Texas at Austin.

Çöl, B. G., İmre, M., & Yıkmış, S. (2023). Virtual reality and augmented reality technologies in gastronomy: A review. *EFood, 4*(3), e84. doi:10.1002/efd2.84

Compton, M., Willis, S., Rezaie, B., & Humes, K. (2018). Food processing industry energy and water consumption in the Pacific northwest. *Innovative Food Science & Emerging Technologies, 47*, 371–383. doi:10.1016/j.ifset.2018.04.001

Cooremans, K., & Geuens, M. (2019). Same but different: Using anthropomorphism in the battle against food waste. *J Publ Pol Market, 38*(2), 232–245. doi:10.1177/0743915619827941

Coşkun, A., & Özbük, R. M. Y. (2020). What influences consumer food waste behavior in restaurants? An application of the extended theory of planned behavior. *Waste Management (New York, N.Y.), 117*, 170–178. doi:10.1016/j.wasman.2020.08.011 PMID:32861079

Coşkun, F., & Gülleroğlu, H. D. (2021). Yapay zekânın tarih içindeki gelişimi ve eğitimde kullanılması. *Ankara University Journal of Faculty of Educational Sciences, 54*(3), 947–966.

Couch, L. M., Wien, M., Brown, J. L., & Davidson, P. (2016). Food nanotechnology: Proposed uses, safety concerns and regulations. *Agro Food Industry Hi-Tech, 27*(1).

Coutinho, N. M., Silveira, M. R., Rocha, R. S., Moraes, J., Ferreira, M. V. S., Pimentel, T. C., Freitas, M. Q., Silva, M. C., Raices, R. S. L., Ranadheera, C. S., Borges, F. O., Mathias, S. P., Fernandes, F. A. N., Rodrigues, S., & Cruz, A. G. (2018). Cold plasma processing of milk and dairy products. *Trends in Food Science & Technology, 74*, 56–68. doi:10.1016/j.tifs.2018.02.008

Creydt, M., & Fischer, M. (2019). Blockchain and more-Algorithm driven food traceability. *Food Control, 105*, 45–51. doi:10.1016/j.foodcont.2019.05.019

Crofton, E. C., Botinestean, C., Fenelon, M., & Gallagher, E. (2019). Potential applications for virtual and augmented reality technologies in sensory science. *Innovative Food Science & Emerging Technologies, 56*, 102178. doi:10.1016/j.ifset.2019.102178

Crosby, M., Pattanayak, P., Verma, S., & Kalyanaraman, V. (2016). Blockchain technology: Beyond bitcoin. *Applied Innovation*, *2*(6-10), 71.

Cui, H., Wu, X., Li, Z., Li, X., & Yang, Y. (2020). A molecularly imprinted polymer-based fluorescent sensor for highly selective and sensitive detection of tetracycline antibiotics in milk. *Food Chemistry*, *310*, 125966.

Cushen, M., Kerry, J., Morris, M., Cruz-Romero, M., & Cummins, E. (2014). Evaluation and simulation of silver and copper nanoparticle migration from polyethylene nanocomposites to food and an associated exposure assessment. *Journal of Agricultural and Food Chemistry*, *62*(6), 6. doi:10.1021/jf404038y PMID:24450547

Dai, Y., Cao, X., Ma, H., He, P., Zhang, Q., & Fang, Y. (2016). Characteristics of nitrogen and phosphorus removal from agricultural runoff using a constructed wetland with biochar/zeolite mixture. *Journal of Environmental Management*, *182*, 90–96.

Dalgleish, D. G. (2004). Food emulsions: their structures and properties. In Journal of Food Emulsions, 4th Edn, eds. Friberg SE, Larsson K, Sjöblom, 1-44. New York: Marcel Dekker.

Damoska Sekuloska, J. & Erceg, A. (2022). Blockchain Technology toward Creating a Smart Local Food Supply Chain. *Computers*. . doi:10.3390/computers11060095

Dang, J. (2017). *How 3D printing could change the way we eat*. CNN. https://www.cnn.com/2017/06/22/health/3d-printing-food/index.html

Daniel, V. L. (2015). *3D Food Printing Creating Shapes and Texture*. Elesvier. http://refhub.elsevier.com/S0924-2244(17)30082-1/sref8

Dankar, I., Haddarah, A., Omar, F. E. L., Sepulcre, F., & Pujola, M. (2018). 3D printing technology: The new era for food customization and elaboration. *Trends in Food Science & Technology*, *75*, 231–242. doi:10.1016/j.tifs.2018.03.018

Datta, A. K., & Anantheswaran, R. C. (2000). *Handbook of microwave technology for food applications*. Marcel Dekker Inc.

David Szanto. (2016). Stories of Eco-Gastronomy: Torino. Retrieved from https://www.youtube.com/watch?v=d43gt6xw4IQ Access Date: 13.06.2023.

David Szanto. (2017) Stories of Eco-Gastronomy: Montreal. Retrieved from https://www.youtube.com/watch?v=hNlyaTHxI1o Access Date: 13.06.2023.

de Dicastillo, C. L., Vidal, C. P., Falco, I., Sanchez, G., Marquez, P., & Escrig, J. (2020). Antimicrobial bilayer nanocomposites based on the incorporation of as-synthetized hollow zinc oxide nanotubes. *Nanomaterials (Basel, Switzerland)*, *10*(3), 3. doi:10.3390/nano10030503 PMID:32168893

Debeaufort F., (2021). *Metal Packaging*. Wiley Online library. doi:10.1002/9781119825081.ch4

Değerli, C., & El, S. N. (2017). Üç boyutlu (3D) Yazıcı Teknolojisi ile Gıda Üretimine Genel Bakış. *Turkish Journal of Agriculture-Food Science and Technology*, *5*(6), 593–599. doi:10.24925/turjaf.v5i6.593-599.1062

Demirbaş, N. (2018). Dünyada ve Türkiye'de Gıda Israfını Önleme Çalışmalarının Değerlendirilmesi. *VIII.* [Plovdiv.]. *IBANESS Kongreler Serisi*, *21*(22), 521–526.

Demirbaş, N. (2019). yi Tarım Uygulamaları ile Meyve Bahçelerinde Ortaya Çıkan Üretim, Hasat ve Hasat Sonrası Kayıpları Azaltılabilir mi? *I. IBANESS Kongreler Serisi, Tekirdağ. Türkiye*, *9*(10), 289–296.

Demir, Y. (2022). *Eğitim Mutfaklarında Gıda İsrafını Belirlemeye Yönelik Bir Araştırma*. İstanbul Üniversitesi Sosyal Bilimler Enstitüsü.

Demir, Y. (2023). Sığır eti üretiminde su ayak izi durumu (Evaluation of water footprint in beef production). *Aydın Gastronomy*, *7*(1), 161–171. doi:10.17932/IAU.GASTRONOMY.2017.016/gastronomy_v07i10011

Dempsey, M. (2022). *AI: Why chefs are turning to artificial intelligence.* BBC. https://www.bbc.com/news/business-59651077

Deshwal, G. K., Panjagari, N. R., & Alam, T. (2019). An overview of paper and paper based food packaging materials: Health safety and environmental concerns. *Journal of Food Science and Technology*, *56*(10), 4391–4403. doi:10.100713197-019-03950-z PMID:31686671

Dickinson, E., & Stainsby, G. (1982). *Colloids in foods*. Elsevier Applied Science Publishers.

Difrancesco, R. M., Meena, P., & Kumar, G. (2022). How blockchain technology improves sustainable supply chain processes: A practical guide [Preprint]. *Operations Management Research*, (0123456789). doi:10.100712063-022-00343-y

DigitalFoodLab. (2010). *NYMBLE LABS—The Smart Kitchen Appliance*. Digital Food Lab. https://www.digitalfoodlab.com/en/ foodtech-database/nymble-labs

Dinu, A. (2022).The impact of food delivery applications on Romanian consumers' behaviour during the COVID-19 pandemic. *Socio-Economic Planning Sciences, 82. Part A*, (August), 101220.

Dirsehan, T., & Cankat, E. (2021). Role of mobile food-ordering applications in developing restaurants' brand satisfaction and loyalty in the pandemic period. *Journal of Retailing and Consumer Services*, *102608*, 102608. Advance online publication. doi:10.1016/j.jretconser.2021.102608

Djekic, I., Jankovic, D., & Rajkovic, A. (2017). Analysis of foreign bodies present in European food using data from Rapid Alert System for Food and Feed (RASFF). *Food Control*, *79*, 143–149. doi:10.1016/j.foodcont.2017.03.047

Dogan, M. E., Goru Dogan, T., & Bozkurt, A. (2023). The use of artificial intelligence (AI) in online learning and distance education processes: A systematic review of empirical studies. *Applied Sciences (Basel, Switzerland)*, *13*(5), 3056. doi:10.3390/app13053056

Dolci, R. (2017, July). IoT solutions for precision farming and food manufacturing: artificial intelligence applications in digital food. In *2017 IEEE 41st Annual Computer Software and Applications Conference (COMPSAC)* (*Vol. 2*, pp. 384-385). IEEE. 10.1109/COMPSAC.2017.157

Donsi, F. (2018). In S. M. Jafari & D. J. McClements (Eds.), *"Applications of nanoemulsions in foods," in Nanoemulsions: Formulation, Applications, and Characterization* (pp. 349–377). Academic Press.

Dooley, D. M., Griffiths, E. J., Gosal, G. S., Buttigieg, P. L., Hoehndorf, R., Lange, M. C., Schriml, L. M., Brinkman, F. S. L., & Hsiao, W. W. L. (2018). FoodOn: A harmonized food ontology to increase global food traceability, quality control and data integration. *NPJ Sci. Food*, *2*(1), 23. doi:10.103841538-018-0032-6 PMID:31304272

Dordevic, D., Buchtova, H., Jancikova, S., Macharackova, B., Jarosova, M., Vitez, T., & Kushkevych, I. (2019). Aluminum contamination of food during culinary preparation: Case study with aluminum foil and consumers' preferences. *Food Science & Nutrition*, *7*(10), 3349–3360. doi:10.1002/fsn3.1204 PMID:31660148

Dossena, C., & Mochi, F. (2020). 'Managing Online Communities and E-WOM: Prosumers' Characteristics and Behaviors in the Food Service Sector. In *Digital Business Transformation* (pp. 205–217). Springer. doi:10.1007/978-3-030-47355-6_14

Drewitt, T. (2013). *Food Waste Prevention in Quick Service Restaurants*. Lund University Master of Science in Environmental Management and Policy.

Duan, J., Zhang, C., Gong, Y., Brown, S., & Li, Z. (2020). A Content-Analysis Based Literature Review in Blockchain Adoption within Food Supply Chain. *International Journal of Environmental Research and Public Health, 17*(5), 1784. doi:10.3390/ijerph17051784 PMID:32182951

Duan, X., Zhang, M., Mujumdar, A. S., & Wang, R. (2010). Trends in microwave-assisted freeze drying of foods. *Drying Technology, 28*(4), 444–453. doi:10.1080/07373931003609666

Durmuş, H. (2022). *Gıda İsrafının Boyutu ve Ekonomik Etkilerinin Tespiti Üzerine Bir Araştırma: Sakarya İli Örneği.* Sakarya: Sakarya Üniversitesi Sosyal Bilimler Enstitüsü.

Dutta, P., Choi, T.-M., Somani, S., & Butala, R. (2020). Blockchain technology in supply chain operations: Applications, challenges and research opportunities. *Transportation Research Part E, Logistics and Transportation Review, 142*, 102067. doi:10.1016/j.tre.2020.102067 PMID:33013183

EcoCircuitos Panama. (2017). Eco-Story Telling: Panama's Gastronomy with Patricia Miranda. Retrieved from https://www.youtube.com/watch?v=O2fL2uugKHc Access Date: 17.06.2023.

Eco-Gastronomy. (2023) Eating Is an Agricultural Act'-Wendell Berry. Retrieved from https://eco-gastronomy.com/ Access Date: 12.05.2023.

Ecotsy (2023). Being Environmentally Friendly vs Eco Friendly: Which is Better? Retrieved from https://ecotsy.com/environmentally-friendly-vs-eco-friendly/ Access Date: 6.06.2023.

Edelman, D. C., & Singer, M. (2015). Competing on Customer Journeys. *Sales and Marketing*. https://hbr.org/2015/11/competing-on-customer-journeys

Eftimov, T., Koroušić Seljak, B., & Korošec, P. (2016). Grammar and dictionary based named-entity linking for knowledge extraction of evidence-based dietary recommendations. In *Proceedings of the 8th international Joint conference on knowledge discovery, knowledge engineering and knowledge management.* SCITEPRESS. 10.5220/0006032401500157

Eftimov, T., Korošec, P., & Koroušić Seljak, B. (2017). StandFood: Standardization of foods using a semi-automatic system for classifying and describing foods according to FoodEx2. *Nutrients, 9*(6), 121–148. doi:10.3390/nu9060542 PMID:28587103

El-Allaly, E., Khoshgoftaar, T. M., & Wald, R. (2021). An Intelligent Recipe Generation Model Using Deep Learning. In *2021 IEEE International Conference on Systems, Man, and Cybernetics (SMC)* (pp. 458-463). IEEE.

Elferink, M., & Schierhorn, F. (2016). Global demand for food is rising. Can we meet it. *Harvard Business Review, 7*(04), 2016.

El-Masri, M., Altrabsheh, N., & Mansour, H. (2017). Successes and challenges of Arabic senti-ment analysis research: A literature review. *Social Network Analysis and Mining, 7*(1), 1–22. doi:10.100713278-017-0474-x

El-Ramady, H. R., Domokos-Szabolcsy, É., Abdalla, N. A., Taha, H. S., & Fári, M. (2015). Postharvest management of fruits and vegetables storage. *Sustainable Agriculture Research, 15*, 65–152.

Epuran, G. (2021). Enhancing Consumer Experience through Development of Implicit Attitudes Using Food Delivery Applications. *Journal of Theoretical and Applied Electronic Commerce Research, 16*(7), 2858–2882. doi:10.3390/jtaer16070157

Erin, B. (2021). *11 Best Restaurants Review Sites.* Available at: https://restaurantclicks.com/restaurant-review-sites/

Escriche, I., Chiralt, A., Moreno, J., & Serra, J. A. (2000). Influence of blanching-osmotic dehydration treatments on volatile fraction of strawberries. *Journal of Food Science, 65*(7), 107. doi:10.1111/j.1365-2621.2000.tb10247.x

Eun-Jung, K., & Loren, G. V. (2003). Consumers' Restaurant Choice Behavior and the Impact of Socio-Economic and Demographic Factors. *Journal of Foodservice Business Research*, *6*(1), 3–24. doi:10.1300/J369v06n01_02

Evans, G. W. (2017). *Artificial intelligence: where we came from, where we are now, and where we are going.*

Evans, D. (2012). Beyond the throwaway society: Ordinary domestic practice and a sociological approach to household food waste. *Sociology*, *2012*(46), 41–56. doi:10.1177/0038038511416150

Ezeh, O., Yusoff, M. M., & Niranjan, K. (2018). Nonthermal processing technologies for fabrication of microstructures to enhance food quality and stability. In *Food Microstructure and Its Relationship with Quality and Stability* (pp. 239–274). Woodhead Publishing. doi:10.1016/B978-0-08-100764-8.00012-5

Falade, K. O., & Igbeka, J. C. (2007). Osmotic dehydration of tropical fruits and vegetables. *Food Reviews International*, *23*(4), 373–405. doi:10.1080/87559120701593814

FAO. (2011). Global food losses and food waste – Extent, causes and prevention. Roma.

FAO. (2013). *Food Wastage Footprint Impacts on Natural Resources.* FAO. https://www.fao.org/3/i3347e/i3347e.pdf

FAO. (2015). Save Food: Global Initiative on Food Loss and Waste Reduction. Roma.

FAO. (2018). *The future of food and agriculture Alternative pathways to 2050.* Licence: CC BY-NC-SA 3.0 IGO. FAO.

FAO. (2019). No one likes labels… unless it comes to food. Retrieved from https://www.fao.org/fao-stories/article/en/c/1177378/ Access Date: 25.06.2023.

FAO. (2020). Calling all foodies: this one's for you. Retrieved from https://www.fao.org/fao-stories/article/en/c/1198076/ Access Date: 10.06.2023.

FAO. (2020). *Grow, Nourish, sustain. Together.* FAO. https://www.fao.org/world-food-day/home/en/

FAO. (2021a). A spicy agricultural heritage: there's more to wasabi than meets the eye. Retrieved from https://www.fao.org/fao-stories/article/en/c/1391113/ Access Date: 17.06.2023.

FAO. (2021b). We are what we eat: Transforming diets to transform agri-food systems. Retrieved from https://www.fao.org/fao-stories/article/en/c/1393055/ Access Date: 19.06.2023.

FAO. (2021c). A cup of tea… or cha? Retrieved from https://www.fao.org/fao-stories/article/en/c/1401152/ Access Date: 10.06.2023.

FAO. (2022). Local food for healthy locals. Retrieved from https://www.fao.org/fao-stories/article/en/c/1481073/ Access Date: 12.06.2023.

FAO. (2022). *Tackling food loss and waste: A triple win opportunity.* FAO. https://www.fao.org/newsroom/detail/FAO-UNEP-agriculture-environment-food-loss-waste-day-2022/en.

FAO. (2023). Freekeh: The ancient dish firing up rural women's futures in Lebanon. Retrieved from https://www.fao.org/fao-stories/article/en/c/1634616/ Access Date: 21.06.2023.

FDA. (2019). How to cut food waste and maintain food safety. FDA. https://www.fda.gov/food/consumers/how-cut-food-waste-andmaintain- food-safety.

Feng, H., Wang, X., Duan, Y., Zhang, J., & Zhang, X. (2020). Applying blockchain technology to improve agri-food traceability: A review of development methods, benefits and challenges. *Journal of Cleaner Production*, *260*, 121031. doi:10.1016/j.jclepro.2020.121031

Fernandes, F. A. N., Gallao, M. I., & Rodrigues, S. (2008). Effect of osmotic dehydration and ultrasound pretreatment on cell structure: Melon dehydration. *Lebensmittel-Wissenschaft + Technologie*, *41*(4), 604–610. doi:10.1016/j.lwt.2007.05.007

Filieri, R., & McLeay, F. (2014). E-WOM and accommodation: An analysis of the factors that influence travelers' adoption of information from online reviews. *Journal of Travel Research*, *53*(1), 44–57. doi:10.1177/0047287513481274

Fito, P., Chiralt, A., Betoret, N., Gras, M., Chafer, M., Martnez-Monzo, J., Andres, A., & Vidal, D. (2001). Vacuum impregnation and osmotic dehydration in matrix engineering applications in functional fresh food development. *Journal of Food Engineering*, *49*(2-3), 175–183. doi:10.1016/S0260-8774(00)00220-X

Flink, J. M. (1979). Dehydrated carrot slices: influence of osmotic concentration on drying behaviour and product quality. In Food Process Engineering, Eds. Linko, P., Malkki, Y. Olkku J. and Larinkari, J., 412-418. Applied Science Publishers: London.

Fonseca, E., Oliveira, I., Lobo, J., Mota, T., Martins, J., & Au-Yong-Oliveira, M. (2019). Kitchen Robots: The Importance and Impact of Technology on People's Quality of Life. In Á. Rocha, H. Adeli, L. Reis, & S. Costanzo (Eds.), *New Knowledge in Information Systems and Technologies. WorldCIST'19 2019. Advances in Intelligent Systems and Computing* (Vol. 931). Springer. doi:10.1007/978-3-030-16184-2_19

Food and Agriculture Organization of the United Nations (FAO). (2018). Sustainable food systems-Concept and framework. Retrieved from https://www.fao.org/3/ca2079en/CA2079EN.pdf Access Date: 12.05.2023.

Food Safety and Standards Regulation. (2011). *Food Safety and Standard Authority of India (FSSAI)*. FSSAI. https://www.fssai.gov.in/cms/food-safety-and-standards-regulations.php

Food Waste Faqs (no date) USDA. https://www.usda.gov/foodwaste/faqs

Ford, S., & Despeisse, M. (2016). Additive manufacturing and sustainability: An exploratory study of the advantages and challenges. *Journal of Cleaner Production*, *137*, 1573–1587. doi:10.1016/j.jclepro.2016.04.150

Fowler, M. R., & Park, J. W. (2015). Effect of salmon plasma protein on Pacific whiting surimi gelation under various ohmic heating conditions. *Lebensmittel-Wissenschaft + Technologie*, *61*(2), 309–315. doi:10.1016/j.lwt.2014.12.049

França, A. S. L., Neto, J. A., Gonçalves, R. F., & Almeida, C. M. V. B. (2020). Proposing the use of blockchain to improve the solid waste management in small municipalities. *Journal of Cleaner Production*, *244*, 118529. doi:10.1016/j.jclepro.2019.118529

Francesc Fusté-Forné. (2021) Robot chefs in gastronomy tourism: What's on the menu? *Tourism Management Perspectives*, *37*. doi:10.1016/j.tmp.2020.100774

Franco-Mariscal, A. J. (2018). Discovering the Chemical Elements in Food. *Journal of Chemical Education*, *95*(3), 403–409. doi:10.1021/acs.jchemed.7b00218

Freeman, C. (1987). *Technology, policy, and economic performance: lessons from Japan*. Burns & Oates.

Frey, C. B., & Osborne, M. A. (2017). The future of employment: How susceptible are jobs to computerisation? *Technological Forecasting and Social Change*, *114*, 254–280. doi:10.1016/j.techfore.2016.08.019

Friedman, N., & Ormiston, J. (2022). Blockchain as a sustainability-oriented innovation?: Opportunities for and resistance to Blockchain technology as a driver of sustainability in global food supply chains. *Technological Forecasting and Social Change*, *175*, 121403. doi:10.1016/j.techfore.2021.121403

Fritz, W., Hadi, R., & Stephen, A. (2022). From tablet to table: How augmented reality influences food desirability. *Journal of the Academy of Marketing Science*, 1–27. doi:10.1007/S11747-022-00919-X/FIGURES/2 PMID:36589144

Future Food Institute. (2020). Food is a Conversation: Sustainable Gastronomy Day - Japanese Session. https://www.youtube.com/watch?v=gr7U5BsSeBM Access Date: 19.06.2023.

Gad, A. G. (2022). Emerging Trends in Blockchain Technology and Applications: A Review and Outlook. *Journal of King Saud University - Computer and Information Sciences*, *34*(9), 6719–6742.

Galiano Conservancy Association. (2021). John Volpe on EcoGastronomy | Galiano Conservancy Association. Retrieved from https://www.youtube.com/watch?v=lqeFMwzxql4 Access Date: 14.06.2023.

Gandini, A. (2016). *The reputation economy: Understanding knowledge work in digital society*. Palgrave Macmillan. doi:10.1057/978-1-137-56107-7

Garcia, A., Mount, J. R., & Davidson, P. M. (2003). Ozone and chlorine treatment of minimally processed lettuce. *Journal of Food Science*, *68*(9), 2747–2751. doi:10.1111/j.1365-2621.2003.tb05799.x

Garcia-Guzman, L., Cabrera-Barjas, G., Soria-Hernández, C. G., Castaño, J., Guadarrama-Lezama, A. Y., & Rodríguez Llamazares, S. (2022). Progress in Starch-Based Materials for Food Packaging Applications. *Polysaccharides*, *3*(1), 1. doi:10.3390/polysaccharides3010007

Garcia-Haro, J. M., Oña, E. D., Hernandez-Vicen, J., Martinez, S., & Balaguer, C. (2021). Service Robots in Catering Applications: A Review and Future Challenges. *Electronics (Basel)*, *10*(1), 47. doi:10.3390/electronics10010047

Garcia-Herrero, I., Hoehn, D., Margallo, M., Laso, J., Bala, A., Batlle-Bayer, L., Fullana, P., Vazquez-Rowe, I., Gonzalez, M. J., Durá, M. J., Sarabia, C., Abajas, R., Amo-Setien, F. J., Quiñones, A., Irabien, A., & Aldaco, R. (2018). On the Estimation of Potential Food Waste Reduction to Support Sustainable Production and Consumption Policies. *Food Policy*, *80*, 24–38. doi:10.1016/j.foodpol.2018.08.007

Garg, V. (2019). The Study of Interest of Consumers In Mobile Food Ordering Apps. *International Journal of Recent Technology and Engineering* (IJRTE), *8* (1).

Garnett, T. (2011). Where are the best opportunities for reducing greenhouse gas emissions in the food system (including the food chain)? *Food Policy*, *36*, S23–S32. doi:10.1016/j.foodpol.2010.10.010

Gastro Tourism Academy. (2021). Eco-Gastronomy: Minjangan Grass. https://www.youtube.com/watch?v=FjISfzcr7Vw Access Date: 10.06.2023.

Gastro Tourism Academy. (2023). Climate Eaters Webinar- Climate Action for Food Sustainability: Eco Gastronomy and Greenpreneurs. Retrieved from https://www.youtube.com/watch?v=Az1NvVpeNaM Access Date: 17.06.2023.

Gaudin, V. (2017). Advances in biosensor development for the screening of antibiotic residues in food products of animal origin - A comprehensive review. *Biosensors & Bioelectronics*, *90*, 363–377. doi:10.1016/j.bios.2016.12.005 PMID:27940240

Gavahian, M., Chu, Y. H., Khaneghah, A. M., Barba, F. J., & Misra, N. N. (2018). A critical analysis of the cold plasma induced lipid oxidation in foods. *Trends in Food Science & Technology*, *77*, 32–41. doi:10.1016/j.tifs.2018.04.009

Gebbers, R., & Adamchuk, V. I. (2010). Precision agriculture and food security. *Science*, *327*(5967), 828–831. doi:10.1126cience.1183899 PMID:20150492

Gemeç, E., & Gürbüz, İ. B. (2022). Gıda İsrafının Önlenmesine Yönelik Tutum ve Davranışların Belirlenmesi; Üniversite Öğrencileri Üzerine Bir İnceleme. *Balkan ve Yakın Doğu Sosyal Bilimler Dergisi*, *8*, 164–175.

Ghaffari Laleh, N., Ligero, M., Perez-Lopez, R., & Kather, J. N. (2023). Facts and hopes on the use of artificial intelligence for predictive immunotherapy biomarkers in cancer. *Clinical Cancer Research*, *29*(2), 316–323. doi:10.1158/1078-0432.CCR-22-0390 PMID:36083132

Ghinoi, S., Silvestri, F., & Steiner, B. (2020). Toward the creation of novel food waste management systems: A network approach. *Journal of Cleaner Production*, *246*, 118987. doi:10.1016/j.jclepro.2019.118987

Gilland, B. (2002). World Population and Food Supply: Can Food Production Keep Pace with Population Growth in the Next Half-century? *Food Policy*, *27*(1), 47–63. doi:10.1016/S0306-9192(02)00002-7

Giri, S. K., & Prasad, S. (2007). Drying kinetics and rehydration characteristics of microwave vacuum and convective hot-air dried mushrooms. *Journal of Food Engineering*, *78*(2), 512–521. doi:10.1016/j.jfoodeng.2005.10.021

Gjerris, M., & Gaiani, S. (2013). Household Food Waste in Nordic Countries: Estimations and Ethical İmplications. *Nordic Journal of Applied Ethics*, *7*(1), 6–23. doi:10.5324/eip.v7i1.1786

Godfray, H. J., Crute, I. R., Haddad, L., Lawrence, D., Muir, J. F., Nisbett, N., Pretty, J., Robinson, S., Toulmin, C., & Whiteley, R. (2010). The Future of the Global Food System. *Philosophical Transactions of the Royal Society of London. Series B, Biological Sciences*, *365*(1554), 2769–2777. doi:10.1098/rstb.2010.0180 PMID:20713383

Godoi, F. C., Prakash, S., & Bhandari, B. R. (2016). 3D Printing Technologies Applied for Food Design: Status and Prospects. *Journal of Food Engineering*, *179*, 44–54. doi:10.1016/j.jfoodeng.2016.01.025

Goel, M., Chakraborty, P., Ponnaganti, V., Khan, M., Tatipamala, S., Saini, A., & Bagler, G. (2022, May). Ratatouille: A tool for Novel Recipe Generation. In *2022 IEEE 38th International Conference on Data Engineering Workshops (ICDEW)* (pp. 107-110). IEEE. 10.1109/ICDEW55742.2022.00022

Gonçalves, H. M., Silva, G. M., & Martins, T. G. (2018). Motivations for posting online reviews in the hotel industry. *Psychology and Marketing*, *35*(11), 807–817. doi:10.1002/mar.21136

Gong, X., Zhang, Y., Tang, J., Zhang, Y., & Li, G. (2017). Fabrication of a Molecularly Imprinted Polymer Film Coated Quartz Crystal Microbalance Sensor for Selective Detection of Nonylphenol. *Analytical Letters*, *50*(12), 1958–1968.

Goody, J. (2018). Industrial food: Towards the development of a world cuisine. In Food and culture. In C. Counihan, P. V. Estenik, & A. P. Julier, Food and Culture (pp. 263-282). Routledge.

Gopalakrishnan, P. K., Hall, J., & Behdad, S. (2021). Cost analysis and optimization of Blockchain-based solid waste management traceability system. *Waste Management (New York, N.Y.)*, *120*, 594–607. doi:10.1016/j.wasman.2020.10.027 PMID:33288397

Gordon, J. S., & Nyholm, S. (2021). Ethics of artificial intelligence. Internet Encyclopedia of Philosophy.

Görkem, O., Bayram, M., Bertan, S., & Bayram, Ü. (2016). Approach of the Young to Eco-Gastronomy: The Case of Turkish Cuisine. Global Issues and Trends in Tourism, 602.

Gossner, C. M.-E., Schlundt, J., Ben Embarek, P., Hird, S., Lo-Fo-Wong, D., Beltran, J. J. O., Teoh, K. N., & Tritscher, A. (2009). The melamine incident: Implications for international food and feed safety. *Environmental Health Perspectives*, *117*(12), 1803–1808. doi:10.1289/ehp.0900949 PMID:20049196

Govindan, K. (2018). Sustainable consumption and production in the food supply chain: A conceptual framework. *International Journal of Production Economics*, *195*, 419–431. doi:10.1016/j.ijpe.2017.03.003

Govindan, K., Nasr, A. K., Karimi, F., & Mina, H. (2022). Circular economy adoption barriers: An extended fuzzy best–worst method using fuzzy DEMATEL and Supermatrix structure. *Business Strategy and the Environment, 31*(4), 1566–1586. doi:10.1002/bse.2970

Graham, D. M. (1997). Use of ozone for food processing. *Food Technology, 6*(51), 72–75.

Graham-Rowe, E., Jessop, D. C., & Sparks, P. (2014). Identifying Motivations and Barriers to Minimising Household Food Waste. *Resources, Conservation and Recycling, 84*, 15–23. doi:10.1016/j.resconrec.2013.12.005

Green Restaurant Association. (2023a). Greening Restaurants Since 1990. Retrieved from https://www.dinegreen.com/about Access Date: 21.06.2023.

Green Restaurant Association. (2023b). Product & Service Certification Standards. Retrieved from https://www.dinegreen.com/product-certification-standards Access Date: 21.06.2023.

Green Restaurant Association. (2023c). Green your restaurant-Become a green vendor-Be a green consumer-Become a green employee. Retrieved from https://www.dinegreen.com/ Access Date: 21.06.2023.

Griffin, M., Sobal, J., & Lyson, T. A. (2009). An Analysis of a Community Food Waste Stream. *Agriculture and Human Values, 26*(1-2), 67–81. doi:10.100710460-008-9178-1

GTHB. (2018). Meyve ve Sebzelerin Depolama ve Nakliye Uygulamaları Eğitim Materyali. In Gıda Zincirindeki Hasat Sonrası Kayıpları Azaltmak için Yenilikçi Yaklaşımlar.

Gu, C., Huang, T., Wei, W., Yang, C., Chen, J., Miao, W., Lin, S., Sun, H., & Sun, J. (2023). The Effect of Using Augmented Reality Technology in Takeaway Food Packaging to Improve Young Consumers’ Negative Evaluations. *Agriculture 2023, 13*(2), 335. doi:10.3390/agriculture13020335

Güneş, S. G. (2019). Eco-gastronomy, Tourism and Sustainability: The Rise of Sustainable Restaurants in the World. *Erasmus International Academic Research Symposium on Educational and Social Sciences*, 5-6 April 2019, Izmir, (pp.67-84).

Gün, S., & Kılıç, G. D. (2022). Eko Gastronomi ve Sürdürülebilir Gastronomi (Eco Gastronomy and Sustainable Gastronomy). *Journal of Tourism and Gastronomy Studies, 10*(2), 999–1015.

Gupta, M., & Wong, W. L. E. (2007). Microwaves and metals. Singapore: John Wiley & Sons. doi:10.1002/9780470822746

Gupta, H., & Barua, M. K. (2016). Identifying enablers of technological innovation for Indian MSMEs using best–worst multi criteria decision making method. *Technological Forecasting and Social Change, 107*, 69–79. doi:10.1016/j.techfore.2016.03.028

Gupta, M. (2019). A Study on Impact of Online Food delivery app on Restaurant Business special reference to zomato and swiggy. *International Journal of Research and Analytical Reviews, 6*(1), 889–893. http://ijrar.com/

Gupta, P., Anand, S., & Gupta, H. (2017). Developing a roadmap to overcome barriers to energy efficiency in buildings using best worst method. *Sustainable Cities and Society, 31*, 244–259. doi:10.1016/j.scs.2017.02.005

Gupta, V., & Duggal, S. (2021). How the consumer's attitude and behavioural intentions are influenced: A case of online food delivery applications in India. *International Journal of Culture, Tourism and Hospitality Research, 15*(1), 77–93. doi:10.1108/IJCTHR-01-2020-0013

Gustavsson, J., Cederberg, C., & Sonesson, U. (2011). *Food Losses and Food Waste: Extent, Causes and Prevention.* FAO. https://www.fao.org/3/i2697e/i2697e.pdf:

Guzel-Seydim, Z. B., Greene, A. K., & Seydim, A. C. (2004). Use of ozone in the food industry. *Lebensm-Wiss-UTechnologie, 37*(4), 453–460. doi:10.1016/j.lwt.2003.10.014

Haji, M. (2020). Roles of Technology in Improving Perishable Food Supply Chains. *Logistics*. . doi:10.3390/logistics4040033

Haken, H. (1991). *Synergetic Computers and Cognition: A Top-Down Approach to Neural Nets*. Springer Series In Synergetics. https://books.google.com.tr/books?hl=en&lr=&id=-e6JfVRtLMwC&oi=fnd&pg=PA1&ots=iLJZne3UIh&sig=qDwqQH_BasNLUU0f1nLbWhlMK1g&redir_esc=y#v=onepage&q&f=false

Halupka, V., Almahr, A., Pan, Y., & Cheok, A. D. (2012). Chop chop: A sound augmented kitchen prototype. Lecture Notes in Computer Science (Including Subseries Lecture Notes in Artificial Intelligence and Lecture Notes in Bioinformatics), 7624 LNCS, 494–497. doi:10.1007/978-3-642-34292-9_43/COVER

Hammond, S. T., Brown, J. H., Burger, J. R., Flanagan, T. P., Fristoe, T. S., Silva, N. M., Nekola, J. C., & Okie, J. G. (2015). Food Spoilage, Storage, and Transport: Implications for a Sustainable Future. *Bioscience*, *65*(8), 758–768. doi:10.1093/biosci/biv081

Hamon, M. O., Oyarzabal, A., & Hong, J. W. (2013). Nanoliter/picoloter scale fluidic systems for food safety. In B. Park & M. Appell (Eds.), *Advances in applied nanotechnology for agriculture* (pp. 145–165). ACS. doi:10.1021/bk-2013-1143.ch008

Han, J. H. (2014). Edible Films and Coatings: A Review, Book Chapter. In *Food Science and Technology, Innovations in Food Packaging* (2nd ed., pp. 213–255). Academic Press. doi:10.1016/B978-0-12-394601-0.00009-6

Harnkarnsujarit, N., & Charoenrein, S. (2011). Influence of collapsed structure on stability of ß carotene in freeze-dried mangoes. *Food Reviews International*, *44*(10), 3188–3194. doi:10.1016/j.foodres.2011.08.008

Hassan, A. B., Mahmood, N. S., Elmamoun, K., Adiamo, Q., Isam, A., & Ahmed, M. (2017). Effects of gamma irradiation on the protein characteristics and functional properties of sesame (*Sesamum indicum* L.) seeds. *Radiation Physics and Chemistry*, *2017*, 144.

Hassanien, A. E. (2022). *Artificial Intelligence: A Real Opportunity in the Food Industry* (Vol. 1000). Springer Nature.

Hassoun, A., Jagtap, S., Trollman, H., Garcia-Garcia, G., Abdullah, N. A., Goksen, G., ... Lorenzo, J. M. (2022). Food processing 4.0: Current and future developments spurred by the fourth industrial revolution. *Food Control*, 109507.

Hawkes, J., & Flink, J. M. (1978). Osmotic concentration of fruit slices prior to freeze dehydration. *Journal of Food Processing and Preservation*, *2*(4), 265–284. doi:10.1111/j.1745-4549.1978.tb00562.x

Hearn, A. (2010). Structuring feeling: Web 2.0, online ranking and rating, and the digital reputation economy. *Ephemera*, *10*, 421–438.

Heck, S., Campos, H., Barker, I., Okello, J. J., Baral, A., Boy, E., Brown, L., & Birol, E. (2020). Resilient agri-food systems for nutrition amidst COVID-19: Evidence and lessons from food-based approaches to overcome micronutrient deficiency and rebuild livelihoods after crises. *Food Security*, *12*(4), 823–830. doi:10.100712571-020-01067-2 PMID:32839664

Heikkila, L., Reinikainen, A., Katajajuuri, J.-M., Silvennoinen, K., & Hartikainen, H. (2016). Elements Affecting Food Waste in the Food Service Sector. *Waste Management (New York, N.Y.)*, *56*, 446–453. doi:10.1016/j.wasman.2016.06.019 PMID:27373724

Hennig-Thurau, T., Gwinner, K. P., Walsh, G., & Gremler, D. D. (2004). Electronic word-of-mouth via consumer-opinion platforms: What motivates consumers to articulate themselves on the internet? *Journal of Interactive Marketing*, *18*(1), 38–52. doi:10.1002/dir.10073

Henson, S., & Caswell, J. (1999). Food safety regulation: An overview of contemporary issues. *Food Policy*, *24*(6), 589–603. doi:10.1016/S0306-9192(99)00072-X

Herman, L. E., Martínez-Sánchez, C. E., Pacheco-Angulo, H., Carmona-García, R., Ruiz-Espinosa, H., & Ruiz-López, I. I. (2013). Mass transfer modeling of equilibrium and dynamic periods during osmotic dehydration of radish in NaCl solutions. *Food and Bioproducts Processing*, *9*(3), 216–224. doi:10.1016/j.fbp.2012.10.001

He, S., & Shi, X. (2021). Microbial Food Safety in China: Past, Present, and Future. *Foodborne Pathogens and Disease*, *18*(8), 510–518. doi:10.1089/fpd.2021.0009 PMID:34242111

Hilken, T., de Ruyter, K., Chylinski, M., Mahr, D., & Keeling, D. I. (2017). Augmenting the eye of the beholder: Exploring the strategic potential of augmented reality to enhance online service experiences. *Journal of the Academy of Marketing Science*, *45*(6), 884–905. doi:10.100711747-017-0541-x

Hinton, T. J., Jallerat, Q., Palchesko, R. N., Park, J. H., Grodzicki, M. S., Shue, H. J., Ramadan, M. H., Hudson, A. R., & Feinberg, A. W. (2015). Three-dimensional printing of complex biological structures by freeform reversible embedding of suspended hydrogels. *Science Advances*, *1*(9), e1500758. doi:10.1126ciadv.1500758 PMID:26601312

HISPACOOP. (2012). *Estudio Sobre el Desperdicio de Alimentos en Los Hogares*. HISPACOOP. http://www.hispacoop.es/home/index.php?option=com_docman&task=doc_view&gi: onfederación Española de Cooperativas de Consumidores y Usuarios.

Hitkova, H., Stoyanova, A., Ivanova, N., Sredkova, M., Popova, V., Iordanova, R., & Bachvarova-Nedelcheva, A. (2012). Study of antibacterial activity of nonhydrolytic synthesized TiO_2 against E. coli, P. aeruginosa and S. aureus. *Journal of Optoelectronics and Biomedical Materials*, *4*(1), 9–17.

Hivos people unlimited (2023). Sustainable Diets For All. Retrieved from https://hivos.org/program/sustainable-diets-4-all/ Access Date: 23.06.2023.

Hof, B. (2021). The turtle and the mouse: How constructivist learning theory shaped artificial intelligence and educational technology in the 1960s. *History of Education*, *50*(1), 93–111. doi:10.1080/0046760X.2020.1826053

Horvath, J., & Cameron, R. (2020). *Mastering 3D Printing: A Guide to Modeling, Printing and Prototyping*. Apress Media. doi:10.1007/978-1-4842-5842-2

Hossain, M. M. (2022). Consumers Buying Intention towards Eco-Friendly Products: A Perspectives of Coronavirus Outbreak. *International Journal of Humanities and Social Science Review*, *8*(1), 11–24.

Howson, P. (2020). Building trust and equity in marine conservation and fisheries supply chain management with blockchain. *Marine Policy*, *115*, 103873. doi:10.1016/j.marpol.2020.103873

Huang, H. W., Hsu, C. P., Yang, B. B., & Wang, C. Y. (2013). Potential utility of high-pressure processing to address the risk of food allergen concerns. *Comprehensive Reviews in Food Science and Food Safety*, *13*(1), 78–90. doi:10.1111/1541-4337.12045 PMID:33412689

Huang, Y. W., & Toledo, R. (1982). Effect of high doses of high and low intensity UV irradiation on surface microbiological counts and storage life of fish. *Journal of Food Science*, *47*(5), 1667–1669. doi:10.1111/j.1365-2621.1982.tb05007.x

Huang, Y., Huang, L., Li, Y., Li, W., Li, X., & Li, Y. (2018). Molecularly imprinted polymers coupled with high-performance liquid chromatography for determination of sulfamethazine and its metabolite residues in food matrices. *Journal of Agricultural and Food Chemistry*, *66*(28), 7503–7510.

Huang, Z., Marra, F., & Wang, S. (2016). A novel strategy for improving radio frequency heating uniformity of dry food products using computational modeling. *Innovative Food Science & Emerging Technologies*, *34*, 100–111. doi:10.1016/j.ifset.2016.01.005

Hudson, S., & Gilbert, D. (2006). The Internet and small hospitality businesses: B&B marketing in Canada. *Journal of Hospitality & Leisure Marketing*, *14*(1), 99–116. doi:10.1300/J150v14n01_06

Hughes, L., Dwivedi, Y. K., Misra, S. K., Rana, N. P., Raghavan, V., & Akella, V. (2019). Blockchain research, practice and policy: Applications, benefits, limitations, emerging research themes and research agenda. *International Journal of Information Management*, *49*, 114–129. doi:10.1016/j.ijinfomgt.2019.02.005

Hung, Y. H., Chen, C. H., & Huang, S. W. (2017). Applying augmented reality to enhance learning: A study of different teaching materials. *Journal of Computer Assisted Learning*, *33*(3), 252–266. doi:10.1111/jcal.12173

Huppertz, T., Hinz, K., Zobrist, M. R., Uniacke, T., Kelly, A. L., & Fox, P. F. (2005). Effects of high pressure treatment on the rennet coagulation and cheese-making properties of heated milk. *Innovative Food Science & Emerging Technologies*, *6*(3), 279–285. doi:10.1016/j.ifset.2005.03.005

Hutto, C., & Gilbert, E. (2014). VADER: A Parsimonious Rule-Based Model for Sentiment Analysis of Social Media Text. *Proceedings of the International AAAI Conference on Web and Social Media*, *8*(1), 216-225. Available at: https://ojs.aaai.org/index.php/ICWSM/article/view/14550

IBM (2020) *Blockchain Component Overview, BM Blockchain Platform for IBM Cloud.* Available at: BM Blockchain Platform for IBM Cloud.

IBM. (2022) *Hyperledger Fabric Access Control System for Internet of Things Layer in Blockchain-Based Applications.* Available at: https://www.ibm.com/products/blockchain-platform-hyperledger-fabric

Ibrišimović, N., Ibrišimović, M., Kesić, A., & Pittner, F. (2015). Microbial biosensor: A new trend in the detection of bacterial contamination. *Monatshefte für Chemie*, *146*(8), 1363–1370. doi:10.100700706-015-1451-6

Idili, A., Amodio, A., Vidonis, M., Feinberg-Somerson, J., Castronovo, M., & Ricci, F. (2014). Folding-upon-binding and signal-on electrochemical DNA sensor with high affinity and specificity. *Analytical Chemistry*, *86*(18), 9013–9019. doi:10.1021/ac501418g PMID:24947124

Iftekhar, A. (2020). In L. Patarata (Ed.), *'Application of Blockchain and Internet of Things to Ensure Tamper-Proof Data Availability for Food Safety', Journal of Food Quality* (p. 5385207). doi:10.1155/2020/5385207

Iftekhar, A., Cui, X., Tao, Q., & Zheng, C. (2021). Hyperledger Fabric Access Control System for Internet of Things Layer in Blockchain-Based Applications. *Entropy (Basel, Switzerland)*, *23*(8), 1054. doi:10.3390/e23081054 PMID:34441194

İlkbahar, A., & Köksel, H. (2016). *Ekmek.* Ankara Halk Ekmek ve Un Fabrikası AŞ.

İlyasov, A. (2017). *Türk Misafirlerin Gıda İsrafı Tutumları Üzerine Beş Yıldızlı Otel İşletmelerinde Bir Araştırma: Alanya Örneği.* Süleyman Demirel Üniversitesi Sosyal Bilimler Enstitüsü.

Inamo. (2022). *Inamo London.* Inamo. http://www.inamorestaurant.com

Inbaraj, B. S., & Chen, B. H. (2016). Nanomaterial-based sensors for detection of foodborne bacterial pathogens and toxins as well as pork adulteration in meat products. *Yao Wu Shi Pin Fen Xi*, *24*(1), 15–28. PMID:28911398

Ines, G. R. M., Juliana, M. C., Araceli, O. M., Laura, S. T., & Olga, M. B. (2015). Long-term stability of food-grade nanoemulsions from high methoxyl pectin containing essential oils. *Food Hydrocolloids*, *52*, 438–446.

Inmon, W. H. (1996). The data warehouse and data mining. *Communications of the ACM*, *39*(11), 49–51. doi:10.1145/240455.240470

Intellidigest (2023). Course Catalog. Retrieved from https://intellidigest.com/courses/ Access Date: 20.06.2023.

Ionel, M. (2016). Hospitality industry. *Ovidius University Annals: Economic Sciences Series*, *1*(1), 187–191.

Islam, M. N., & Flink, J. N. (1982). Dehydration of potato II. Osmotic concentration and its effect on air drying behavior. *Journal of Food Technology*, *17*(3), 387–403. doi:10.1111/j.1365-2621.1982.tb00194.x

Ismail, S., Reza, H., Salameh, K., Kashani Zadeh, H., & Vasefi, F. (2023). Toward an Intelligent Blockchain IoT-Enabled Fish Supply Chain: A Review and Conceptual Framework. *Sensors (Basel)*, *23*(11), 5136. doi:10.339023115136 PMID:37299864

Ispirova, G., Eftimov, T., Koroušić Seljak, B., & Korošec, P. (2017). Mapping food composition data from various data sources to a domain-specific ontology. In *Proceedings of the 9th international Joint conference on knowledge discovery, knowledge engineering and knowledge management*. SCITEPRESS. 10.5220/0006504302030210

Ispirova, G., Eftimov, T., Korošec, P., & Koroušić Seljak, B. (2019). MIGHT: Statistical methodology for missing-data imputation in food composition databases. *Applied Sciences (Basel, Switzerland)*, *9*(19), 4111. doi:10.3390/app9194111

Ivanov, S., Gretzel, U., Berezina, K., Sigala, M., & Webster, C. (2019). Progress on robotics in hospitality and tourism: A review on the literature. *Journal of Hospitality and Tourism Technology*, *10*(4), 489–521. doi:10.1108/JHTT-08-2018-0087

Ivanov, S., & Webster, C. (2019). Perceived Appropriateness and Intention to Use Service Robots in Tourism. In J. Pesonen & J. Neidhardt (Eds.), *Information and Communication Technologies in Tourism*. Springer. (Original work published 2019), doi:10.1007/978-3-030-05940-8_19

Jagtap, S., Saxena, P., & Salonitis, K. (2021). Food 4.0: Implementing augmented reality systems in the food industry. *Procedia CIRP*, *104*, 1137–1142. doi:10.1016/j.procir.2021.11.191

Jain, P. K., Pamula, R., & Srivastava, G. (2021). *A systematic literature review on machine learning applications for consumer sentiment analysis using online reviews*. Elsevier BV. doi:10.1016/j.cosrev.2021.100413

Jakob, S. J., & Hansen, F. (2005). New Chemical and Biochemical Hurdles. In D. Sun (Ed.), *Emerging technologies for Food Technology* (pp. 387–418). Elsivier Ltd.

Jampilek, J., Kos, J., & Kralova, K. (2019). Potential of nanomaterial applications in dietary supplements and foods for special medical purposes. *Nanomaterials (Basel, Switzerland)*, *9*(2), 296. doi:10.3390/nano9020296 PMID:30791492

Jamroz, E., Kulawik, P., & Kopel, P. (2019). The effect of nanofillers on the functional properties of biopolymer-based films: A review. *Polymers*, *11*(4), 4. doi:10.3390/polym11040675 PMID:31013855

Javaid, M. (2021). Blockchain technology applications for Industry 4.0: A literature-based review. *Blockchain: Research and Applications*, *2*(4), 100027.

Javaid, M., Haleem, A., Singh, R. P., & Suman, R. (2022). Enhancing smart farming through the applications of Agriculture 4.0 technologies. *International Journal of Intelligent Networks*, *3*, 150–164. doi:10.1016/j.ijin.2022.09.004

Jayas, D. S. (2017). The role of sensors and bio-imaging in monitoring food quality. *Resource Magazine*, *2017*, 12.

Jayashankar, P., Johnston, W., & Nilakanta, S. (2020, July). A Conceptual Introduction to an AI-Based Food Marketing System: Navigating Assortment Expansion from Farm to Fork. In *the 45th Annual Macromarketing Conference* (*Vol. 35*).

Jee-Won, K., & Young, N. (2019). The role of personalization on continuance intention in food service mobile apps: A privacy calculus perspective. *International Journal of Contemporary Hospitality Management*, *31*(2), 734–752. doi:10.1108/IJCHM-12-2017-0783

Jeong, E., & Jang, S. S. (2011). Restaurant experiences triggering positive electronic word-of-mouth (eWOM) motivations. *International Journal of Hospitality Management*, *30*(2), 356–366. doi:10.1016/j.ijhm.2010.08.005

Jerath, K., Sajeesh, S., & John Zhang, Z. (2016). A model of unorganized and organized retailing in emerging economies. *Marketing Science*, *35*(5), 756–778. doi:10.1287/mksc.2015.0962

Jin, C., Sun, S., Yang, D., Sheng, W., Ma, Y., He, W., & Li, G. (2021). Anaerobic digestion: An alternative resource treatment option for food waste in China. *The Science of the Total Environment*, *779*, 146397. doi:10.1016/j.scitotenv.2021.146397 PMID:33743457

Joardder, M. U., & Masud, M. H. (2019). *Food Preservation in Developing Countries: Challenges and Solutions*. Cham.

Johnson, L., Levine, A., Smith, R., & Stone, S. (2010). The 2010 Horizon Report. In *New Media Consortium*. New Media Consortium. http://www.nmc.org

Juan, M. C., Charco, J. L., García-García, I., & Mollá, R. (2019). An Augmented Reality App to Learn to Interpret the Nutritional Information on Labels of Real Packaged Foods. *Frontiers of Computer Science*, *1*, 1. doi:10.3389/fcomp.2019.00001

Julio, N.-M., & Claudia, S.-S. (2019). Progress in wireless technologies in hospitality and tourism. *Journal of Hospitality and Tourism Technology*, *10*(4), 587–599. doi:10.1108/JHTT-07-2018-0061

KabaQ. (2021). *Augmented Reality Food*. kabaq.io

Kader, A. A. (2002). *Postharvest Technology of Horticultural Crops*. University of California Agriculture and Natural Resources.

Kader, A. A., & Watkins, C. B. (2000). Modified atmosphere packaging-toward 2000 and beyond. *HortTechnology*, *10*(3), 483–486. doi:10.21273/HORTTECH.10.3.483

Kakani, V., Nguyen, V. H., Kumar, B. P., Kim, H., & Pasupuleti, V. R. (2020). A critical review on computer vision and artificial intelligence in food industry. *Journal of Agriculture and Food Research*, *2*, 100033. doi:10.1016/j.jafr.2020.100033

Kamble, S., Gunasekaran, A., & Arha, H. (2019). Understanding the Blockchain technology adoption in supply chains-Indian context. *International Journal of Production Research*, *57*(7), 2009–2033. doi:10.1080/00207543.2018.1518610

Kamilaris, A., Fonts, A., & Prenafeta-Boldύ, F. X. (2019). The rise of blockchain technology in agriculture and food supply chains. *Trends in Food Science & Technology*, *91*, 640–652. doi:10.1016/j.tifs.2019.07.034

Kane, P. (2020). *Jan Smink: How 3D printing is transforming fine dining*. CNN. https://www.cnn.com/style/article/3d-printing-fine-dining-jan-smink/index.html

Kang, J. (2018). Effective marketing outcomes of hotel Facebook pages: The role of active participation and satisfaction. *Journal of Hospitality and Tourism Insights*, *1*(2), 106–120. doi:10.1108/JHTI-10-2017-0003

Kaplan, A. M., & Haenlein, M. (2010). Users of the world, unite! The challenges and opportunities of Social Media. *Business Horizons*, *53*(1), 59–68. doi:10.1016/j.bushor.2009.09.003

Karam, M. C., Petit, J., Zimmer, D., Djantou, E. B., & Scher, J. (2016). Effects of drying and grinding in production of fruit and vegetable powders: A review. *Journal of Food Engineering*, *188*, 32–49. doi:10.1016/j.jfoodeng.2016.05.001

Kasuriya S. and Atong D. (2004). Rapid Drying of Ceramic and Efficient Food Processing with a Continuous Microwave Belt Furnace, 27-29.

Katajajuuri, J.-M., Silvennoinen, K., Hartikainen, H., Heikkilä, L., & Reinikainen, A. (2014). Food Waste in the Finnish Food Chain. *Journal of Cleaner Production*, *73*, 322–329. doi:10.1016/j.jclepro.2013.12.057

Kaul, V., Enslin, S., & Gross, S. A. (2020). History of artificial intelligence in medicine. *Gastrointestinal Endoscopy*, *92*(4), 807–812. doi:10.1016/j.gie.2020.06.040 PMID:32565184

Kaur, A., Singh, G., Kukreja, V., Sharma, S., Singh, S., & Yoon, B. (2022). Adaptation of IoT with Blockchain in Food Supply Chain Management: An Analysis-Based Review in Development, Benefits and Potential Applications. *Sensors (Basel)*, *22*(21), 8174. doi:10.339022218174 PMID:36365871

Kaur, P., Dhir, A., Talwar, S., & Ghuman, K. (2021). The value proposition of food delivery apps from the perspective of theory of consumption value. *International Journal of Contemporary Hospitality Management*, *33*(4), 1129–1159. doi:10.1108/IJCHM-05-2020-0477

Kawashima, N., Yagi, T., & Kojima, K. (2019). How Do Bioplastics and Fossil-Based Plastics Play in a Circular Economy? *Macromolecular Materials and Engineering*, *304*(9), 1900383. doi:10.1002/mame.201900383

Kayikci, Y., Subramanian, N., Dora, M., & Bhatia, M. S. (2022). Food supply chain in the era of Industry 4.0: Blockchain technology implementation opportunities and impediments from the perspective of people, process, performance, and technology. *Production Planning and Control*, *33*(2-3), 301–321. doi:10.1080/09537287.2020.1810757

Kayıkcı, Y., Usar, D. D., & Aylak, B. L. (2021). Using blockchain technology to drive operational excellence in perishable food supply chains during outbreaks. *International Journal of Logistics Management*.

Keogh, J. G., Dube, L., Rejeb, A., Hand, K. J., Khan, N., & Dean, K. (2020). The future food chain: digitization as an enabler of Society 5.0. *Building the Future of Food Safety Technology*, 11–38. doi:10.1016/B978-0-12-818956-6.00002-6

Khadre, M. A., & Yousef, A. E. (2001). Decontamination of a multilaminated aseptic food packaging material and stainless steel by ozone. *Journal of Food Safety*, *21*(1), 1–13. doi:10.1111/j.1745-4565.2001.tb00304.x

Khaliq, A. (2022). Determinants of customers' intention to use online food delivery application through smart phone in Malaysia. *British Food Journal*, *124*(3), 732–753. doi:10.1108/BFJ-01-2021-0075

Khamidovich, Z. K. (2021). Pandemic and trends in the restaurant business in 2021. *Central Asian Journal of Innovations on Tourism Management and Finance*, *2*(6), 54–61. doi:10.47494/cajitmf.v2i6.129

Khan, A. A., Badshah, S., Liang, P., Waseem, M., Khan, B., Ahmad, A., & Akbar, M. A. (2022, June). Ethics of AI: A systematic literature review of principles and challenges. In *Proceedings of the 26th International Conference on Evaluation and Assessment in Software Engineering* (pp. 383-392). 10.1145/3530019.3531329

Khan, M. R., Volpe, S., Valentino, M., Miele, N. A., Cavella, S., & Torrieri, E. (2021). Active Casein Coatings and Films for Perishable Foods: Structural Properties and Shelf-Life Extension. *Coatings*, *11*(8), 899. doi:10.3390/coatings11080899

Khan, S., Kaushik, M. K., Kumar, R., & Khan, W. (2022). Investigating the barriers of blockchain technology integrated food supply chain: A BWM approach. *Benchmarking*.

Kickbusch, I., Allen, L., & Franz, C. (2016). The commercial determinants of health. *The Lancet. Global Health*, *4*(12), e895–e896. doi:10.1016/S2214-109X(16)30217-0 PMID:27855860

Kiddon, C., Zettlemoyer, L., & Choi, Y. (2016, November). Globally coherent text generation with neural checklist models. In Proceedings of the 2016 conference on empirical methods in natural language processing (pp. 329-339). ACL Anthology. 10.18653/v1/D16-1032

Kılınç Şahin, S., & Bekar, A. (2018). Küresel Bir Sorun "Gıda Atıkları": Otel İşletmelerindeki Boyutları. *Journal of Tourism and Gastronomy Studies*, *6*(4), 1039–1061.

Kim, H., Jung, S., & Ryu, G. (2020). A study on the restaurant recommendation service app based on AI chatbot using personalization information. *International Journal of Advanced Culture Technology (IJACT), 8*(4), 263-270. doi:https://www.earticle.net/Article/A388080

Kim, B., & Velthuis, O. (2021). From reactivity to reputation management: Online consumer review systems in the restaurant industry. *Journal of Cultural Economics, 14*(6), 675–693. doi:10.1080/17530350.2021.1895280

Kim, B., Yun, H., Jung, S., Jung, Y., Jung, H., Choe, W., & Jo, C. (2011). Effect of atmospheric pressure plasma on inactivation of pathogens inoculated onto bacon using two different gas compositions. *Food Microbiology, 28*(1), 9–13. doi:10.1016/j.fm.2010.07.022 PMID:21056769

Kim, H. W., Bae, H., & Park, H. J. (2017). Classification of The Printability of Selected Food for 3D Printing: Development of An Assessment Method Using Hydrocolloids As Reference Material. *Journal of Food Engineering, 215*, 23–32. doi:10.1016/j.jfoodeng.2017.07.017

Kim, J. G., Yousef, A. E., & Dave, S. (1999). Application of ozone for enhancing the microbiological safety and quality of foods: A review. *Journal of Food Protection, 62*(9), 1071–1087. doi:10.4315/0362-028X-62.9.1071 PMID:10492485

Knijnenburg, B. P. (2022). *Modern Socio-Technical Perspectives on Privacy*. Modern Socio-Technical Perspectives on Privacy. doi:10.1007/978-3-030-82786-1

Knorr, D., Augustin, M. A., & Tiwari, B. (2020). Advancing the Role of Food Processing for Improved Integration in Sustainable Food Chains. *Frontiers in Nutrition, 7*(April), 1–8. doi:10.3389/fnut.2020.00034 PMID:32309287

Knorr, D., Eshtiaghi, M. N., & Rastogi, N. K. (1999). Accelerated mass transfer during osmotic dehydration of high intensity electrical field pulse pretreated carrots. *Journal of Food Science, 64*(6), 1020–1023. doi:10.1111/j.1365-2621.1999.tb12272.x

Knorr, D., Froehling, A., Jaeger, H., Reineke, K., Schlueter, O., & Schoessler, K. (2011). Emerging technologies in food processing. *Annual Review of Food Science and Technology, 2*(1), 203–235. doi:10.1146/annurev.food.102308.124129 PMID:22129381

Kocacık, F. (1998). *Tüketim Eğilimleri Ve Sorunları*. Cumhuriyet Üniversitesi Yayınları.

Kocira, A., Kozłowicz, K., Panasiewicz, K., Staniak, M., Szpunar-Krok, E., & Hortyńska, P. (2021). Polysaccharides as Edible Films and Coatings: Characteristics and Influence on Fruit and Vegetable Quality—A Review. *Agronomy (Basel), 11*(5), 813. doi:10.3390/agronomy11050813

Kodandaram, R. D., Waghray, K., & Sathyanarayana, S. V. (2020). Infrared Heating - A New Green Technology for Process Intensification in Drying of Purslane Leaves to Reduce the Thermal Losses. In *International Conference on Emerging Trends in Engineering*. Springer Nature: Switzerland. 10.1007/978-3-030-24314-2_56

Köhler, S., & Pizzol, M. (2020). Technology assessment of blockchain-based technologies in the food supply chain. *Journal of Cleaner Production, 269*, 122193. doi:10.1016/j.jclepro.2020.122193

Komaiko, J. S., & McClements, D. J. (2016). Formation of food-grade nanoemulsions using low-energy preparation methods: A review of available methods. *Comprehens Rev Food Sci Safety, 15*(2), 331–352. doi:10.1111/1541-4337.12189 PMID:33371595

Korachi, M., Ozen, F., Aslan, N., Vannini, L., Maria, E. G., Davide, G., & Fatma, Y. E. (2015). Biochemical changes to milk following treatment by a novel, cold atmospheric plasma system. *International Dairy Journal, 42*, 64–69. doi:10.1016/j.idairyj.2014.10.006

Kosseva, M. R. (2013). Introduction: causes and challenges of food wastage. *Food Industry Wastes,* xv-xxiv.

Kouhizadeh, M., Saberi, S., & Sarkis, J. (2021). Blockchain technology and the sustainable supply chain: Theoretically exploring adoption barriers. *International Journal of Production Economics*, *231*, 107831. doi:10.1016/j.ijpe.2020.107831

Kounavis, C. D., Kasimati, A. E., & Zamani, E. D. (2012). Enhancing the Tourism Experience through Mobile Augmented Reality: Challenges and Prospects. *International Journal of Engineering Business Management*, *4*, 1–6. doi:10.5772/51644

Kovačević, D. B., Putnik, P., Dragović-Uzelac, V., Pedisić, S., Jambrak, A. R., & Herceg, Z. (2016). Effects of cold atmospheric gas phase plasma on anthocyanins and color in pomegranate juice. *Food Chemistry*, *190*, 317–323. doi:10.1016/j.foodchem.2015.05.099 PMID:26212976

Kowalska, H. (2006). Kinetykaosmotycznegoodwadnianiadyni. *Żywność. Nauka. Technologia. Jakość*, *2*(47), 135–142.

Kozinets, R. V. (2002). The field behind the screen: using netnography for marketing research in outline communities. *Journal of Marketing Research.*

Krippendorff, K. (1980). Validity in content analysis. In E. Mochmann (Ed.), *Computerstrategien fur die kommunikations analyse* [Computer strategies for communication analysis]. (pp. 69–112). Campus.

Krishnamurthy, K., Khurana, H. K., Soojin, J., Irudayaraj, J., & Demirci, A. (2008). Infrared heating in food processing: An overview. *Comprehensive Reviews in Food Science and Food Safety*, *7*(1), 2–13. doi:10.1111/j.1541-4337.2007.00024.x

Kshetri, N. (2018). 1 Blockchain's roles in meeting key supply chain management objectives. *International Journal of Information Management*, *39*, 80–89. doi:10.1016/j.ijinfomgt.2017.12.005

Kujur, Rubi.(2019). Satisfaction of consumers by using online food services. *International Journal of Humanities and Social Sciences* (IJHSS) ISSN (P): 2319-393X; ISSN(E): 2319-3948 8(4) Jun - Jul 2019; 35-44.

Kumar, A., Liu, R., & Shan, Z. (2020). Is blockchain a silver bullet for supply chain management? Technical challenges and research opportunities. *Decision Sciences*, *51*(1), 8–37. doi:10.1111/deci.12396

Kumar, I., Rawat, J., Mohd, N., & Husain, S. (2021). Opportunities of artificial intelligence and machine learning in the food industry. *Journal of Food Quality*, *2021*, 1–10. doi:10.1155/2021/4535567

Kumari, L., Narsaiah, K., Grewal, M. K., & Anurag, R. K. (2015). Application of RFID in agri-food sector. *Trends in Food Science & Technology*, *43*(2), 144–161. doi:10.1016/j.tifs.2015.02.005

Kumar, M., Choubey, V. K., Raut, R. D., & Jagtap, S. (2023). Enablers to achieve zero hunger through IoT and blockchain technology and transform the green food supply chain systems. *Journal of Cleaner Production*, *405*, 136894. doi:10.1016/j.jclepro.2023.136894

Kumar, R. L., Khan, F., Kadry, S., & Rho, S. (2022). A Survey on blockchain for industrial Internet of Things. *Alexandria Engineering Journal*, *61*(8), 6001–6022. doi:10.1016/j.aej.2021.11.023

Kumar, S., Raut, R. D., Agrawal, N., Cheikhrouhou, N., Sharma, M., & Daim, T. (2022). Integrated blockchain and internet of things in the food supply chain: Adoption barriers. *Technovation*, *118*, 102589. doi:10.1016/j.technovation.2022.102589

Kumar, V. V., & Anthony, S. P. (2016). *Antimicrobial studies of metal and metal oxide nanoparticles. Surface Chemistry of Nanobiomaterials: Applications of Nanobiomaterials*. William Andrew Publishing., doi:10.1016/B978-0-323-42861-3.00009-1

Kummu, M., de Moel, H., Porkka, M., Siebert, S., Varis, O., & Ward, P. J. (2012). Lost Food, Wasted Resources: Global Food Supply Chain Losses and Their İmpacts on Freshwater, Cropland, and Fertiliser Use. *The Science of the Total Environment*, *438*, 477–488. doi:10.1016/j.scitotenv.2012.08.092 PMID:23032564

Kuo, T. C., & Smith, S. (2018). A systematic review of technologies involving eco-innovation for enterprises moving towards sustainability. *Journal of Cleaner Production, 192*, 207–220. doi:10.1016/j.jclepro.2018.04.212

Kurt, Y. (2019). *Yiyecek İçecek Personelinin Gıda İsrafı Hakkında Bilgi, Görüş ve Davranışları: Eskişehir İli Örneği.* Gazi Üniversitesi Sosyal Bilimler Enstitüsü.

Kutyauripo, I., Rushambwa, M., & Chiwazi, L. (2023). Artificial intelligence applications in the agrifood sectors. *Journal of Agriculture and Food Research, 11*, 100502. doi:10.1016/j.jafr.2023.100502

Kyaw, K. S., Adegoke, S. C., Ajani, C. K., Nwabor, O. F., & Onyeaka, H. (2022). Toward in-process technology-aided automation for enhanced microbial food safety and quality assurance in milk and beverages processing. *Critical Reviews in Food Science and Nutrition*, 1–21. doi:10.1080/10408398.2022.2118660 PMID:36066463

Lagnika, C., Huang, J., Jiang, N., Li, D., Liu, C., Song, J., Wei, Q., & Zhang, M. (2018). Ultrasound-assisted osmotic process on quality of microwave vacuum drying sweet potato. *Drying Technology, 36*(11), 1367–1379. doi:10.1080/07373937.2017.1402786

Lai, I. K. W., Liu, Y., & Lu, D. (2021b). The effects of tourists' destination culinary experience on electronic word-of-mouth generation intention: The experience economy theory. *Asia Pacific Journal of Tourism Research, 26*(3), 231–244. doi:10.1080/10941665.2020.1851273

Lai, X., Wang, F., & Wang, X. (2021a). Asymmetric relationship between customer sentiment and online hotel ratings: The moderating effects of review characteristics. *International Journal of Contemporary Hospitality Management, 33*(6), 2137–2156. doi:10.1108/IJCHM-07-2020-0708

Lai, Y., Wu, S., Cheng, S., Chen, Y., Zhou, J., & Lai, H. (2020). Selective and sensitive detection of gliadin in wheat flour by molecularly imprinted polymers. *Food Chemistry, 309*, 125705.

Lambert, F. (2019). *3D printed food: A comprehensive review.* Science Direct. https://www.sciencedirect.com/science/article/pii/S0924224419305242

Law, R., Chan, I. C., & Wang, L. (2018). A comprehensive review of mobile technology use in hospitality and tourism. *Journal of Hospitality Marketing & Management, 27*(6), 626–648. doi:10.1080/19368623.2018.1423251

Lawrence, R. L., & Ken, M. W. (1988). The Effect of Multiple Measures of Age In Segmenting Hotel Markets. *Journal of Hospitality & Tourism Research (Washington, D.C.), 12*(2), 91–98. doi:10.1177/109634808801200210

Le-Bail, A., Maniglia, B. C., & Le-Bail, P. (2020). Recent advances and future perspective in additive manufacturing of foods based on 3D printing. *Current Opinion in Food Science, 35*, 54–64. doi:10.1016/j.cofs.2020.01.009

LeCun, Y., Bottou, L., Bengio, Y., & Haffner, P. (1998). Gradient-based learning applied to document recognition. *Proceedings of the IEEE, 86*(11), 2278–2324. doi:10.1109/5.726791

Lee, S. W. Y., Tsai, C. C., Wu, Y. T., Tsai, M. J., Liu, T. C., Hwang, F. K., Lai, C. H., Liang, J. C., Wu, H. C., & Chang, C. Y. (2011). *Internet-based Science Learning: A review of journal publications.* doi:10.1080/09500693.2010.536998

Lee, H., Kim, J. E., Chung, M. S., & Min, S. C. (2015). Cold plasma treatment for the microbiological safety of cabbage, lettuce, and dried figs. *Food Microbiology, 51*, 74–80. doi:10.1016/j.fm.2015.05.004 PMID:26187830

Lee, H., Van Dolen, W., & Kolk, A. (2013). On the role of social media in the 'responsible' food business: Blogger buzz on health and obesity issues. *Journal of Business Ethics, 118*(4), 695–707. doi:10.100710551-013-1955-0

Lee, J. Y., & Deniniger, R. A. (1999). Survival of bacteria after ozonation. *Ozone Science and Engineering, 22*(1), 65–75. doi:10.1080/01919510008547229

Lee, M., Kwon, W., & Back, K. J. (2021). Artificial intelligence for hospitality big data analytics: Developing a prediction model of restaurant review helpfulness for customer decision-making. *International Journal of Contemporary Hospitality Management*, *33*(6), 2117–2136. doi:10.1108/IJCHM-06-2020-0587

Lee, S. (2020). A study on the user experience of unmanned payment kiosk systems in fast food restaurants. *Int. J. Smart Bus. Technol*, *8*(1), 1–6. doi:10.21742/IJSBT.2020.8.1.01

Lee, W. L., Liu, C. H., & Tseng, T. W. (2022). The multiple effects of service innovation and quality on transitional and electronic word-of-mouth in predicting customer behavior. *Journal of Retailing and Consumer Services*, *64*, 102791. doi:10.1016/j.jretconser.2021.102791

Lei, S., & Law, R. (2015). Content analysis of TripAdvisor reviews on restaurants: A case study of Macau. *Journal of Tourism*, *16*(1), 17-28.

Leonard, A., Blacher, S., Nimmol, C., & Devahastin, S. (2008). Effect of far-infrared radiation assisted drying on microstructure of banana slices: An illustrative use of X-ray microtomography in microstructural evaluation of a food product. *Journal of Food Engineering*, *85*(1), 154–162. doi:10.1016/j.jfoodeng.2007.07.017

Le, S., Bazger, W., Hill, A. R., & Wilcock, A. (2014). Awareness and perceptions of food safety of artisan cheese makers in Southwestern Ontario: A qualitative study. *Food Control*, *41*(1), 158–167. doi:10.1016/j.foodcont.2014.01.007

Leung, X. Y., & Wen, H. (2021). How emotions affect restaurant digital ordering experiences: A comparison of three ordering methods. *Journal of Hospitality and Tourism Technology*, *12*(3), 439–453. doi:10.1108/JHTT-05-2020-0117

Lewicki, P. P., & Lenart, A. (2015). Osmotic dehydration of fruits and vegetables. In A. S. Mujumdar (Ed.), *Handbook of industrial drying* (4th ed., pp. 661–674). CRC Press.

Lewis, T., & Phillipov, M. (2018). Food/media: Eating, cooking, and provisioning in a digital world. *Communication Research and Practice*, *4*(3), 207–211. doi:10.1080/22041451.2018.1482075

Lezoche, M., Hernandez, J. E., Díaz, M. D. M. E. A., Panetto, H., & Kacprzyk, J. (2020). Agri-food 4.0: A survey of the supply chains and technologies for the future agriculture. *Computers in Industry*, *117*, 103187. doi:10.1016/j.compind.2020.103187

Li, D., Wang, X., Chan, H. K., & Manzini, R. (2014). Sustainable food supply chain management. *International Journal of Production Economics*, *152*, 1–8. doi:10.1016/j.ijpe.2014.04.003

Li, H., Meng, F., & Pan, B. (2020). How does review disconfirmation influence customer online review behavior? A mixed-method investigation. *International Journal of Contemporary Hospitality Management*, *32*(11), 3685–3703. https://www.researchgate.net/profile/Bing-Pan-5/publication/349913300_An_exploratory_study_on_Chinese_tourists%27_visitation_to_a_US_National_Park/links/62758c473a23744a72693115/An-exploratory-study-on-Chinese-tourists-visitation-to-a-US-National-Park.pdf. doi:10.1108/IJCHM-03-2020-0234

Li, K., Lee, J. Y., & Gharehgozli, A. (2021). Blockchain in food supply chains: A literature review and synthesis analysis of platforms, benefits and challenges. *International Journal of Production Research*, 1–20.

Lille, M., Nurmela, A., Nordlund, E., Metsa-Kortelainen, S., & Sozer, N. (2018). Applicability of protein and fiber-rich food materials in extrusion-based 3D printing. *Journal of Food Engineering, 220*, 20–27. . JFOODENG.2017.04.034 doi:10.1016/J

Lipinski, B., Hanson, C., Waite, R., & Searchinger, T. (2013). *Reducing Food Loss and Waste.* (World Resources Institute Working Paper, 1-40). World Resources Institute Working Paper. https://files.wri.org/d8/s3fs-public/reducing_food_loss_and_waste.pdf

Lipton, J. I., Cutler, M., Nigl, F., Cohen, D., & Lipson, H. (2015). Additive Manufacturing for The Food Industry. *Trends in Food Science & Technology*, *43*(1), 114–123. doi:10.1016/j.tifs.2015.02.004

Lipton, J., Arnold, D., Nigl, F., Lopez, N., Cohen, D., Norén, N., & Lipson, H. (2010, September). Multi-material Food Printing with Complex Internal Structure Suitable for Conventional Post-Processing. In *2010 International Solid Freeform Fabrication Symposium*. University of Texas at Austin.

Litvin, S. W., Goldsmith, R. E., & Pan, B. (2008). Electronic word-of-mouth in hospitality and tourism management. *Tourism Management*, *29*(3), 458–468. doi:10.1016/j.tourman.2007.05.011

Liu, F., Zhang, X., Xie, F., Liu, J., Jiang, X., Luo, L., & Huang, X. (2017). Highly sensitive detection of Escherichia coli O157: H7 using molecularly imprinted electrochemical sensor based on PAMAM dendrimers. *Biosensors & Bioelectronics*, *92*, 235–241. doi:10.1016/j.bios.2017.02.041 PMID:28285201

Liu, W., & Lanier, T. C. (2016). Rapid (microwave) heating rate effects on texture, fat/water holding, and microstructure of cooked comminuted meat batters. *Food Research International*, *81*, 108–113. doi:10.1016/j.foodres.2016.01.005

Liu, Z., Zhang, M., Bhandari, B., & Wang, Y. (2017). 3D printing: Printing precision and application in food sector. *Trends in Food Science & Technology*, *69*, 83–94. doi:10.1016/j.tifs.2017.08.018

Li, Y.-H., Zheng, J., Yue, S.-T., & Fan, Z.-P. (2021). Capturing and analyzing e-WOM for travel products: A method based on sentiment analysis and stochastic dominance. *Kybernetes*. Advance online publication. doi:10.1108/K-02-2021-0146

Loaharanu, P., & Ahmad, M. (1991). Advantages and disadvantages of the use of irradiation for Food Preservation. *Journal of Agricultural & Environmental Ethics*, *4*(1), 14–30. doi:10.1007/BF02229144

Logaldo, M. (2014). *Promoting Augmented Reality in the Food Industry: A Linguistic Analysis of Web-Based Communication*. Food and Identity in Literature and Culture.

Lohr, S., & Markoff, J. (2011). Computer Wins on 'Jeopardy!': Trivial, It's Not. *The New York Times*. https://www.nytimes.com/2011/02/17/science/17jeopardy-watson.html

Lo, I. M., & Woon, K. S. (2016). Food waste collection and recycling for value-added products: Potential applications and challenges in Hong Kong. *Environmental Science and Pollution Research International*, *23*(8), 7081–7091. doi:10.100711356-015-4235-y PMID:25772864

Lopez-Sanchez, A., Luque-Badillo, A. C., Orozco-Nunnely, D., Alencastro-Larios, N. S., Ruiz-Gomez, J. A., Garcia-Cayuela, T., & Gradilla-Hernandez, M. S. (2021). Food Loss in the Agricultural Sector of a Developing Country: Transitioning to a More Sustainable Approach. The Case of Jalisco, Mexico. *Environmental Challenges*, *5*, 1–16. doi:10.1016/j.envc.2021.100327

Lu, C. H., Yang, H. H., & Zhu, C. L. (2019). Molecularly imprinted polymers for the analysis of food contaminants. *Trends in Analytical Chemistry*, *111*, 61–70.

Lukanova, G., & Ilieva, G. (2019). Robots, Artificial Intelligence, and Service Automation in Hotels. Ivanov, S. and Webster, C. (Ed.) Robots, Artificial Intelligence, and Service Automation in Travel, Tourism and Hospitality. Emerald Publishing Limited, Bingley. doi:10.1108/978-1-78756-687-320191009

Lundberg, D. E. (1989). *The hotel and restaurant business* (Vol. 5). Van Nostrand Reinhold.

Luque, A., Peralta, M. E., de las Heras, A., & Córdoba, A. (2017). State of the Industry 4.0 in the Andalusian food sector. *Procedia Manufacturing*, *13*, 1199–1205. doi:10.1016/j.promfg.2017.09.195

Macarthur, S., & Hemmings, F. J. (2017). Fibres, Yarns and Fabrics: An Introduction to Production, Structure and Properties. In Forensic Examination of Fibres, 3rd Edition. CRC Press.

Maftei, A. A., Lavric, A., Petrariu, A. I., & Popa, V. (2023). Massive Data Storage Solution for IoT Devices Using Blockchain Technologies. *Sensors (Basel)*, *23*(3), 1570. doi:10.339023031570 PMID:36772609

Mahalik, N. P., & Nambiar, A. N. (2010). Trends in food packaging and manufacturing systems and technology. *Trends in Food Science & Technology*, *21*(3), 117–128. doi:10.1016/j.tifs.2009.12.006

Mahapatra, A. K., Muthukumarappan, K., & Julson, I. J. (2007). Applications of ozone, bacteriosins and irradiation in Food processing: A review. *Critical Reviews in Food Science and Technology*, *45*(6), 447–461. doi:10.1080/10408390591034454

Mahdi, J. S., Yinghe, H., & Bhandari, B. (2007). Production of sub-micron emulsions by ultrasound and microfluidization techniques. *Journal of Food Engineering*, *82*(4), 478–488. doi:10.1016/j.jfoodeng.2007.03.007

Mairhofer, J., Roppert, K., & Ertl, P. (2009). Microfluidic systems for pathogen sensing: A review. *Sensors (Basel)*, *9*(6), 4804–4823. doi:10.339090604804 PMID:22408555

Malek, J., & Desai, T. N. (2019). Prioritization of sustainable manufacturing barriers using Best Worst Method. *Journal of Cleaner Production*, *226*, 589–600. doi:10.1016/j.jclepro.2019.04.056

Maloney, N., & Harrison, M. (2016). Advanced heating technologies for food processing A2. In C. E. Leadley (Ed.), *Innovation and Future Trends in Food Manufacturing and Supply Chain Technologies* (pp. 203–256). Woodhead Publishing Series in Food Science, Technology and Nutrition. doi:10.1016/B978-1-78242-447-5.00008-3

Mandić, A., Pivčević, S., & Petrić, L. (2020). Restaurant online reputation and destination competitiveness: insight into TripAdvisor data. In *Gastronomy for Tourism Development* (pp. 155–184). Emerald Publishing Limited. doi:10.1108/978-1-78973-755-420201011

Mangaraj, S., & Goswami, T. (2009). Modified atmosphere packaging of fruits and vegetables for extending shelf-life: A review. *Fresh Produce*, *3*(1), 1–31.

Mangla, S. K., Kazançoğlu, Y., Yıldızbaşı, A., Öztürk, C., & Çalık, A. (2022). A conceptual framework for blockchain-based sustainable supply chain and evaluating implementation barriers: A case of the tea supply chain. *Business Strategy and the Environment*, *31*(8), 3693–3716. doi:10.1002/bse.3027

Mao, X., & Huang, T. J. (2012). Microfluidic diagnostics for the developing world. *Lab on a Chip*, *12*(8), 1412–1416. doi:10.1039/c2lc90022j PMID:22406768

Marazuela, M. D., Moreno-Bondi, M. C., & Orellana, G. (2018). Molecularly imprinted polymers: Powerful tools in biosensor design. *Biosensors & Bioelectronics*, *117*, 720–733.

María-Eugenia, R.-M., Irene, G.-S., & Gloria, B.-C. (2014). Information and Communication Technology as a Differentiation Tool in Restaurants. *Journal of Foodservice Business Research*, *17*(5), 410–428. doi:10.1080/15378020.2014.967639

Marina, B., Liudmila, L., Alexander, S., Olga, B., & Natalyia, A. (2022). Innovative Engineering Developments of Equipment and Vending Machines. *Networked Control Systems for Connected and Automated Vehicles*, 1675–1682. doi:10.1007/978-3-031-11058-0_170

Marın, J., Biswas, A., Ofli, F., Hynes, N., Salvador, A., Aytar, Y., Weber, I., & Torralba, A. (2021). Recipe1m+: A dataset for learning cross-modal embeddings for cooking recipes and food images. *IEEE Transactions on Pattern Analysis and Machine Intelligence*, *43*(1), 187–203. doi:10.1109/TPAMI.2019.2927476 PMID:31295105

Marlowe, D. E. (2005). Nanotechnology and the U.S. Food and Drug Administration. *ASTM Standardization News*, *33*, 29–31. PMID:16237872

Marra, F. (2013). *Fighting Food Loss and Food Waste in Japan.* Leiden University: M. A. in Japanese Studies –. *Asian Studies (Quezon City, Philippines)*, 2011–2013.

Marra, F., Zhang, L., & Lyng, J. (2009). Radio frequency treatment of foods: Review of recent advances. *Journal of Food Engineering*, *91*(4), 497–508. doi:10.1016/j.jfoodeng.2008.10.015

Mashayekhy, Y. (2022). *Impact of Internet of Things (IoT) on Inventory Management: A Literature Survey*. Logistics. doi:10.3390/logistics6020033

Masoud, E. M., Liu, L., & Peng, B. (2020). Synthesis, Characterization, and Applications of Polymer Nanocomposites, *Journal of Nanomaterials. Article ID*, *5439136*, 1–2. *Advance online publication.* doi:10.1155/2020/5439136

Masud, M. H., Karim, A., Ananno, A. A., & Ahmed, A. (2020). *Sustainable food drying techniques in developing countries: prospects and challenges*. Springer International Publishing. doi:10.1007/978-3-030-42476-3

Mateo-Babiano, I., & Palipane, K. (2020). Reimagining place through the sandbox studio pedagogy: An introduction. Placemaking sandbox. *Emergent approaches, techniques and practices to create more thriving places*, 1-14.

Mathath, A., & Fernando, Y. (2015). Robotic transformation and its business applications in food industry. In Z. Luo (Ed.), *Robotics, Automation, and Control in Onustrial and Service Settings* (pp. 281–305). IGI Global. doi:10.4018/978-1-4666-8693-9.ch011

Matranga, V., & Corsi, I. (2012). Toxic effects of engineered nanoparticles in the marine environment: Model organisms and molecular approaches. *Marine Environmental Research*, *76*, 32–40. doi:10.1016/j.marenvres.2012.01.006 PMID:22391237

Mavani, N. R., Ali, J. M., Othman, S., Hussain, M. A., Hashim, H., & Rahman, N. A. (2022). Application of artificial intelligence in food industry—A guideline. *Food Engineering Reviews*, *14*(1), 134–175. doi:10.100712393-021-09290-z

McAdams, B., von Massow, M., Gallant, M., & Hayhoe, M. A. (2019). A cross industry evaluation of food waste in restaurants. *Journal of Foodservice Business Research*, *22*(5), 449–466. doi:10.1080/15378020.2019.1637220

McClements, D. J. (2019). *Future Foods: How Modern Science Is Transforming the way We Eat.* Switerzerland, Springer Scientific. doi:10.1007/978-3-030-12995-8

McClements, D. J., Decker, E. A., & Weiss, J. (2007). Emulsion-based delivery systems for lipophilic bioactive components. *Journal of Food Science*, *72*(8), R109–R124. doi:10.1111/j.1750-3841.2007.00507.x PMID:17995616

McClements, D. J., & Gunasekaran, S. (1997). Ultrasonic characterization of foods and drinks: Principles, methods, and applications. *Critical Reviews in Food Science and Nutrition*, *37*(1), 1–46. doi:10.1080/10408399709527766 PMID:9067087

McClements, D. J., & Rao, J. (2011). Food-grade nanoemulsions: Formulation, fabrication, properties, performance, biological fate, and potential toxicity. *Critical Reviews in Food Science and Nutrition*, *51*(4), 285–330. doi:10.1080/10408398.2011.559558 PMID:21432697

McGrew, S., Ortega, T., Breakstone, J., & Wineburg, S. (2017). The Challenge That's Bigger than Fake News: Civic Reasoning in a Social Media Environment. *American Educator*, *41*(3), 4.

McKinley, J. (2014). With farm robotics, the cows decide when it's milking time. *The New York Times*.

Mello, L. D., & Kubota, L. T. (2002). Review of the use of biosensors as analytical tools in the food and drink industries. *Food Chemistry*, *77*(2), 237–256. doi:10.1016/S0308-8146(02)00104-8

Memon, R. A., Li, J. P., & Ahmed, J. (2019). Simulation model for blockchain systems using queuing theory. *Electronics (Basel)*, *8*(2), 234. doi:10.3390/electronics8020234

Mena, C., Terry, L. A., Williams, A., & Ellram, L. (2014). Causes of waste across multi-tier supply networks: Cases in the UK food sector. *International Journal of Production Economics*, *152*, 144–158. doi:10.1016/j.ijpe.2014.03.012

Mengucci, C., Ferranti, P., Romano, A., Masi, P., Picone, G., & Capozzi, F. (2022). Food structure, function and artificial intelligence. *Trends in Food Science & Technology*, *123*, 251–263. doi:10.1016/j.tifs.2022.03.015

Meral, H., & Demirdöven, A. (2022). Gıdanın Özelleştirilmesi ve Detaylandırılmasında Üç Boyutlu (3D) Yazıcının Kullanımı. *Gida*, *47*(1), 91–106. doi:10.15237/gida.GD21112

Mete, H. (2012). Peynir Altı Suyunun Ekmekçilikte Değerlendirilmesi ve Ekonomik Önemi. *Tekirdağ S.M.M.M Odası Sosyal Bilimler Dergisi, 1*, 1-10.

Michelin Guide Asia. (2020). MICHELIN Introduces New Sustainable Gastronomy Icon. Retrieved from https://www.youtube.com/watch?v=BzsQ4Uouyvs Access Date: 13.06.2023.

Michelin Guide. (2020). MICHELIN Debuts New Sustainable Gastronomy Icon. Retrieved from https://Guide.michelin.com/sg/en/article/news-and-views/michelin-new-sustainable-gastronomy-icon Access Date: 20.06.2023.

Michelin Guide. (2021a). The MICHELIN Guide and sustainable gastronomy. Retrieved from https://mb.cision.com/Public/55/3413982/85252c3e3c870f10.pdf Access Date: 10.06.2023.

Michelin Guide. (2021b). What is a MICHELIN Green Star? Retrieved from https://Guide.michelin.com/kr/en/article/features/what-is-a-michelin-green-star-kr Access Date: 19.06.2023.

Michelin Guide. (2022). What is a Michelin Green Star and How Do You Earn One? Retrieved from https://Guide.michelin.com/ie/en/article/features/what-is-the-michelin-green-star-and-how-do-you-earn-one Access Date: 10.06.2023.

Michelin Guide. (2023a). Sustainability Leads the Way at Le Manoir aux Quat'Saisons, A Belmond Hotel. https://Guide.michelin.com/gb/en/article/sustainable-gastronomy/sustainability-leads-the-way-at-le-manoir-aux-quat-saisons-a-belmond-hotel Access Date: 15.06.2023.

Michelin Guide. (2023b). World Bee Day–20th May. https://Guide.michelin.com/ie/en/article/sustainable-gastronomy/world-bee-day Access Date: 20.06.2023.

Michelin Guide. (2023c). Green Star Focus: Apricity, London. Retrieved from https://Guide.michelin.com/ie/en/article/sustainable-gastronomy/green-star-focus-apricity-london Access Date: 20.06.2023.

Michelin Guide. (2023d). Sustainable Gastronomy Day–18th June. Retrieved from https://Guide.michelin.com/gb/en/article/features/sustainable-gastronomy-day---18th-june Access Date:20.06.2023.

Michelin Guide. (2023e). Chantelle Nicholson's Green Star Recipe. Retrieved from https://Guide.michelin.com/ie/en/article/dining-in/chantelle-nicholson-s-green-star-recipe Access Date: 20.06.2023.

Mihalca, V., Kerezsi, A. D., Weber, A., Gruber-Traub, C., Schmucker, J., Vodnar, D. C., Dulf, F. V., Socaci, S. A., Fărcaş, A., Mureşan, C. I., Suharoschi, R., & Pop, O. L. (2021). Protein-Based Films and Coatings for Food Industry Applications. *Polymers*, *13*(5), 769. doi:10.3390/polym13050769 PMID:33801341

Milgram, P., & Kishino, F. (1994). A Taxanomy of Mixed Reality Visual Displays. *IEICE Transactions on Information and Systems*, *12*.http://vered.rose.utoronto.ca/people/paul_dir/IEICE94/ieice.html

Miller, K. B., Eckberg, J. O., Decker, E. A., & Marinangeli, C. P. F. (2021). Role of Food Industry in Promoting Healthy and Sustainable Diets. *Nutrients*, *13*(8), 2740. doi:10.3390/nu13082740 PMID:34444900

Min-jing, P., & Bo, L. (2013). An Augmented Reality Based Disclosure Model of Information on Food Safety and Its Application. *International Journal of U- and e- Service. Science and Technology*, *6*(4), 69–80.

Misra, N. N., & Jo, C. (2017). Applications of cold plasma technology for microbiological safety in meat industry. *Trends in Food Science & Technology*, *64*, 74–86. doi:10.1016/j.tifs.2017.04.005

Misra, N. N., Patil, S., Moiseev, T., Bourke, P., Mosnier, J. P., Keener, K. M., & Cullen, P. J. (2014). In-package atmospheric pressure cold plasma treatment of strawberries. *Journal of Food Engineering*, *125*, 131–138. doi:10.1016/j.jfoodeng.2013.10.023

Mohammad, Z. H., Ahmad, F., Ibrahim, S. A., & Zaidi, S. (2022). Application of nanotechnology in different aspects of the food industry. *Discover Food*, *2*(1), 12. *Advance online publication.* doi:10.100744187-022-00013-9

Moley Robotics. (2015). *MK1—The World's First Robotic Kitchen.* Moley Robotics. https://www.moley.com/ (accessed on 10 December 2020).

Momtaz, M., Bubli, S. Y., & Khan, M. S. (2023). Mechanisms and Health Aspects of Food Adulteration: A Comprehensive Review. *Foods (Basel, Switzerland)*, *12*(1), 199. doi:10.3390/foods12010199 PMID:36613416

Morgan, K., & Murdoch, J. (2000). Organic vs. conventional agriculture: Knowledge, power and innovation in the food chain. *Geoforum*, *31*(2), 159–173. doi:10.1016/S0016-7185(99)00029-9

Mor, R. S., Kumar, D., & Singh, A. (Eds.). (2022). *Agri-Food 4.0: Innovations, Challenges and Strategies.* Emerald Publishing Limited. doi:10.1108/S1877-6361202227

Morrison, R. (2022). Fancy a byte to eat? Chinese restaurant chain is forced to use ROBOT waiters priced at £ 14,500 each after struggling with staff shortages during the Covid pandemic. *Daily Mail Online.* https://www.dailymail.co.uk/sciencetec/ article-10402701/Chinese-restaurant-chain-forced-use-ROBOT-waiters-Covid-pandemic.html

Morrissey, S. (2006). Nanotechnology in food and agriculture. *Chemical and Engineering News*, *84*, 31–31.

Mostafaeipour, A., Alvandimanesh, M., Najafi, F., & Issakhov, A. (2021). Identifying challenges and barriers for development of solar energy by using fuzzy best-worst method: A case study. *Energy*, *226*, 120355. doi:10.1016/j.energy.2021.120355

Moyer, J. C., & Stotz, E. (1947). The blanching of vegetables by electronics. *Food Technology*, *1*, 252–257.

Muhammad Nizam, S. S., Lam, M. C., Arshad, H., & Suwadi, N. A. (2018). A Scoping Review on Tangible and Spatial Awareness Interaction Technique in Mobile Augmented Reality-Authoring Tool in Kitchen. *Advances in Multimedia*, *2018*, 1–14. doi:10.1155/2018/5320984

Munir, K. (2022). Sustainable food waste management strategies by applying practice theory in hospitality and food services-a systematic literature review. *Journal of Cleaner Production*, *331*, 129991. doi:10.1016/j.jclepro.2021.129991

Murphy, R. (2020). *Local Consumer Review Survey.* Available in: https://www.brightlocal.com/research/local-consumer-review-survey/

Murphy, G., Corcoran, C., Tatlow-Golden, M., Boyland, E., & Rooney, B. (2020). See, like, share, remember: Adolescents' responses to unhealthy-, healthy-and non-food advertising in social media. *International Journal of Environmental Research and Public Health*, *17*(7), 2181. doi:10.3390/ijerph17072181 PMID:32218252

Murphy, J., Hofacker, C., & Gretzel, U. (2017). Dawning of the Age of Robots in Hospitality and Tourism: Challenges for Teaching and Research. *European Journal of Tourism Research*, *15*, 104–111. doi:10.54055/ejtr.v15i.265

Müßigmann, B., von der Gracht, H., & Hartmann, E. (2020). Blockchain technology in logistics and supply chain management—A bibliometric literature review from 2016 to January 2020. *IEEE Transactions on Engineering Management, 67*(4), 988–1007. doi:10.1109/TEM.2020.2980733

Muthukrishnan, N., Maleki, F., Ovens, K., Reinhold, C., Forghani, B., & Forghani, R. (2020). Brief history of artificial intelligence. *Neuroimaging Clinics, 30*(4), 393–399. doi:10.1016/j.nic.2020.07.004 PMID:33038991

Muthukumarappan, K., Halaweish, F., & Naidu, A. S. (2000). Ozone. In A. S. Naidu (Ed.), *Natural food antimicrobial systems* (pp. 783–800). CRC Press. doi:10.1201/9781420039368.ch29

Nachal, N., Moses, J. A., Karthik, P., & Anandharamakrishnan, C. (2019). Applications of 3D Printing in Food Processing. *Food Engineering Reviews, 11*(3), 123–141. doi:10.100712393-019-09199-8

Nag, S., Das, D., & Roy, R. B. (2023). Voltammetry Application of Molecularly Imprinted Polyacrylamide as Vanillin Receptor in Desserts. *IEEE Sensors Journal, 23*(4), 3346–3452. doi:10.1109/JSEN.2023.3235933

Nag, S., Pradhan, S., Das, D., Tudu, B., Bandyopadhyay, R., & Roy, R. B. (2021). Fabrication of a Molecular Imprinted Polyacrylonitrile Engraved Graphite Electrode for Detection of Formalin in Food Extracts. *IEEE Sensors Journal, 22*(1), 42–49. doi:10.1109/JSEN.2021.3128520

Naik, L., Sharma, R., Rajput, Y. S., & Manju, G. (2013). Application of High Pressure Processing Technology for Dairy Food Preservation - Future Perspective: A Review. *J. Anim Prod Adv, 3*(8), 232–241. doi:10.5455/japa.20120512104313

Nalarobotics, (n.d.) *Nalarobotics-The Wingman.* Nalarobotics. https://nalarobotics.com/the-wingman.html

Nalarobotics. (n.d.) *Nalarobotics-Nala Chef.* Nalarobotics. https://nalarobotics.com/nala-chef.html

Nam, S. W., & Lee, D. H. (2000). The effect of Mn on the mechanical behavior of Al alloys. *Metals and Materials, 6*(1), 13–16. doi:10.1007/BF03026339

Nathakaranakule, A., Jaiboon, P., & Soponronnarit, S. (2010). Farinfrared radiation assisted drying of longan fruit. *Journal of Food Engineering, 100*(4), 662–668. doi:10.1016/j.jfoodeng.2010.05.016

Naumov, N. (2019). The impact of robots, artificial intelligence, and service automation on service quality and service experience in hospitality. In S. Ivanov & C. Webster (Eds.), *Robots, artificial intelligence, and service automation in travel, tourism, and hospitality* (pp. 123–133). Emerald Publishing Limited. doi:10.1108/978-1-78756-687-320191007

Neethirajan, S., Kobayashi, I., Nakajima, M., Wu, D., Nandagopal, S., & Lin, F. (2011). Microfluidics for food, agriculture and biosystems industries. *Lab on a Chip, 11*(9), 1574–1586. doi:10.1039/c0lc00230e PMID:21431239

Nespresso (2021). Sustainable Gastronomy with Adam Handling | Edition 28. Retrieved from https://www.youtube.com/watch?v=qzewMPs8jZ0 Access Date: 10.06.2023.

New York Botanical Garden. (2019). Creating a More Eco-friendly and Sustainable Dining Experience. Retrieved from https://www.youtube.com/watch?v=FBA_QNq2MTI Access Date: 10.06.2023.

Nguyen, T. (2013). *Innovation Insight: Augmented Reality Will Become an Important Workplace Tool.* Gartner Research.

Nic, M., Jirat, J., & Kosata, B. (2006). IUPAC compendium of chemical terminology, the "Gold Book" 2. Blackwell Scientific Publications: Oxford: Oliveira, M.E.C. & Franca, A.S. (2002). Microwave heating of foodstuff. *Journal of Food Engineering, 53*, 347–359.

Nico Dingemans. (2023). Sustainable Gastronomy Day-18.06.2023-From Farm to Fork Dinner in Amman. Retrieved from https://www.youtube.com/watch?v=IsiXsILPqNo Access Date: 10.06.2023.

Nihalani, J. C., Narang, R. H., Chandrasekaran, G., Eshraghi, M. Z., & Bhatia, S. (2010). *Artificial Intelligence in Gaming*. Academic Press.

Nižetić, S., Šolić, P., López-de-Ipiña González-de-Artaza, D., & Patrono, L. (2020). Internet of Things (IoT): Opportunities, issues and challenges towards a smart and sustainable future. *Journal of Cleaner Production*, *274*, 122877. doi:10.1016/j.jclepro.2020.122877 PMID:32834567

Noort, M., Van Bommel, K., & Renzetti, S. (2017). 3D-printed Cereal Foods. *Cereal Foods World*, *62*(6), 272–277. doi:10.1094/CFW-62-6-0272

Nosratabadi, S., Khazami, N., Abdallah, M. B., Lackner, Z., Band, S. S., Mosavi, A., & Mako, C. (2020). Social capital contributions to food security: a comprehensive literature review. *Foods*, *9*(11). Home (2023) Numenta. https://www.numenta.com/

Nosratabadi, S., Mosavi, A., & Lakner, Z. (2020). Food Supply Chain and Business Model Innovation. *Foods*, *9*(2), 132. doi:10.3390/foods9020132 PMID:32012751

NPR. (2011). *Science diction: the origin of the word "Robot."* NPR. https://www.npr.org/2011/04/22/135634400/science-diction-the-origin-of-the-word-robot

Nunes, C. A., Ribeiro, M. N., de Carvalho, T. C. L., Ferreira, D. D., de Oliveira, L. L., & Pinheiro, A. C. M. (2023). Artificial intelligence in sensory and consumer studies of food products. *Current Opinion in Food Science*, *50*, 101002. doi:10.1016/j.cofs.2023.101002

Öcal, D. (2021). *Yiyecek İçecek İşletmelerinde Gıda İsrafı Üzerine Bir Araştırma: Kırıkkale Örneği.* Sakarya: Sakarya Üniversitesi Lisansüstü Eğitim Enstitüsü.

Okorie, O., Russell, J., Jin, Y., Turner, C., Wang, Y., & Charnley, F. (2022). Removing Barriers to Blockchain use in Circular Food Supply Chains: Practitioner Views on Achieving Operational Effectiveness. *Cleaner Logistics and Supply Chain*, 100087.

Oliveira, F. M., Cereda, M. P., & Mattoso, L. H. (2019). Molecularly imprinted polymers for the recognition of food proteins: A review. *Trends in Food Science & Technology*, *90*, 24–36.

Oral, Z. (2015). *Dünya'da ve Türkiye'de Gıda İsrafı ve Önlenmesine Yönelik Uygulamalar.* Ankara: AB Uzmanlık Tezi, TC Gıda Tarım ve Hayvancılık Bakanlığı Avrupa Birliği ve Dış İlişkiler Müdürlüğü.

Oronsky, C. R., & Chathoth, P. K. (2007). An exploratory study examining information technology adoption and implementation in full-service restaurant firms. *International Journal of Hospitality Management*, *26*(4), 941–956. doi:10.1016/j.ijhm.2006.04.001

Osmólska, E., Stoma, M., & Starek-Wójcicka, A. (2022). Application of Biosensors, Sensors, and Tags in Intelligent Packaging Used for Food Products—A Review. *Sensors (Basel)*, *22*(24), 9956. Advance online publication. doi:10.339022249956 PMID:36560325

Otero, C. (2018). *Flippy, un automata ´ que hace hamburguesas a media jornada.* BeTech.

Ottenbacher, M. C., & Harrington, R. J. (2009). The product innovation process of quick-service restaurant chains. *International Journal of Contemporary Hospitality Management*, *21*(5), 523–541. doi:10.1108/09596110910967782

Özdal, Z. N. (2019). *Türkiye'de Tüketicilerin Gıda İsrafını Engellemeye Yönelik Tutumlarının Analizi ve Mobil Uygulama Tasarımı.* Çukurova Üniversitesi Fen Bilimleri Enstitüsü.

Özdemir, G., & Güçer, E. (2018). Food Waste Management within Sustainability Perspective: A Study on Five Star Chain Hotels. *Journal of Tourism and Gastronomy Studies*, *6*(1), 280–299. doi:10.21325/jotags.2018.187

Ozkoc, S. O., Sumnu, G., Sahin, S., & Turabi, E. (2009). Investigation of physicochemical properties of breads baked in microwave and infrared-microwave combination ovens during storage. *European Food Research and Technology*, *228*(6), 883–893. doi:10.100700217-008-1001-0

Öztürk, İ. (2006). Türkiye'de Perakende Sektörü. *Çağ Üniversitesi Sosyal Bilimler Dergisi*, *3*(1), 69-81.

Öztürk, K., & Şahin, M. E. (2018). Yapay sinir ağları ve yapay zekâ'ya genel bir bakış. *Takvim-i Vekayi*, *6*(2), 25–36.

Paddle, B. M. (1996). Biosensors for chemical and biological agents of defence interest. *Biosensors & Bioelectronics*, *11*(11), 1079–1113. doi:10.1016/0956-5663(96)82333-5 PMID:8828162

Pallathadka, H., Ramirez-Asis, E. H., Loli-Poma, T. P., Kaliyaperumal, K., Ventayen, R. J. M., & Naved, M. (2023). Applications of artificial intelligence in business management, e-commerce and finance. *Materials Today: Proceedings*, *80*, 2610–2613. doi:10.1016/j.matpr.2021.06.419

Pamuru, V., Khern-Am-Nuai, W., & Kannan, K. (2021). The Impact of an Augmented-Reality Game on Local Businesses: A Study of Pokémon Go on Restaurants. *Information Systems Research*, *32*(3), 950–966. doi:10.1287/isre.2021.1004

Pandey, S. (2021, June). Artificial Intelligence and Food Industry. *In Proceedings of the 22nd International RAIS Conference on Social Sciences and Humanities* (pp. 72-76). Scientia Moralitas Research Institute.

Pandey, V., Pant, M., & Snasel, V. (2022). Blockchain technology in food supply chains: Review and bibliometric analysis. *Technology in Society*, *69*, 101954. doi:10.1016/j.techsoc.2022.101954

Pandiselvam, R., Manikantan, M. R., Subhashree, N., Mathew, A. C., Balasubramanian, D., Shameena Beegum, P. P., Ramesh, S. V., Niral, V., Ranjini, T. N., & Hebbar, K. B. (2019). Correlation and principal component analysis of physical properties of tender coconut (Cocos nucifera L.) in relation to the development of trimming machine. *Journal of Food Process Engineering*, *42*(6), e13217. doi:10.1111/jfpe.13217

Pankaj, S. K., Bueno-Ferrer, C., Misra, N. N., Milosavljević, V., O'Donnell, C. P., Bourke, P., Keener, K. M., & Cullen, P. J. (2014). Applications of cold plasma technology in food packaging. *Trends in Food Science & Technology*, *35*(1), 5–17. doi:10.1016/j.tifs.2013.10.009

Pankaj, S. K., & Keener, K. M. (2017). Cold plasma applications in food packaging. In *Reference Module in Food Science*. Elsevier. doi:10.1016/B978-0-08-100596-5.21417-0

Parfitt, J., Barthel, M., & Macnaughton, S. (2010). Food Waste Within Food Supply Chains: Quantification and Potential for Change to 2050. *Phil. Trans. Biol. Sci.*, *365*(1554), 3065–3081. doi:10.1098/rstb.2010.0126 PMID:20713403

Patidar, R., & Sharma, L. (2011). Credit card fraud detection using neural network. [IJSCE]. *International Journal of Soft Computing and Engineering*, *1*, 32–38.

Patterson, M. F. (2005). Microbiology of pressure treated foods. *Journal of Applied Microbiology*, *98*(6), 1400–1409. doi:10.1111/j.1365-2672.2005.02564.x PMID:15916652

Paul Pairet. (2022). *uvbypp.cc*. http://paulpairet.com/

Pccw Solutions. (2022). *Pizza Hut: Enhancing Customer Engagement with Augmented Reality Game - Lenovo PCCW Solutions*. PCCW. https://www.pccwsolutions.com

Penco, L., Serravalle, F., Profumo, G., & Viassone, M. (2021). Mobile augmented reality as an internationalization tool in the "Made In Italy" food and beverage industry. *The Journal of Management and Governance*, *25*(4), 1179–1209. doi:10.100710997-020-09526-w

Peng, L. (2021). Privacy preservation in permissionless blockchain: A survey. *Digital Communications and Networks*, *7*(3), 295–307.

Pereira, R. N., Martins, J., Mateus, C., Teixeira, J. A., & Vicente, A. A. (2007). Death kinetics of in goat milk and in cloudberry jam treated by ohmic heating. *Chemical Papers*, *61*(2), 121–126. doi:10.247811696-007-0008-5

Pereira, R. N., Martins, R. C., & Vicente, A. A. (2008). Goat milk free fatty acid characterization during conventional and ohmic heating pasteurization. *Journal of Dairy Science*, *91*(8), 2925–2937. doi:10.3168/jds.2007-0873 PMID:18650269

Phan, V. T., Seung, P., & Choo, Y. (2010). A Study on Tangible AR for interior design. *International Journal of CAD/CAM*, *10*(1), 21–28. https://scholar.google.com/scholar_lookup?title=A

Philips, C. A. (1996). Review: Modified atmosphere packaging and its effects on the microbiological quality and safety of produce. *International Journal of Food Science & Technology*, *31*(6), 463–479. doi:10.1046/j.1365-2621.1996.00369.x

Pigłowski, M. (2020). Food hazards on the European Union market: The data analysis of the Rapid Alert System for Food and Feed. *Food Science & Nutrition*, *8*(3), 1603–1627. doi:10.1002/fsn3.1448 PMID:32180969

Pine, B. J., & Gilmore, J. H. (1998). *Welcome to the experience economy*. Harvard Business Review.

Pinel, F. (2015). What's cooking with Chef Watson? an interview with Lav Varshney and James briscione. *IEEE Pervasive Computing*, *14*(4), 58–62. doi:10.1109/MPRV.2015.66

Pingen, S., Sudhaus, N., Becker, A., Krischek, C., & Klein, G. (2016). High pressure as an alternative processing step for ham production. *Meat Science*, *118*, 22–27. doi:10.1016/j.meatsci.2016.03.014 PMID:27019422

Pinna, C., Ramundo, L., Sisca, F. G., Angioletti, C. M., Taisch, M., & Terzi, S. (2016 September). Additive manufacturing applications within food industry: An actual overview and future opportunities. *XXI Summer School "Francesco Turco" - Industrial Systems Engineering*.

Pirani, S. I., & Arafat, H. A. (2014). Solid Waste Management in the Hospitality Industry: A Review. *Journal of Environmental Management*, *146*, 320–336. doi:10.1016/j.jenvman.2014.07.038 PMID:25194519

Pirim, A. G. H. (2006). Yapay zeka. *Yaşar Üniversitesi E-Dergisi*, *1*(1), 81–93.

Piyasena, P., Mohareb, E., & McKellar, R. C. (2003). Inactivation of microbes using ultrasound: A review. *International Journal of Food Microbiology*, *87*(3), 207–216. doi:10.1016/S0168-1605(03)00075-8 PMID:14527793

PlateJoy. (n.d.). *Personalized Meal Plans and Recipes*. PlateJoy. https://www.platejoy.com/

Plestenjak, A., Pozrl, T., Hribar, J., Unuk, T., & Vidrih, R. (2008). Regulation of metabolic changes in shredded cabbage by modified atmosphere packaging. *Food Technology and Biotechnology*, *46*(4), 427–433.

Plotkina, D., Munzel, A., & Pallud, J. (2020). Illusions of truth - Experimental insights into human and algorithmic detections of fake online reviews. *Journal of Business Research*, *109*, 511–523. https://www.elsevier.com/open-access/userlicense/1.0/. doi:10.1016/j.jbusres.2018.12.009

Popa, A., Hnatiuc, M., Paun, M., Geman, O., Hemanth, D., Dorcea, D., Son, L., & Ghita, S. (2019). An Intelligent IoT-Based Food Quality Monitoring Approach Using Low-Cost Sensors. *Symmetry*, *11*(3), 374. doi:10.3390ym11030374

Popovski, G., Seljak, B., & Eftimov, T. (2019). FoodOntoMap: linking food concepts across different food ontologies. In *Proceedings of the 11th international Joint conference on knowledge discovery, knowledge engineering and knowledge management.* SCITEPRESS. 10.5220/0008353201950202

Porter, M. E., & Heppelmann, J. E. (2014). How Smart, Connected Products Are Transforming Competition - Article - Faculty & Research - Harvard Business School. *Harward Business Review, 92*(11), 64–88. https://www.hbs.edu/faculty/Pages/item.aspx?num=48195

Poushneh, A., & Vasquez-Parraga, A. Z. (2017). Discernible impact of augmented reality on retail customer's experience, satisfaction and willingness to buy. *Journal of Retailing and Consumer Services, 34*, 229–234. doi:10.1016/j.jretconser.2016.10.005

Pozelli Sabio, R., & Spers, E. E. (2022). 'Consumers' Expectations on Transparency of Sustainable Food Chains'. *Frontiers in Sustainable Food Systems, 6*(April), 1–8. doi:10.3389/fsufs.2022.853692

Pradhan, N., Singh, S., Ojha, N., Srivastava, A., Barla, A., & Rai, V. (2015). Facets of nanotechnology as seen in food processing, packaging, and preservation industry. *BioMed Research International, 2015*, 365672. doi:10.1155/2015/365672 PMID:26613082

Prayogo, R. R., & Kusumawardhani, A. (2017). Examining relationships of destination image, service quality, e-WOM, and revisit intention to Sabang Island, Indonesia. *APMBA, 5*(2), 89–102.

Prof. M. THANGA DARWIN (2019). Sustainable Gastronomy Day 18 JUNE 2023. Retrieved from https://www.youtube.com/watch?v=6xGczlEcqMI Access Date: 10.06.2023.

Pufahl, L. (2021). Enabling Financing in Agricultural Supply Chains Through Blockchain: Interorganizational Process Innovation Through Blockchain. Business Process Management Cases Vol. 2: Digital Transformation-Strategy, Processes and Execution. Springer. doi:10.1007/978-3-662-63047-1_4

Quested, T. E., Parry, A. D., Easteal, S., & Swannell, R. (2011). Food and Drink Waste From Households in the UK. *Nutrition Bulletin, 36*(4), 460–467. doi:10.1111/j.1467-3010.2011.01924.x

Racz, A., Marchesi, V. V., & Crnković, I. (2018). Economical, Environmental and Ethical Impact of Food Wastage in Hospitality and Other Global Industries. *Jahr: Europski časopis za bioetiku, 9(1)*, 25-42.

Raguseo, E., & Vitari, C. (2017). The effect of brand on the impact of e-WOM on hotels' financial performance. *International Journal of Electronic Commerce, 21*(2), 249–269. doi:10.1080/10864415.2016.1234287

Rahul Soni, K. (2022). Ponappa, Puneet Tandon,A review on customized food fabrication process using Food Layered Manufacturing. *Lebensmittel-Wissenschaft + Technologie, 161*, 113411. doi:10.1016/j.lwt.2022.113411

Ramaswamy, H. S., Marcotte, M., Sastry, S. K., & Abdelrahim, K. (2016). Ohmic heating. In food processing, Eds. Abdelrahim K, pp. 93–102. CRC Press: Boca Raton, USA.

Ramesh, R., Nayak, B., Amirbahman, A., Tripp, C. P., & Mukhopadhyay, S. (2016). Application of ultraviolet light assisted titanium dioxide photocatalysis for food safety: A review. *Innovative Food Science & Emerging Technologies, 38*, 105–115. doi:10.1016/j.ifset.2016.09.015

Rana, S. (2021). Blockchain-Based Model to Improve the Performance of the Next-Generation Digital Supply Chain. *Sustainability*. . doi:10.3390/su131810008

Rana, R. L., Tricase, C., & De Cesare, L. (2021). Blockchain technology for a sustainable agri-food supply chain. *British Food Journal, 123*(11), 3471–3485. doi:10.1108/BFJ-09-2020-0832

Raskin, M. (2016). The law and legality of smart contracts. *Geo. L. Tech. Rev.*, *1*, 305.

Rastogi, N. K. (2010). Opportunities and challenges in non-thermal processing of foods. In Innovation in Food Engineering, 3–58. New Techniques and Products, CRC Press.

Rastogi, N. (2013). Introduction. In *Recent Developments in High Pressure Processing of Foods* (pp. 1–7). Springer. doi:10.1007/978-1-4614-7055-7_1

Rastogi, N. K. (2012). Recent trends and developments in infrared heating in food processing. *Critical Reviews in Food Science and Nutrition*, *52*(9), 737–760. doi:10.1080/10408398.2010.508138 PMID:22698266

Rastogi, N. K. (2015). Infrared heating of foods and its combination with electron beam processing. In *Electron Beam Pasteurization and Complementary Food Processing Technologies* (pp. 61–82). Woodhead Publishing Series in Food Science, Technology and Nutrition. doi:10.1533/9781782421085.2.61

Rastogi, N. K., Raghavarao, K. S. M. S., Niranjan, K., & Knorr, D. (2002). Recent developments in osmotic dehydration: Methods to enhance mass transfer. *Trends in Food Science & Technology*, *2*(13), 48–59. doi:10.1016/S0924-2244(02)00032-8

Rastogi, N. K., Raghavarao, K. S., Balasubramaniam, M. S., Niranjan, V. M., & Knorr, D. (2007). Opportunities and challenges in high pressure processing of foods. *Critical Reviews in Food Science and Nutrition*, *47*(1), 69–112. doi:10.1080/10408390600626420 PMID:17364696

Rauschnabel, P. A., Babin, B. J., tom Dieck, M. C., Krey, N., & Jung, T. (2022). What is augmented reality marketing? Its definition, complexity, and future. *Journal of Business Research*, *142*, 1140–1150. doi:10.1016/j.jbusres.2021.12.084

Raut, R. D., Gardas, B. B., Narwane, V. S., & Narkhede, B. E. (2019). Improvement in the food losses in fruits and vegetable supply chain-a perspective of cold third-party logistics approach. *Operations Research Perspectives*, *6*, 100117. doi:10.1016/j.orp.2019.100117

Rawson, K., & Shore, E. (2019). *Dining out: A global history of restaurants*. Reaktion Books.

Redlingshöfer, B., Coudurier, B., & Georget, M. (2017). Quantifying Food Loss During Primary Production and Processing in France. *Journal of Cleaner Production*, *164*, 703–714. doi:10.1016/j.jclepro.2017.06.173

Reiley, L., & Powell, L. 2022. The robots are here. And they are making you fries. *The Washington Post*. https://www.washingtonpost.com/business/20 22/09/20/robots-automating-restaurant-industry/

Rejeb, A., Rejeb, K., & Keogh, J. G. (2021). Enablers of augmented reality in the food supply chain: A systematic literature review. *Journal of Foodservice Business Research*, *24*(4), 415–444. doi:10.1080/15378020.2020.1859973

Rejeb, A., Rejeb, K., Keogh, J. G., & Zailani, S. (2022). Barriers to blockchain adoption in the circular economy: A fuzzy Delphi and best-worst approach. *Sustainability*, *14*(6), 3611. doi:10.3390u14063611

Rejeb, A., Suhaiza, Z., Rejeb, K., Seuring, S., & Treiblmaier, H. (2022). The Internet of Things and the circular economy: A systematic literature review and research agenda. *Journal of Cleaner Production*, *350*, 131439. doi:10.1016/j.jclepro.2022.131439

Report, C. (2006). *Nanotechnologies in the Food Industry*. Cientifica. http://www.cientifica.com/www/details.php?id47

Restaino, L., Frampton, E. W., Hemphil, J. B., & Palnikar, P. (1995). Efficacy of ozonated water against various food-related microorganisms. *Applied and Environmental Microbiology*, *61*(9), 3471–3475. doi:10.1128/aem.61.9.3471-3475.1995 PMID:7574656

Reyna, A., Martín, C., Chen, J., Soler, E., & Díaz, M. (2018). On blockchain and its integration with IoT. Challenges and opportunities. *Future Generation Computer Systems*, *88*, 173–190. doi:10.1016/j.future.2018.05.046

Reynolds, C. J., Mirosa, M., & Clothier, B. (2016). New Zealand's Food Waste: Estimating the Tonnes, Value, Calories and Resources Wasted. *Agriculture*, *6*(1), 1–15. doi:10.3390/agriculture6010009

Rezaei, J. (2015). Best-worst multi-criteria decision-making method. *Omega*, *53*, 49–57. doi:10.1016/j.omega.2014.11.009

Rezaei, J. (2016). Best-worst multi-criteria decision-making method: Some properties and a linear model. *Omega*, *64*, 126–130. doi:10.1016/j.omega.2015.12.001

Rezaei, J., Wang, J., & Tavasszy, L. (2015). Linking supplier development to supplier segmentation using Best Worst Method. *Expert Systems with Applications*, *42*(23), 9152–9164. doi:10.1016/j.eswa.2015.07.073

Ribeiro, F. W. P., Barroso, M. F., Morais, S., Viswanathan, S., de Lima-Neto, P., Correia, A. N., Oliveira, M. B. P. P., & Delerue-Matos, C. (2014). Simple laccase-based biosensor for formetanate hydrochloride quantification in fruits. *Bioelectrochemistry (Amsterdam, Netherlands)*, *95*, 957–14. doi:10.1016/j.bioelechem.2013.09.005 PMID:24161938

Richard, H. P., Adams, L., & Walter, R. (2012). *Hotel Design, Planning and Development* (2nd ed.). Routledge., doi:10.4324/9780080967004

Richins, M. L. (1983). Negative word-of-mouth by dissatisfied consumers: A pilot study. *Journal of Marketing*, *47*(1), 68–78. doi:10.1177/002224298304700107

Robinson, M. (2017). This robot-powered restaurant is one step closer to putting fast-food workers out of a job. *Business Insider*. https://www.businessinsider.com/momentum-machines-funding-robot-burger-restaurant-2017-6?r=US&IR=T

Robinson, S., & Markowitz, D. (2021). *Just desserts: Baking with AI-made recipes*. https://blog.google/products/google-cloud/just-desserts-baking-ai-made-recipes

Roday, S., Biwal, A., & Joshi, V. (2009). *Tourism Opeartions and Management*. Oxford Publication.

Rød, S. K., Hansen, F., Leipold, F., & Knøchel, S. (2012). Cold atmospheric pressure plasma treatment of ready-to-eat meat: Inactivation of Listeria innocua and changes in product quality. *Food Microbiology*, *30*(1), 233–238. doi:10.1016/j.fm.2011.12.018 PMID:22265306

Roe, B. E., Qi, D., Apolzan, J. W., & Martin, C. K. (2020). Selection, intake, and plate waste patterns of leftover food items among U.S. consumers: A pilot study. *PLoS One*, *2020*(15), e0238050. doi:10.1371/journal.pone.0238050 PMID:32903260

Rogerson, M., & Parry, G. C. (2020). Blockchain: Case studies in food supply chain visibility. *Supply Chain Management*, *25*(5), 601–614. doi:10.1108/SCM-08-2019-0300

Rohini, C., Geetha, P. S., Vijayalakshmi, R., Mini, M. L., & Pasupathi, E. (2020). Global Effects of Food Waste. *Journal of Pharmacognosy and Phytochemistry*, *9*(2), 690–699.

Rojas-Graü, M. A., Soliva-Fortuny, R., & Martín-Belloso, O. (2009). Edible coatings to incorporate active ingredients to fresh-cut fruits: A review. *Trends in Food Science & Technology*, *20*(10), 438–447. doi:10.1016/j.tifs.2009.05.002

Rolle, R. S. (2006). *Improving Postharvest Management and Marketing in the Asia-Pacific Region: Issues and Challenges Trends.* Semantic Scholar.

Ross, M. M. (2022). *3D printing of dairy-based ingredients and investigation into Irish consumer acceptance of 3D food printing applications*. [PhD Thesis, University College Cork].

Rotariu, L., Lagarde, F., Jaffrezic-Renault, N., & Bala, C. (2016). Electrochemical biosensors for fast detection of food contaminants – trends and perspective. *Trends in Analytical Chemistry*, *2016*, 7980–7987. doi:10.1016/j.trac.2015.12.017

Russell, S., & Norvig, P. (2010). *Artificial Intelligence: A Modern Approach* (3rd ed.). Prentice Hall.

Ryu, K., Jarumaneerat, T., Promsivapallop, P., & Kim, M. (2023). What influences restaurant dining out and diners' self-protective intention during the COVID-19 pandemic: Applying the Protection Motivation Theory. *International Journal of Hospitality Management*, *109*, 103400. doi:10.1016/j.ijhm.2022.103400 PMID:36439897

Sadowska-Rociek, A., Śliwińska, M., & Koczoń, P. (2019). Molecularly imprinted polymers as recognition materials in sensors for detection of food additives. *Chemosensors (Basel, Switzerland)*, *7*(2), 19.

Sagar, V. R., & Suresh Kumar, P. (2010). Recent advances in drying and dehydration of fruits and vegetables: A review. *Journal of Food Science and Technology*, *47*(1), 15–26. doi:10.100713197-010-0010-8 PMID:23572596

Şahin, K., & Turan, B. O. (2018). Üç Boyutlu Yazıcı Teknolojilerinin Karşılaştırmalı Analizi. *Stratejik ve Sosyal Araştırmalar Dergisi*, *2*(2), 97–116. doi:10.30692isad.441648

Sahni, V., Srivastava, S., & Khan, R. (2021). Modelling techniques to improve the quality of food using artificial intelligence. *Journal of Food Quality*, *2021*, 1–10. doi:10.1155/2021/2140010

Saifullah, M., Ahsan, A., & Shishir, M. R. I. (2016). Production, stability and application of micro- and nanoemulsion in food production and the food processing industry. In A. M. Grumezescu (Ed.), *Nanotechnology in the Agri-Food Industry, Emulsions* (pp. 405–442). Academic Press. doi:10.1016/B978-0-12-804306-6.00012-X

Salawi, A. (2022). Pharmaceutical Coating and Its Different Approaches, a Review. *Polymers*, *14*(16), 3318. doi:10.3390/polym14163318 PMID:36015575

Salazar-Gonzalez, C., San Martin-Gonzalez, M. F., Lopez-Malo, A., & Sosa-Morales, M. E. (2012). Recent studies related to microwave processing of fluid foods. *Food and Bioprocess Technology*, *5*(1), 31–46. doi:10.100711947-011-0639-y

Saldaña, E., Siche, R., Luján, M., & Quevedo, R. (2013). Review: Computer vision applied to the inspection and quality control of fruits and vegetables. *Brazilian Journal of Food Technology*, *16*(4), 254–272. doi:10.1590/S1981-67232013005000031

Saleem, R., Khurshid, M., & Ahmed, S. (2018). Laccases, Manganese Peroxidases and Xylanases Used for the Bio-bleaching of Paper Pulp: An Environmental Friendly Approach. *Protein and Peptide Letters*, *25*(2), 180–186. doi:10.2174/0929866525666180122100133 PMID:29359649

Salvador, A., Drozdzal, M., Giró-i-Nieto, X., & Romero, A. (2019). Inverse cooking: Recipe generation from food images. In *Proceedings of the IEEE/CVF Conference on Computer Vision and Pattern Recognition* (pp. 10453-10462). IEEE. 10.1109/CVPR.2019.01070

Salvador, A., Hynes, N., Aytar, Y., Marin, J., Ofli, F., Weber, I., & Torralba, A. (2017). Learning cross-modal embeddings for cooking recipes and food images. In *Proceedings of the IEEE conference on computer vision and pattern recognition* (pp. 3020-3028). IEEE. 10.1109/CVPR.2017.327

Salvagno, M., Taccone, F. S., & Gerli, A. G. (2023). Can artificial intelligence help for scientific writing? *Critical Care (London, England)*, *27*(1), 1–5. PMID:36597110

Salvia-Trujillo, L., Rojas-Graü, M. A., Soliva-Fortuny, R., & Martín-Belloso, O. (2015). Use of antimicrobial nanoemulsions as edible coatings: Impact on safety and quality attributes of fresh-cut Fuji apples. *Postharvest Biology and Technology*, *105*, 8–16. doi:10.1016/j.postharvbio.2015.03.009

Salwa, H. N., Sapuan, S. M., Mastura, M. T., & Zuhri, M. Y. M. (2019), Green Bio composites For Food Packaging. *International Journal of Recent Technology and Engineering* (IJRTE) *8*, 2S4, https://courseware.cutm.ac.in/wp-content/uploads/2021/06/Composites-for-food-Packaging.pdf

Sandhya, K. V. (2010). Modified atmosphere packaging of fresh produce: Current status and future needs. *Lebensmittel-Wissenschaft + Technologie*, *43*(3), 381–392. doi:10.1016/j.lwt.2009.05.018

Santos, K. (2019). *Domino's ordering app to acquire further customer insight with new augmented reality feature*. QSR Media Australia. https://qsrmedia.com.au/technology/news/dominos-ordering-app-acquire-further-customer-insight-new-augmented-reality-feature

Saravacos, G., & Kostaropoulos, A. E. (2016). *Equipment for novel food processes handbook of food processing equipment*. Springer. doi:10.1007/978-3-319-25020-5

Sarfraz, J., Gulin-Sarfraz, T., Nilsen-Nygaard, J., & Pettersen, M. K. (2021). Nanocomposites for food packaging applications: An overview. *Nanomaterials (Basel, Switzerland)*, *11*(1), 10. doi:10.3390/nano11010010 PMID:33374563

Sastry, S. (2008). Safety of Foods Processed Using Four Alternative Processing Technologies. *National Integrated Food Safety Initiative* USDA, 1-2.

Sastry, S. K., Datta, A. K., & Worobo, R. W. (2008). Ultraviolet light. *Journal of Food Science*, (Supplement), 90–92.

Sau, S., Datta, P., Sarkar, T., & Sarkar, S. (2018). Impact of low doses of gamma irradiation on off- season guava at ambient storage conditions. *International Journal of Current Microbiology and Applied Sciences*, *1*(1), 295–307. doi:10.20546/ijcmas.2018.701.032

Savastano, M., Amendola, C., & D'Ascenzo, F. (2018). How Digital Transformation is Reshaping The Manufacturing Industry Value Chain: The New Digital Manufacturing Ecosystem Applied to A Case Study From The Food Industry. In *Network, Smart and Open: Three Keywords for Information Systems Innovation* (pp. 127–142). Springer International Publishing. doi:10.1007/978-3-319-62636-9_9

Segre, A., & Falasconi, L. (2011). *Il Libro Nero Dello Spreco in Italia: Il Cibo*. Edizioni: Ambiente.

Segre, A., Falasconi, L., Politano, A., & Vittuari, M. (2014). *SAVE FOOD: Global Initiative on Food Loss and Waste Reduction Background Paper on the Economics of Food Loss and Waste*. Food and Agriculture Organization of the United Nations.

Sejnowski, T. J., & Rosenberg, C. R. (1985). NETtalk: A parallel network that learns to read aloud. *Artificial Intelligence*, *134*(1–2), 57–83. doi:10.1016/S0004-3702(01)00129-1

Şeker, İ. T., & Karaman, E. E. (2020). Dijital Gastronomi. Abdullah Karaman ve Kürşad Şahin (Ed.) Dijital Turizm: Sektörün Yeni Geleceği içinde (189-196). İstanbul: Eğitim Yayınevi.

Selcuk, N., & Erkan, M. (2015). The effects of modified and palliflex controlled atmosphere storage on postharvest quality and composition of 'Istanbul' medlar fruit. *Postharvest Biology and Technology*, *99*, 9–19. doi:10.1016/j.postharvbio.2014.07.004

Şener, A., & Ünal, Ü. M. (2008). Gıda Sanayii Atıklarının Biyoteknolojik Yöntemlerle Değerlendirilmesi. Türkiye 10. Gıda Kongresi; 21-23 Mayıs 2008, (pp. 21-23). Erzurum.

Senturk Parreidt, T., Müller, K., & Schmid, M. (2018). Alginate-Based Edible Films and Coatings for Food Packaging Applications. *Foods*, *17*(10), 170. doi:10.3390/foods7100170 PMID:30336642

Seo, E. J., Park, J. W., & Choi, Y. J. (2020). The effect of social media usage characteristics on e-WOM, trust, and brand equity: Focusing on users of airline social media. *Sustainability*, *12*(4), 1691. doi:10.3390u12041691

Seo, K. H. (2020). A study on the application of kiosk service as the workplace flexibility: The determinants of expanded technology adoption and trust of quick service restaurant customers. *Sustainability*, *12*(21), 8790. doi:10.3390u12218790

Sera, T., Izukura, S., Hashimoto, I., Motegi, T., & Motohashi, Y. (2020, April). A case study of Food Production Using Artificial Intelligence. In *Proceedings of the 2020 Symposium on Emerging Research from Asia and on Asian Contexts and Cultures* (pp. 33-36). IEEE. 10.1145/3391203.3391211

Serizawa, R., Shitara, M., Gong, J., Makino, M., Kabir, M. H., & Furukawa, H. (2014). 3D jet printer of edible gels for food creation, 2014 Behavior and Mechanics of Multifunctional Materials and Composites, 9058. *Article, 90580A*, 90580A. doi:10.1117/12.2045082

Serrano, M., Martinez-Romero, D., Guillen, F., Castillo, S., & Valero, D. (2006). Maintenance of broccoli quality and functional properties during cold storage as affected by modified atmosphere packaging. *Postharvest Biology and Technology, 39*(1), 61–68. doi:10.1016/j.postharvbio.2005.08.004

Sethi, P., & Sarangi, S. R. (2017). In R. Khanna (Ed.), *'Internet of Things: Architectures, Protocols, and Applications', Journal of Electrical and Computer Engineering* (p. 9324035)., doi:10.1155/2017/9324035

Sevenich, R., Bark, F., Crews, C., Anderson, W., Pye, C., Riddellova, K., Hradecky, J., Moravcova, E., Reineke, K., & Knorr, D. (2013). Effect of high pressure thermal sterilization on the formation of food processing contaminants. *Innovative Food Science & Emerging Technologies, 20*, 42–50. doi:10.1016/j.ifset.2013.07.006

Shabani, N., Munir, A., & Hassan, A. (2019). E-Marketing via Augmented Reality: A Case Study in the Tourism and Hospitality Industry. *IEEE Potentials, 38*(1), 43–47. doi:10.1109/MPOT.2018.2850598

Shamim, S., Yang, Y., Zia, N. U., & Shah, M. H. (2021). Big data management capabilities in the hospitality sector: Service innovation and customer generated online quality ratings. *Computers in Human Behavior, 121*, 121. doi:10.1016/j.chb.2021.106777

Sharma, A., Goud, K. Y., Hayat, A., Bhand, S., & Marty, J. L. (2017). Recent advances in electrochemical-based sensing platforms for aflatoxins detection. *Chemosensors (Basel, Switzerland), 5*(1), 1. doi:10.3390/chemosensors5010001

Sharma, G. P., & Prasad, S. (2001). Drying of garlic cloves by microwave–hot air combination. *Journal of Food Engineering, 50*(2), 99–105. doi:10.1016/S0260-8774(00)00200-4

Sharma, H. B., Vanapalli, K. R., Cheela, V. R. S., Ranjan, V. P., Jaglan, A. K., Dubey, B., Goel, S., & Bhattacharya, J. (2020). Challenges, opportunities, and innovations for effective solid waste management during and post COVID-19 pandemic. *Resources, Conservation and Recycling, 162*, 105052. doi:10.1016/j.resconrec.2020.105052 PMID:32834486

Shelake, A. S. (2020). IoT Based Food Inventory Tracking System for Domestic and Commercial Kitchens. *International Journal of Scientific Research and Engineering Development, 3*(4). www.ijsred.com

Shelbizleee (2022). What a "climatarian" actually eats in a day|most eco-friendly food. Retrieved from https://www.youtube.com/watch?v=KY8WDiGqAEI Access Date: 11.06.2023.

Shi, J., Pan, Z., McHugh, T. H., Wood, D., Hirschberg, E., & Olson, D. (2008). Drying and quality characteristics of fresh and sugarinfused blueberries dried with infrared radiation heating. *Lebensmittel-Wissenschaft + Technologie, 41*(10), 1962–1972. doi:10.1016/j.lwt.2008.01.003

Shi, S., He, D., Li, L., Kumar, N., Khan, M. K., & Choo, K.-K. R. (2020). Applications of blockchain in ensuring the security and privacy of electronic health record systems: A survey. *Computers & Security, 97*, 101966. doi:10.1016/j.cose.2020.101966 PMID:32834254

Sierra, I., Vidal-Valverde, C., & Lopez-Fandino, R. (2000). Effect of high pressure on the Vitamin B1 and B6 content in milk. *Milchwissenschaft. Milk Science International, 55*, 365–367.

Simeone, M., & Scarpato, D. (2020). Sustainable consumption: How does social media affect food choices? *Journal of Cleaner Production, 277*, 124036. doi:10.1016/j.jclepro.2020.124036

Sina, E., Boakye, D., Christianson, L., Ahrens, W., & Hebestreit, A. (2022). Social Media and Children's and Adolescents' Diets: A Systematic Review of the Underlying Social and Physiological Mechanisms. *Advances in Nutrition, 13*(3), 913–937. doi:10.1093/advances/nmac018 PMID:35218190

Singh, A. (2020). Consumer Perception and Attitude towards Online Food Ordering. International Journal of Research in Engineering, Science and Management, 3(5), May-2020 www.ijresm.com | ISSN (Online): 2581-5792.

Singh, T., Shukla, S., Kumar, P., Wahla, V., Bajpai, V. K., & Rather, I. A. (2017). Application of nanotechnology in food science: Perception and overview. *Frontiers in Microbiology, 8*, 1501. doi:10.3389/fmicb.2017.01501 PMID:28824605

Sinha Ray, S., & Okamoto, M. (2003). Polymer/layered silicate nanocomposites: A review from preparation to processing. *Progress in Polymer Science (Oxford), 28*(11), 11. doi:10.1016/j.progpolymsci.2003.08.002

Slovenia. (2023a). A sustainability label to delight and Guide all foodies who care about the environment. Retrieved from https://www.tasteslovenia.si/en/taste-slovenia/european-region-of-gastronomy-2021/slovenia-green-cuisine/ Access Date: 20.06.2023.

Slovenia. (2023b). News from Slovenia-The first International Day of Zero Waste is here. Retrieved from https://www.slovenia.info/en/press-centre/news-of-the-tourism-press-agency/23006-the-first-international-day-of-zero-waste-is-here Access Date: 20.06.2023.

Smith, P. R., & Zook, Z. (2016). *Marketing Communications: Offline and Online Integration, Engagement and Analytics*. Kogan Page.

Soares, S. (2011). *Insects au Gratin*. Sunsan Soares. http://www. susanasoares.com/index.php?id=79

Soares, S., & Forkers, A. (2014). Insects Au Gratin - an investigation into the experiences of developing a 3D printer that uses insect protein-based flour as a building medium for the production of sustainable food. In E. Bohemia, A. Eger, W. Eggink, A. Kovacevic, B. Parkinson, & W. Wits (Eds.), *DS 78: Proceedings of the E&PDE 2014 16th international Conference on Engineering and product design*, (pp. 426–431). The Design Society & Institution of Engineering Designers. https://www.designsociety.org/p ublication/35727/ds_78_proceedings_of_the_e_pde_2014_16th_international_conference_on_engineering_and_product_design_university_of_twente_the_netherlands.

Soltani, M., Alimardani, R., Mobli, H., & Mohtasebi, S. (2015). Modified atmosphere packaging: A progressive technology for shelf-life extension of fruits and vegetables. *Journal of Applied Packaging Research, 7*, 2.

Songür, A. N., & Çakıroğlu, F. P. (2016). Gıda Kayıpları ve Atık Yönetimi. *Türkiye Klinikleri Nutrition And Dietetics-Special Topics, 2*(3), 21–26.

Soon, M. J. (2020). Consumers' Awareness and Trust Toward Food Safety News on Social Media in Malaysia. *Journal of Food Protection, 83*(3), 452–459. doi:10.4315/0362-028X.JFP-19-415 PMID:32065648

Sopher, C. D., Graham, D. M., Rice, R. G., & Strasser, J. H. (2002). Studies on the use of ozone in production agriculture and food processing. In *Proceedings of the International Ozone Association, Pan American Group, Raleigh-Durham,* NC, USA, 18–22 May, 1–15.

Soria, A. C., & Villamiel, M. (2010). Effect of ultrasound on the technological properties and bioactivity of food: A review. *Trends in Food Science & Technology, 21*(7), 323–331. doi:10.1016/j.tifs.2010.04.003

Spence, C., Okajima, K., Cheok, A. D., Petit, O., & Michel, C. (2016). Eating with our eyes: From visual hunger to digital satiation. *Brain and Cognition, 110*, 53–63. doi:10.1016/j.bandc.2015.08.006 PMID:26432045

Spink, J., Ortega, D. L., Chen, C., & Wu, F. (2017). Food fraud prevention shifts the food risk focus to vulnerability. *Trends in Food Science & Technology*, *62*, 215–220. doi:10.1016/j.tifs.2017.02.012

Sputnik. (2019). *Türkiye Yılda 214 Milyar Liralık Gıda İsraf Ediyor*. Sputnik. https://sputniknews.com.tr/20190929/turkiye-yilda-214-milyar-liralik-gida-israf-ediyor-1040278150.html

Spyce. (2018). *Culinary Excellence Elevated by Technology*. Spyce. https://www.spyce.com/

Statista. (2023) *Food - Worldwide*. Statista. https://www.statista.com/outlook/cmo/food/worldwide

Steinhart, H. (2006). Novel foods and novel processing techniques as threats and challenges to a hypersensitive world. *Allergy Matters: New Approaches to Allergy Prevention and Management.*, *10*, 65–77. doi:10.1007/1-4020-3897-6_8

Stenger, T. (2014). Social media and online reputation management as practice: First steps towards social CRM? *International Journal of Technology and Human Interaction*, *10*(4), 49–64. doi:10.4018/ijthi.2014100104

Stranieri, S., Riccardi, F., Meuwissen, M. P., & Soregaroli, C. (2021). Exploring the impact of blockchain on the performance of agri-food supply chains. *Food Control*, *119*, 107495. doi:10.1016/j.foodcont.2020.107495

Strong, A. I. (2016). Applications of artificial intelligence & associated technologies. *Science [ETEBMS-2016]*, *5*(6).

Stuart, T. (2009). *Waste: Uncovering the Global Food Scandal*. WW Norton & Company.

Suave, J., Dall'agnol, E. C., Pezzin, A. P. T., Silva, D. A. K., Meier, M. M., & Soldi, V. (2006). Microencapsulação: Inovaçãoemdiferentesáreas. *RevistaSaúde e Ambiente*, *7*, 12–20.

Sun, J., Peng, Z., Zhou, W., Fuh, J. Y. H., Hong, G. S., & Chiu, A. (2015). A review on 3D printing for customized food fabrication. *Procedia Manufacturing*, *1*, 308–319. doi:10.1016/j.promfg.2015.09.057

Sun, J., Zhou, W., Yan, L., Huang, D., & Lin, L. (2018). Extrusion-based food printing for digitalized food design and nutrition control. *Journal of Food Engineering*, *220*, 1–11. doi:10.1016/j.jfoodeng.2017.02.028

Sun, Y., Li, X., Xu, J., Li, W., & Li, D. (2020). Detection of 2, 4-dinitrophenol by fluorescence quenching of carbon dots based on the inner filter effect. *Analytical Methods*, *12*(6), 706–710. doi:10.1039/c9ay02590k

Sustainable Gastronomy Initiatives Map. (2023). Retrieved from https://www.google.com/maps/d/u/0/viewer?mid=10_pkm4ERhUZrsZDLQDeL6icHGgesnxVf&ll=-3.81666561775622e-14%2C0&z=1 Access Date: 21.06.2023.

Swan, M. (2015). *Blockchain: Blueprint for a new economy*. O'Reilly Media, Inc.

Symons, M. (2013). The rise of the restaurant and the fate of hospitality. *International Journal of Contemporary Hospitality Management*, *25*(2), 247–263. https://www.emerald.com/insight/content/doi/10.1108/09596111311301621/full/html. doi:10.1108/09596111311301621

T.C. Tarım ve Orman Bakanlığı. (2020). *Türkiye'nin Gıda Kayıpları ve İsrafının Önlenmesi, Azaltılması ve Yönetimine İlişkin Ulusal Strateji Belgesi ve Eylem Planı*. Birleşmiş Milletler Gıda ve Tarım Örgütü.

Taherdoost, H. (2023). Smart Contracts in Blockchain Technology: A Critical Review. *Information (Basel)*, *14*(2), 117. doi:10.3390/info14020117

Talens, P., Martínez-Navarrete, N., Fito, P., & Chiralt, A. (2002). Changes in optical and mechanical properties during osmodehydrofreezing of kiwi fruit. *Innovative Food Science & Emerging Technologies*, *3*(2), 191–199. doi:10.1016/S1466-8564(02)00027-9

Tan, W. C., & Sidhu, M. S. (2022). Review of RFID and IoT integration in supply chain management. *Operations Research Perspectives*, *9*, 00229.

Tappi, S., Gozzi, G., Vannini, L., Berardinelli, A., Romani, S., Ragni, L., & Rocculi, P. (2016). Cold plasma treatment of fresh cut melon stabilization. *Innovative Food Science & Emerging Technologies*, *33*, 225–233. doi:10.1016/j.ifset.2015.12.022

Taylor, P., Steenmans, K., & Steenmans, I. (2020). Blockchain technology for sustainable waste management. *Frontiers in Political Science, 15.*

TDK. (2020). *İsraf.* TDK. https://sozluk.gov.tr/

Team, K. (no date) *Keras Documentation: Character-level text generation with LSTM.* Keras. https://keras.io/examples/generative/lstm_character_level_text_generation/

Tekiner, İ. H., Mercan, N. N., Kahraman, A., & Özel, M. (2021). Dünya ve Türkiye'de Gıda İsrafı ve Kaybına Genel Bir Bakış. *İstanbul Sabahattin Zaim Üniversitesi Fen Bilimleri Enstitüsü Dergisi, 3*(2), 123-128.

Tekiner, E. (2023). *Gıda İsrafı ve Çevre Koruma Bilinci: Kocaeli İli'nde Bir Uygulama.* Uludağ Üniversitesi Fen Bilimleri Enstitüsü.

Teoh, C. W., Koay, K. Y., & Chai, P. S. (2022). The role of social media in food waste prevention behaviour. *British Food Journal, 124*(5), 1680–1696. doi:10.1108/BFJ-04-2021-0368

Termeer, C. J., Drimie, S., Ingram, J., Pereira, L., & Whittingham, M. J. (2018). A Diagnostic Framework for Food System Governance Arrangements: The Case of South Africa. *NJAS Wageningen Journal of Life Sciences*, *84*(1), 85–93. doi:10.1016/j.njas.2017.08.001

Thakur, M. S., & Ragavan, K. V. (2013). Biosensors in food processing. *Journal of Food Science and Technology*, *50*(4), 625–641. doi:10.100713197-012-0783-z PMID:24425965

The Atlantic (2020). The Struggle of a Zero-Waste Restaurant. Retrieved from https://www.youtube.com/watch?v=Gj38m08QLoE Access Date: 13.06.2023.

Thelwall, M. (2019). *Sentiment analysis for tourism. Big Data and Innovation in Tourism, Travel, and Hospitality*. Springer.

Thévenot, D. R., Toth, K., Durst, R. A., & Wilson, G. S. (2001). Electrochemical biosensors: Recommended definitions and classification1international union of pure and applied chemistry: Physical chemistry division, commission i.7 (biophysical chemistry); analytical chemistry division, commission v.5 (electroanalytical chemistry)1. *Biosensors & Bioelectronics*, *16*(1), 121–131. doi:10.1016/S0956-5663(01)00115-4 PMID:11261847

Thorat, S. B., Nayak, S. K., & Dandale, J. P. (2010). *Facial recognition technology: An analysis with scope in India.* arXiv preprint arXiv:1005.4263.

Thyberg, K. L. (2015). *Food Waste and Sustainability: Quantifying Food Waste Disposal and Evaluating the Environmental Impacts of Technologies and Policies.* [Thesis, Stony Brook University].

Ticaret Bakanlığı, T. C. (2018). *Türkiye İsraf Raporu.* Tüketicinin Korunması ve Piyasa Gözetimi Genel Müdürlüğü.

Tillemann, T., Price, A., Tillemann-Dick, G., & Knight, A. (2019). *The blueprint for blockchain and social innovation.* New America.

Tirkey, K. (2011). *Effect of Harvest Time and Postharvest Operations on the Quality and Quantity of Sesame (Sesamum indicum L.).* [MSc Agriculture Thesis: Jawaharlal Nehru Krishi Vishwa Vidyalaya, Jabalpur, College of Agriculture].

Tiwari, B. K., Donnell, C. P. O., Muthukumarappan, K., & Cullen, P. J. (2009). Ascorbic acid degradation kinetics of sonicated orange juice during storage and comparison with thermally pasteurized juice. *Lebensmittel-Wissenschaft + Technologie*, *42*(3), 700–704. doi:10.1016/j.lwt.2008.10.009

Tiwari, R. B. (2005). Application of osmo-air dehydration for processing of tropical fruits in rural areas. *Indian Food Ind, 24*(6), 62–69.

Todorović, V., Milić, N., & Lazarević, M. (2019). Augmented Reality in Food production traceability - Use case. *EUROCON 2019 - 18th International Conference on Smart Technologies.* IEEE10.1109/EUROCON.2019.8861734

Tokkozhina, U., Martins, A. L., & Ferreira, J. C. (2022). Use of Blockchain Technology to Manage the Supply Chains: Comparison of Perspectives between Technology Providers and Early Industry Adopters. *Journal of Theoretical and Applied Electronic Commerce Research, 17*(4), 1616–1632. doi:10.3390/jtaer17040082

Tomasevic, I., Putnik, P., Valjak, F., Pavlic, B., Sojic, B., Markovinovic, A. B., & Kovacevic, D. B. (2021). 3D printing as novel tool for fruit-based functional food production. *Current Opinion in Food Science, 41,* 138–145. . COFS.2021.03.015 doi:10.1016/J

Tönnissen, S., & Teuteberg, F. (2020). Analysing the impact of blockchain-technology for operations and supply chain management: An explanatory model drawn from multiple case studies. *International Journal of Information Management, 52*, 101953. doi:10.1016/j.ijinfomgt.2019.05.009

Tribhuvan, Aditya (2020). A Study on Consumers Perception on Food Apps.2020 *International Journal of Advance Research and Innovative Ideas in Education,* ISSN (O)-2395-4396. 6(4).

Tribhuvan, A. (2020). Analysis of customer attitude towards Electronic Food Ordering. *International Journal of Advance Research and Innovative Ideas in Education, 6*(4), 36.

Tribst, A. A. L., de Castro Leite Júnior, B. R., de Oliveira, M. M., & Cristianini, M. (2016). High pressure processing of cocoyam, Peruvian carrot and sweet potato: Effect on oxidative enzymes and impact in the tuber color. *Innovative Food Science & Emerging Technologies, 34*, 302–309. doi:10.1016/j.ifset.2016.02.010

Troitino, C. (2018, June 21). Meet the world's first fully automated burger robot: Creator debuts the Big Mac killer. *Forbes.* https://www.forbes.com/sites/christinatroitino/2018/06/21/meet-the-worlds-first-fully-automated-burger-robot-creator-debuts-the-big-mac-killer/#7959e2926a89

Tsoukas, V., Gkogkidis, A., Kampa, A., Spathoulas, G., & Kakarountas, A. (2022). Enhancing Food Supply Chain Security through the Use of Blockchain and TinyML. *Information (Basel), 13*(5), 213. Advance online publication. doi:10.3390/info13050213

Tuncer, I. (2020). Customer Experience in the Restaurant Industry: Use of Smart Technologies. Handbook of Research on Smart Technology Applications in the Tourism Industry, 254–272. doi:10.4018/978-1-7998-1989-9.ch012

Tuomi, A., Tussyadiah, I. P., & Steinmetz, J. (2020). Applications and implications of service robots in hospitality. *Cornell Hospitality Quarterly*, 1–16.

Turner, A. P. F., Karube, I., & Wilson, G. S. (1987). *Biosensors, fundamentals and applications.* Oxford University Press.

Tzima, S., Styliaras, G., Zaragas, H., & Bassounas, A. (2018). The Effects of an Educational Augmented Reality Application on the Interpretation and Communication of Cultural Assets: The "Greek Type" Watermill's Case. *Proceedings of EdMedia: World Conference on Educational Media and Technology*, (pp. 909–925).

Ubbink, J., & Kruger, J. (2006). Physical approaches for the delivery of active ingredients in foods. *Trends in Food Science & Technology, 17*(5), 244–254. doi:10.1016/j.tifs.2006.01.007

UDA. (2014). *K.* Social Media, Its Impact on Food Culture in the World.

Uddin, M. A. (2021). A survey on the adoption of blockchain in IoT: challenges and solutions. *Blockchain: Research and Applications, 2*(2), 100006.

UNEP. (2021). *UNEP Food Waste Index Report 2021.* UNEP. https://www.unep.org/resources/report/unep-food-waste-index-report-2021

UNH Paul College of Business and Economics. (2020). Inside the Classroom: Ecogastronomy. Retrieved from https://www.youtube.com/watch?v=QuKyDeljE8Q Access Date: 13.06.2023.

United Nations. (2017) *World population projected to reach 9.8 billion in 2050, and 11.2 billion in 2100No Title, Department of Economic and Social Affairs.* UN. https://www.un.org/development/desa/en/news/population/world-population-prospects-2017.html (Accessed: 12 June 2023).

United Nations. (2023) Sustainable Gastronomy Day 18 June. Retrieved from https://www.un.org/en/observances/sustainable-gastronomy-day#:~:text=Sustainable%20gastronomy%2C%20therefore%2C%20means%20cuisine,and%20eventually%20to%20our%20plates. Access Date: 15.06.2023.

UNtoday. (2023). Sustainable Gastronomy Day June 18, 2023. Retrieved from https://nationaltoday.com/sustainable-gastronomy-day/ Access Date: 20.06.2023.

US Embassy BSB. (2022a). Sustainable Gastronomy Ep. 1. Retrieved from https://www.youtube.com/watch?v=j1BBpBznugU Access Date: 17.06.2023.

US Embassy BSB. (2022b). Sustainable Gastronomy Ep.2. Retrieved from https://www.youtube.com/watch?v=JRzgeRCw7Pk&list=PLjWN0y_JL94pQWTXt-Qq_EYeIN-dNKJCo&index=2 Access Date: 17.06.2023.

US Embassy BSB. (2022c). Sustainable Gastronomy Ep.3. Retrieved from https://www.youtube.com/watch?v=jy1iUOv4L0w&list=PLjWN0y_JL94pQWTXt-Qq_EYeIN-dNKJCo&index=3 Access Date: 17.06.2023.

Uyar, R., Bedane, T. F., Erdogdu, F., Palazoglu, T. K., Farag, K. W., & Marra, F. (2015). Radio-frequency thawing of food products—A computational study. *Journal of Food Engineering, 146*, 163–171. doi:10.1016/j.jfoodeng.2014.08.018

van Erp, M., Reynolds, C., Maynard, D., Starke, A., Ibáñez Martín, R., Andres, F., Leite, M. C. A., Alvarez de Toledo, D., Schmidt Rivera, X., Trattner, C., Brewer, S., Adriano Martins, C., Kluczkovski, A., Frankowska, A., Bridle, S., Levy, R. B., Rauber, F., Tereza da Silva, J., & Bosma, U. (2021). Using Natural Language Processing and Artificial Intelligence to Explore the Nutrition and Sustainability of Recipes and Food. *Front. Artif. Intell., 3*, 621577. doi:10.3389/frai.2020.621577 PMID:33733227

van Hilten, M., Ongena, G., & Ravesteijn, P. (2020). Blockchain for Organic Food Traceability: Case Studies on Drivers and Challenges. *Frontiers in Blockchain, 3*(September), 1–13. doi:10.3389/fbloc.2020.567175

Vanderroost, M., Ragaert, P., Devlieghere, F., & De Meulenaer, B. (2014). Intelligent food packaging: The next generation. *Trends in Food Science & Technology, 39*(1), 47–62. doi:10.1016/j.tifs.2014.06.009

Vanderroost, M., Ragaert, P., Verwaeren, J., De Meulenaer, B., De Baets, B., & Devlieghere, F. (2017). The digitization of a food package's life cycle: Existing and emerging computer systems in the logistics and post-logistics phase. *Computers in Industry, 87*, 15–30. doi:10.1016/j.compind.2017.01.004

Vasile, C. (2018). Polymeric nanocomposites and nanocoatings for food packaging: A review. *Materials (Basel), 11*(10), 1834. doi:10.3390/ma11101834 PMID:30261658

Veebrant. (2022). *Dining at Sublimotion in Ibiza Is a Mind-Blowing Multisensory Experience.* veebrant.com.

Vejdan, A., Ojagh, S. M., Adeli, A., & Abdollahi, M. (2016). Effect of TiO_2 nanoparticles on the physico-mechanical and ultraviolet light barrier properties of fish gelatin/agar bilayer film. *Lebensmittel-Wissenschaft + Technologie*, *71*(2), 88–95. doi:10.1016/j.lwt.2016.03.011

Verdouw, C., Wolfert, S., & Tekinerdogan, B. (2016). Internet of Things in agriculture. CABI Reviews, 1–12. doi:10.1079/PAVSNNR201611035

Verdouw, C. N., Wolfert, J., Beulens, A. J. M., & Rialland, A. (2016). Virtualization of food supply chains with the internet of things. *Journal of Food Engineering*, *176*, 128–136. doi:10.1016/j.jfoodeng.2015.11.009

Vicente, A. A., Pereira, R. N., Penna, T. C. V., & Knirsch, M. (2016). Electricity effects on microorganisms and enzymes. In H. S. Ramaswamy, M. Marcotte, S. Sastry, & K. Abdelrahim (Eds.), *Ohmic heating in food processing* (pp. 93–102). CRC Press.

Vignali, G., Bertolini, M., Bottani, E., Di Donato, L., Ferraro, A., & Longo, F. (2018). Design and testing of an augmented reality solution to enhance operator safety in the food industry. *International Journal of Food Engineering*, *14*(2). doi:10.1515/ijfe-2017-0122

Vilarino, M. V., Franco, C., & Quarrington, C. (2017). Food Loss and Waste Reduction as an Integral Part of a Circular Economy. *Frontiers in Environmental Science*, *5*, 1–5. doi:10.3389/fenvs.2017.00021

Vladitsi, M., Nikolaou, C., Kalogiouri, N. P., & Samanidou, V. F. (2022). Analytical Methods for Nanomaterial Determination in Biological Matrices. *Methods and Protocols*, *5*(4), 61. doi:10.3390/mps5040061 PMID:35893587

Volo, S. (2017). Emotions in Tourism: From Exploration to Design. In D. Fesenmaier & Z. Xiang (Eds.), *Design Science in Tourism. Tourism on the Verge* (pp. 31–40). Springer. doi:10.1007/978-3-319-42773-7_3

Vu, N., Ghadge, A., & Bourlakis, M. (2021). Blockchain adoption in food supply chains: A review and implementation framework. *Production Planning and Control*, 1–18.

Walker, J. R. (2021). *The restaurant: from concept to operation*. John Wiley & Sons.

Wang, H. Y. (2011). Exploring the factors of gastronomy blogs influencing readers' intention to taste. *International Journal of Hospitality Management*, *30*(3), 503–514. doi:10.1016/j.ijhm.2010.07.009

Wang, H., Lin, G., Hoi, S. C., & Miao, C. (2022). Learning structural representations for recipe generation and food retrieval. *IEEE Transactions on Pattern Analysis and Machine Intelligence*, 1–15. doi:10.1109/TPAMI.2022.3181294 PMID:35687622

Wang, H., Ma, B., Cudjoe, D., Farrukh, M., & Bai, R. (2023). What influences students' food waste behaviour in campus canteens? *British Food Journal*, *125*(2), 381–395. doi:10.1108/BFJ-10-2021-1103

Wang, P., Zhang, X., Li, J., Zhang, L., Liu, Y., & Wang, J. (2019). Molecularly imprinted polymer modified electrode for the determination of atrazine in lake water. *Journal of Electroanalytical Chemistry (Lausanne, Switzerland)*, *847*, 113261.

Wang, Q. J., Mielby, L. A., Thybo, A. K., Bertelsen, A. S., Kidmose, U., Spence, C., & Byrne, D. V. (2019). Sweeter together? Assessing the combined influence of product-related and contextual factors on perceived sweetness of fruit beverages. *Journal of Sensory Studies*, *34*(3), e12492. doi:10.1111/joss.12492

Wang, Y., Li, X., Sun, G., Dong, L., & Pan, Z. (2014). A comparison of dynamic mechanical properties of processing-tomato peel as affected by hot lye and infrared radiation heating for peeling. *Journal of Food Engineering*, *126*, 27–34. doi:10.1016/j.jfoodeng.2013.10.032

Watson, C. (2021). *IBM Chef Watson*. IBM Research. https://www.research.ibm.com/artificial-intelligence/project-chef-watson.shtml

Watzke, H., & German, J. (2011). Personalizing foods. In An integrated approach to new food product development, Ed. Howard, R., Moskowitz, I. Sam, S., Tim, S., Boca, R., 133-173. CRC Press: US.

Webrazzi. (2023). https://webrazzi.com/2023/06/23/yapay-zeka-tarifleriyle-hazirlanan-icecekler-mcdonalds-turkiyede/

Weizenbaum, J. (1966). ELIZA - A Computer Program For the Study of Natural Language Communication Between Man And Machine. *Communications of the ACM*, *9*(1), 36–45. doi:10.1145/365153.365168

Weking, A. N., Suyoto, & Santoso, A. J. (2020). *A Development of Augmented Reality Mobile Application to Promote the Traditional Indonesian Food.*

Wen, H., Pookulangara, S., & Josiam, B. M. (2022). A comprehensive examination of consumers' intentions to use food delivery apps. *British Food Journal*, *124*(5), 1737–1754. doi:10.1108/BFJ-06-2021-0655

Whitesides, G. M. (2006). The origins and the future of microfluidics. *Nature*, *442*(7101), 368–373. doi:10.1038/nature05058 PMID:16871203

Wilkins, K., & Mackenzie, S. G. (2007). Work injuries. *Health Reports*, *18*(3), 25–42. https://pubmed.ncbi.nlm.nih.gov/17892250/ PMID:17892250

Wolfert, S., Ge, L., Verdouw, C., & Bogaardt, M.-J. (2017). Big data in smart farming–a review. *Agricultural Systems*, *153*, 69–80. doi:10.1016/j.agsy.2017.01.023

Wongsa-Ngasri, P., & Sastry, S. K. (2016). Tomato peeling by ohmic heating: Effects of lye-salt combinations and post-treatments on weight loss, peeling quality and firmness. *Innovative Food Science & Emerging Technologies*, *34*, 148–153. doi:10.1016/j.ifset.2016.01.013

World Tourism Organization (UNWTO). (2023). Gastronomy. Retrieved from https://www.unwto.org/gastronomy Access Date: 12.04.2023.

WRAP. (2009). *Household Food and Drink Waste in the UK*. https://wrap.org.uk/sites/default/files/2020-12/Household-Food-and-Drink-Waste-in-the-UK-2009.pdf:

Writer, S. (2022, February 17). *McDonald's & Sonic Self Order Kiosk Ai Technology*. Kiosk Industry. https://kioskindustry.org/mcdonalds-sonic-optimizing-with-self-order-ai-technology/

Wu, H. K., Lee, S. W. Y., Chang, H. Y., & Liang, J. C. (2013). Current status, opportunities and challenges of augmented reality in education. *Computers & Education*, *62*, 41–49. doi:10.1016/j.compedu.2012.10.024

Wunderlich, S., & Martinez, N. M. (2018). Conserving natural resources through food loss reduction: Production and consumption stages of the food supply chain. *International Soil and Water Conservation Research*, *6*(4), 331–339. doi:10.1016/j.iswcr.2018.06.002

Wu, S., Lai, E. P., Zhang, Y., & Zhao, Y. (2019). Development of a highly sensitive and selective electrochemical sensor for bisphenol A detection using molecularly imprinted polymer modified graphene oxide nanocomposites. *Talanta*, *197*, 135–142.

Wu, W., Zhang, A., van Klinken, R. D., Schrobback, P., & Muller, J. M. (2021). Consumer Trust in Food and the Food System: A Critical Review. *Foods (Basel, Switzerland)*, *10*(10), 2490. doi:10.3390/foods10102490 PMID:34681539

Wu, Y., Ngai, E. W., Wu, P., & Wu, C. (2020). Fake online reviews: Literature review, synthesis, and directions for future research. *Decision Support Systems*, *132*, 113280. doi:10.1016/j.dss.2020.113280

Xiao, C., & Zhao, L. (2022). Robotic chef versus human chef: The effects of anthropomorphism, novel cues, and cooking difficulty level on food quality prediction. *International Journal of Social Robotics*, *14*(28), 1697–1710. https://www.mckinsey.com/featured-insights/future-of-work/jobs-lost-jobs-gained-what-the-future-of-work-will-mean-for-jobs-skills-and-wages. doi:10.100712369-022-00896-9 PMID:35910296

Xie, J., Chai, J. J. K., O'Sullivan, C., & Xu, J. L. (2022). Trends of Augmented Reality for Agri-Food Applications. *Sensors (Basel)*, *22*(8333), 8333. doi:10.339022218333 PMID:36366030

Xu, X. (2019). Case study: AgriDigital: Blockchain technology in the trade and finance of agriculture supply chains. *Architecture for blockchain applications*, 239–255.

Xu, J., Jia, Q., Liu, Y., Zhang, H., Wang, S., & Yang, F. (2021). A rapid and highly sensitive electrochemical sensor for detecting patulin based on molecularly imprinted polymers modified copper oxide nanosheets. *Food Chemistry*, *337*, 127730.

Yadav, N., Luthra, S., & Garg, D. (2023). Blockchain Technology for Sustainable Supply Chains : A Network Cluster Analysis for Future Research Propositions. *Environmental Science and Pollution Research International*, *30*(24), 64779–64799. doi:10.100711356-023-27049-3 PMID:37086323

Yadav, V. S., Singh, A. R., Raut, R. D., & Govindarajan, U. H. (2020). Blockchain technology adoption barriers in the Indian agricultural supply chain: An integrated approach. *Resources, Conservation and Recycling*, *161*, 104877. doi:10.1016/j.resconrec.2020.104877

Yağcı, S., Altan, A., Göğüs, F., & Maskan, M. (2006). Gıda Atıklarının Alternatif Kullanım Alanları. Türkiye 9. Gıda Kongresi: 24-26 Mayıs 2006, (pp. 24-26). Bolu.

Yang, C., Yu, L., Wang, L., Guan, Q., & Gao, M. (2020). Preparation of a surface molecularly imprinted polymer and its application in the selective separation and detection of theophylline in food samples. *Food Analytical Methods*, *13*(8), 1599–1611.

Yang, F., Zhang, M., & Bhandari, B. (2017). Recent development in 3D food printing. *Critical Reviews in Food Science and Nutrition*, *57*(14), 3145–3153. doi:10.1080/10408398.2015.1094732 PMID:26479080

Yang, H., Liu, Y., Zhang, R., & Zhang, Z. (2019). Molecularly imprinted polymer for the selective recognition of ovalbumin in egg white. *Journal of the Science of Food and Agriculture*, *99*(6), 2816–2822.

Yan, X., Wang, J., & Chau, M. (2015). Customer revisit intention to restaurants: Evidence from online reviews. *Information Systems Frontiers*, *17*(3), 645–657. doi:10.100710796-013-9446-5

Yarış, A. & Sezgin, A. (2017). Food Packaging: Glass And Plastic. *Researches On Science And Art In 21st Century*.

Yiğitoğlu, V., & Cerrah, S. (2021). Yiyecek Artıklarıyla İlgili Yasal Düzenlemelerin Değerlendirilmesi. In G. S. Ergün (Ed.), *Multidisipliner Turizm Çalışmaları* (pp. 105–115). Çizgi Kitabevi Yayınları.

Yılmaz, M. (2022). *Konaklama İşletmelerinde Helal Konsept Uygulamalarının Gıda Atığı ve İsrafı Üzerindeki Etkisi*. Ankara Hacı Bayram Veli Üniversitesi Lisansüstü Eğitim Enstitüsü.

Yim, M. Y. C., & Yoo, C. Y. (2022). Are Digital Menus Really Better than Traditional Menus? The Mediating Role of Consumption Visions and Menu Enjoyment. doi:10.1016/j.intmar.2020.01.001

Yin, Z., Hoffmann, M., & Jiang, S. (2018). Sludge disinfection using electrical thermal treatment: The role of ohmic heating. *The Science of the Total Environment*, *615*, 262–271. doi:10.1016/j.scitotenv.2017.09.175 PMID:28972902

Young, I., & Waddell, L. (2016). Barriers and Facilitators to Safe Food Handling among Consumers: A Systematic Review and Thematic Synthesis of Qualitative Research Studies. *PLoS One*, *11*(12), e0167695. doi:10.1371/journal.pone.0167695 PMID:27907161

Yousuf, B., Sun, Y., & Wu, S. (2022). Lipid and Lipid-containing Composite Edible Coatings and Films. *Food Reviews International*, *38*(sup1), 574–597. doi:10.1080/87559129.2021.1876084

Yu, C. E., & Zhang, X. (2020). The embedded feelings in local gastronomy: A sentiment analysis of online reviews. *Journal of Hospitality and Tourism Technology*, *11*(3), 461–478. doi:10.1108/JHTT-02-2019-0028

Yucel, U. (2016). *Intelligent Packaging. Reference Module in Food Science*. Elsevier., doi:10.1016/B978-0-08-100596-5.03374-6

Yusaf, T., & Al-Juboori, R. A. (2014). Alternative methods of microorganism disruption for agricultural applications. *Applied Energy*, *114*, 909–923. doi:10.1016/j.apenergy.2013.08.085

Yu, Z., Zang, H., & Wan, X. (2020, November). Routing enforced generative model for recipe generation. In *Proceedings of the 2020 Conference on Empirical Methods in Natural Language Processing (EMNLP)* (pp. 3797-3806). IEEE. 10.18653/v1/2020.emnlp-main.311

Zhang, H., Bai, W., Zhang, M., & Fang, Y. (2016). Development of an immunochromatographic assay for the rapid detection of chloramphenicol in milk. *Food Chemistry*, *199*, 216–220.

Zhang, H., Wu, L., Cai, X., & Lin, X. (2018). Selective detection of malic acid in apple juice using a molecularly imprinted polymer-coated surface plasmon resonance sensor. *Food Chemistry*, *245*, 1125–1131.

Zhang, J., Zhang, X., Liu, W., Ji, M., & Mishra, A. R. (2022). Critical success factors of blockchain technology to implement the sustainable supply chain using an extended decision-making approach. *Technological Forecasting and Social Change*, *182*, 121881. doi:10.1016/j.techfore.2022.121881

Zhang, M., Tang, J., Mujumdar, A. S., & Wang, S. (2006). Trends in microwave-related drying of fruits and vegetables. *Trends in Food Science & Technology*, *17*(10), 524–534. doi:10.1016/j.tifs.2006.04.011

Zhang, Y., Yang, L., Xu, W., Li, H., & Li, J. (2020). Molecularly imprinted electrochemical sensor for the selective detection of glucose in honey. *Analytical Methods*, *12*(4), 453–459.

Zhang, Y., Zhao, L., & Qian, C. (2017). Modeling of an IoT-enabled supply chain for perishable food with two-echelon supply hubs. *Industrial Management & Data Systems*, *117*(9), 1890–1905. doi:10.1108/IMDS-10-2016-0456

Zhang, Z., Ye, Q., Law, R., & Li, Y. (2010). The impact of e-word-of-mouth on the online popularity of restaurants: A comparison of consumer reviews and editor reviews. *International Journal of Hospitality Management*, *29*(4), 694–700. doi:10.1016/j.ijhm.2010.02.002

Zoran, A., & Coelho, M. (2011). Cornucopia: The Concept of Digital Gastronomy. *Leonardo*, *44*(5), 425–431. doi:10.1162/LEON_a_00243

Zylberberg, T. (1989). THESEUS initiative in management. *Industry and Higher Education*, *3*(4), 243–244. doi:10.1177/095042228900300420

About the Contributors

Şule Aydın was born in Artvin. She graduated from Mersin University, Faculty of Tourism, Department of Tourism Management in 1994. After completing her graduate education in Department of Tourism Management at Çanakkale Onsekiz Mart University between 1996-1998. She completed doctorate education at Dokuz Eylül University in Department of Tourism Management between 1998-2004. Prof. Dr. Aydın was a faculty member at Çanakkale Onsekiz Mart University (2004-2011), Trakya University (2012) and Batman University (2015-2016). Since 2017, Prof. Dr. Şule AYDIN has been an on-duty dean Faculty of Tourism at Nevşehir Hacı Bektaş Veli University. Her research interest includes hotel management, organizational behavior, human research management

Eda Özgül Katlav pursued her undergraduate, graduate, and doctoral studies at Nevşehir Hacı Bektaş Veli University's Institute of Social Sciences, specializing in Tourism Management. Following her studies, she undertook postdoctoral research at the University of South Florida's M3 Center for Hospitality Technology & Innovation from 2021 to 2022. She currently holds a position as an associate professor in the Department of Tourism Management at the Faculty of Tourism. Her research interests include employee behavior, rural tourism, and tourism technologies.

Koray Çamlıca born in İzmir. He completed his undergraduate degree at Gazi University's Faculty of Commerce and Tourism Education, specializing in Tourism Management and Hospitality Management. He also achieved a Bachelor's in Business Administration and Tourism Guidance. His master's degree, completed in 2014, was from Hacettepe University's Social Sciences Institute in Tourism Management. He earned his PhD from Nevşehir Hacı Bektaş Veli University's Institute of Social Sciences, majoring in Tourism Management, in 2017. He began his academic journey as a Research Assistant at Bitlis Eren University's School of Tourism and Hotel Management in 2013. From 2014 to 2021, he served at Nevşehir Hacı Bektaş Veli University's Faculty of Tourism. Since 2021, he has been a Doctor Lecturer in the Department of Tourism Guidance at the same institution. As of 2023, he attained his associate professorship. His scholarly pursuits revolve around tourism marketing, sustainable tourism, cultural heritage, and consumer behavior in the realm of "Tourism" Science.

Firdevs Yönet Eren born in Mersin, Turkey, in 1994. She graduated first in her class from Nevşehir Hacı Bektaş Veli University's Gastronomy and Culinary Arts Department in 2016. She joined the same institution as a research assistant in 2017. She earned her master's from Gazi University's Gastronomy and Culinary Arts Department in 2018. In 2022, she achieved her doctoral degree from Nevşehir Hacı Bektaş Veli University, specializing in Gastronomy and Culinary Arts. Over the years, she has participated in numerous workshops, courses, and trainings related to gastronomy. Her research areas encompass culinary culture, gastronomic tourism, cultural heritage and gastronomy, viticulture, local products, and Turkish cuisine.

***

Salih Aka is Assistant Professor in Erzincan Binali Yıldırım University. He is interested in production management, multicriteria decision-making, artificial neural networks, simulation, social manufacturing and food supply chain.

Mitra Amini is an Assisitant professor of Marketing in the School of Management, Bennett University

Feridun Aydınlı graduated from Selçuk University, Department of Tourism Management. He is doing his master degree in Tourism Management at Selcuk University (TR). He has worked on local delicacies. He is currently participating in research on food safety. Areas of interest in the Department of Tourism Management are: Food Safety in Service Services, Local Tastes, Destination Marketing. Research Subject: Food Safety, Traditional Foods

Serkan Bertan works as a Research Assistant at the Pamukkale University, and then attended the PhD Program at Dokuz Eylul University in 2002 and completed his PhD studies within four years. After his PhD degree he joined to Pamukkale University at Department of Tourism and Hotel Management. In 2009 Dr. Serkan was invited as a fellow researcher to Department of Hotel, Restaurant and Tourism Management University of South Carolina for his Post PhD Studies. Assoc. Prof. Dr., Faculty of Tourism, Pamukkale University at Denizli, Turkey.

Birsen Bulut-Solak graduated from Ankara University with a BSc in Dairy Technology. She obtained a MSc and a PhD in Food Engineering from Selcuk University (TR). She was awarded an International Post Doctoral Research Fellowship by the Scientific and Technical Research Council of Turkey (TUBITAK). She fulfilled her post-doctoral research in Food and Nutrition Sciences at University College Cork (IRL) under the supervision of Dr. James Anthony O'Mahony between 2013 and 2015. She was a university lecturer in the Programme of Dairy Technology at Selcuk University between 2002 and 2013. Dr. Bulut-Solak is currently an Associate Professor in the Department of Gastronomy and Culinary Arts at Selcuk University, Konya (TR). She has also been an ambassador in Global Harmonization Initiative for Turkey since 2016.She is currently involved in three significant researches that are concerned with food safety, consumer acceptance of traditional foods and the determinant of food choice. She also works closely with others internationally (e.g. University College Cork and University of Illinois).

Vijayabanu Chidambaram is a Professor belonging to the School of Management in SASTRA University.

Chinna Gopi Simhadri received his B.Tech degree in Computer Science and Engineering from JNTU Kakinada University, India in 2010. He completed his M.Tech, in Computer Science from the KL University, India in 2012. Currently, he is pursuing Doctor of Philosophy (Ph.D.) in the field of

Computer Science and Engineering, VIT-AP University, Andhra Pradesh, India. His research interests include Digital Image Processing, Machine Learning and Deep Learning.

Yeliz Demir graduated from Akdeniz University, Faculty of Tourism/Food and Beverage Management (2014). She received his master's degree from Akdeniz University, Department of Tourism and Hotel Management (2016). She received his PhD degree from the Department of Tourism Management at Istanbul University (2022). She worked as a lecturer at Istanbul Gelişim University\Development Vocational School\Cookery Department (2017-2019). She then worked as a lecturer at Istinye University School\Hotel Restaurant Catering Services\Cookery Department (2019-2021). She worked as a lecturer in the Department of Gastronomy and Culinary Arts, Faculty of Tourism, Dokuz Eylul University between 2021-2022. She has been working as an Assistant Professor at the Faculty of Tourism at Isparta University of Applied Sciences since 2022. Basic fields of study; Gastronomy, Turkish Culinary Culture, Food Waste, Food and Beverage Businesses.

Rajni Gautam is working as Assistant Professor in Department of Physics, School of Basic and Applied Sciences, K.R. Mangalam University, Gurugram, Haryana, India. She completed her PhD from Semiconductor Device Reserach Lab, Department of Electronic Science, University of Delhi in 2014. She has total 35 publications in reputed Journals and conference proceedings. She has H-index of 10 and more than 300 citations. Her research interests include Synthesis and Characterization of doped metal based nanoparticles and their applications.

Kanwal Gul is a PhD student in the Department of Economics and Management at the University of Naples Federico II. Her main research interests are related to Sustainability, Food Waste Consciousness, Artificial Intelligence Food Supply Chain, and Blockchain Innovation in Sustainable Development. She has served as a Lecturer in the Management Department at the Institute of Business Management and Visiting Faculty in various educational Institutions in Pakistan.

Melike İmre is with the Department of Nutrition and Dietetics.

Hari Kishan Kondaveeti is working as an Associate Professor in the School of Computer Science Engineering (SCOPE), acting as Deputy Director of Engineering Clinics at VIT-AP University, Andhra Pradesh, India. He served as a Core Member of Innovation and Incubation Center (IIC). He is an IBM Certified Data Science Professional and NASSCOM Associate Analyst certified faculty and handling multiple Data Analytics courses. He worked as Research Associate in the Naval Science & Technological Laboratory, Andhra Pradesh, India. He is interested in teaching, scientific article writing, developing IoT solutions. He has filed several patents, and authored technical articles, conference papers, journals, and book chapters. He is acting as a reviewer in SCIE journals IEEE Sensors Journal, IET Image Processing, IET Electronics Letters, Frontiers of Computer Science, Journal of Super Computing, International Journal of Medical Informatics, IEEE Access and ESCI Journals International Journal of Information Technology Research (JITR), Journal of Information Retrieval Research (IJIRR). He acted as Convener for ICCIDE2022 and served in the committees of several conferences such as AISP-2021, ICCIDE-2021, ICECCME 2021, ISCDA-2020, AISP-2020, ICCIDE-2020, RACSPA-2019, ICACCI-2017, ICCSA-2017, etc. Dr. Kishan attended, organized, and acted as a resource person in several FDPs, workshops, refresher courses, international guest lectures, and technical talks in his domain and expertise. He received the Best

Guru Male Award 2020 by GISR Foundation, and Best Research with a patent award (2021), Beyond and Above Award (2019) by VIT-AP University, Amaravati, and Top 50 Tech-savvy Academicians in Higher Education Across India for the year 2019. He is a Life member in international professional bodies such as ISTE, SCRS, CSI, IUPRAI, ISSIP, IRED, ISRS, ISSE, SSI, CEGR, ISRD. Member in EAI, CSTA, IPRG, IAENG, SDIWC, ICST, ICTA, IEI, INFORMS, DSA, SCIEI, ICSES, IAASSE, IETE, ACM and IEEE. Fellow Member, International Scientific Research Organization for Science, Engineering, and Technology (ISROSET), India. Associate Member, Universal Association of Computer and Electronics Engineers (UACEE), USA. He is acting as Honorary Member of Asian Council of Science Editors.

Shanthala Mallikarjunaiah obtained her masters and doctorate degree in the field of Zoology from Kuvempu university. she has 16 years of teaching experience and published many international papers in the peer reviewed journals and currently working as teaching faculty in Bangalore university, Karnataka, India

Swapnil Morande is a data scientist with the apex policy think-tank of the Government of India. He holds a doctorate from the University of Naples, Italy and a Masters from Colorado State University, USA. Swapnil's decade-long global work exposure counts in the field of Information Systems with an exemplary record working on technologically advanced initiatives. He has served as Chairman of E-Commerce at the University of Mumbai, India with International publications encompassing Artificial Intelligence, Cloud computing and Blockchain. He believes in the 'Medici Effect' that catapults Design Thinking and aligned with the ecosystemic nature of digitization, seeks Informatics as a research interest.

Sandilyan Pagaldiviti is a PhD in Tourism and Hospitality from IGNOU and earned his Masters in Tourism also from the same University. He has also completed M. Phil in Tourism Management from Madurai Kamaraj University. Dr. Sandilyan is a Tourism and Hospitality professional with over 25 years of experience. In 1995, he started his career with the Taj Group of Hotels as an operations management trainee. In his career he has worked with various prestigious organisation like P&O cruises, Jaypee Group of Hotels, as well as leading academic Institutions like IHM – Hyderabad, Bradford-Rex School of Hospitality & Management, Singapore etc. Having started as a Chef, Dr. Sandilyan is amongst a handful of chefs in the country to hold a PhD degree. Dr. Sandilyan has undergone several advanced learning programmes from Ecole Hoteliere, Lausanne and is also certified by EHL as a "Qualified Learning Facilitator". Dr. Sandilyan has held various senior management positions like HOD for Hotel Management at KLEF – deemed to be University, Vice Principal at SRMIHM. In the industry he has been a General Manager for a QSR chain based at Bangalore as well as GM for SKHL.

Ananiyasri Raghavan is currently pursuing her dual degree of bachelors in Law and Business Administration in SASTRA University.

Célia M.Q. Ramos is a professor at ESGHT at the University of Algarve, Portugal, where she lectures Information Systems, Business Intelligence and Information and Communication Technology, was Director of the undergraduate Course in Marketing and President of the Pedagogical Council. She has authored book chapters, conference papers and journal articles. She is a reviewer of several international scientific journals. At the level of applied research, she has participated in several funded projects. She also is an integrated researcher of CinTurs – Research Centre for Tourism, Sustainability and Well-Being. (Orcid: 0000-0002-3413-4897).

Birendra Kishore Roy is a hospitality professional with over 10 years industry and academic experience.

Partho Pratim Seal has more than two decades of experience in academics and hospitality. An avid wordsmith who believes that writing is the most influential and resourceful medium to disseminate information has authored four books Computers in Hotels: Concepts and Application, Food & Beverage Management, and How to Succeed in Hotel Management Job Interviews and Food Anthropology in India. His interest is in Marketing, Education, and Culinary. He has published research papers in indexed international journals and is also a reviewer of international journals.

G. Kusuma Shanthi is pursuing her Bachelor of Technology in Electronics and Communication from VIT-AP University, Andhra Pradesh, India.

Deepa Sharma has done Ph.D., MBA in Human Resource Management, currently working as Assistant Professor at MMIM, Maharishi Markandeshwar (Deemed to be University), NAAC Accredited Grade 'A++' University, Mullana, Ambala (Haryana). She has more than 7 years of experience in academics. She has written 45+ research papers for various Scopus and UGC listed referred National & International Journals. And has published two book chapters out of which 1 book chapter is published in Taylor & Francis and has earned 45 citations and an h-index of 4 credentials that may be verified on various research platforms like Google Scholar, SSRN, LinkedIn, and Research gate and 5 papers appear in the Scopus preview. She has actively participated in national and international conferences.

Majid Shoeibi is a data analyst and a researcher interested in Supply Chain, Additive Manufacturing, 3DP Reviewer Journal of technology analysis & strategic management MS.C Industrial Management B.A Administration Management WebPage: /in/majid-shoeibi.213253dc-06a0-4d1e-a1c8-bc63a30cd1ff

Ganti Lakshmi Yasaswini is pursuing B.Tech degree in Electronics and Communication from VIT-AP University, Andhra Pradesh, India.

Seydi Yıkmış was born in 1982 in Adıyaman. In 2005, he graduated from Inönü University, Faculty of Engineering, Department of Food Engineering. In 2013, he graduated from Süleyman Demirel University, Institute of Science, Department of Food Engineering. In 2018, he finished his PhD education at Istanbul University, Institute of Health Sciences, Department of Food Hygiene and Technology. He is still working as a Assoc. Prof. Dr. in Department of Food Technology of Namık Kemal University.

Index

H

I

M

N

O

P

Q

R

S

T

U

www.ingramcontent.com/pod-product-compliance
Lightning Source LLC
LaVergne TN
LVHW080636011026
849160LV00014B/1236
9781668490952